# The Cultural Landscapes of Port au Choix

INTERDISCIPLINARY CONTRIBUTIONS TO ARCHAEOLOGY



For more information about this series, please visit the Series Homepage at:
www.springer.com/series/6090

M.A.P. Renouf
Editor

# The Cultural Landscapes of Port au Choix

## Precontact Hunter-Gatherers of Northwestern Newfoundland

Springer

*Editor*
M.A.P. Renouf
Department of Archaeology
Memorial University
St. John's, NL
Canada, A1C 5S7
mapr@mun.ca

ISBN 978-1-4419-8323-7 e-ISBN 978-1-4419-8324-4
DOI 10.1007/978-1-4419-8324-4
Springer New York Dordrecht Heidelberg London

Library of Congress Control Number: 2011924683



Printed on acid-free paper

Springer is part of Springer Science+Business Media (www.springer.com)

*This book is dedicated to*
*Elmer Harp, Jr. and Elaine Groves Harp,*
*together they started what we have*
*continued.*

# Acknowledgements

A number of people greatly helped me with this book. I am indebted to Charles Conway, cartographer with Memorial University Geography Department, for drafting the figures and working tirelessly to ensure their high quality. I am grateful to Chris Hammond, photographer with Memorial University Department of Marketing and Communications, who took many of the artefact photographs, and to Dominique Lavers and Patty Wells, Port au Choix Archaeology Project, who took additional artefact photographs. Larry Nolan of Data Solutions designed our digital data collection and management system, without which the Port au Choix Archaeology Project could not function. Thanks to Dominique Lavers who calibrated the radiocarbon dates for the Appendix and copy-edited the text of the entire book. A special thanks to Robert Anstey, Port au Choix Archaeology Project, who was my primary copy-editor and who made sure that all the bibliographies were stylistically consistent and error-free. I appreciate the time and dedication of many colleagues who reviewed various chapters of this book. Thanks to all the funders of the Port au Choix Archaeology Project, without whose support there would be nothing to write about: Canada Research Chairs Program; Canada Foundation for Innovation; Memorial University; Parks Canada; Provincial Archaeology Office, Government of Newfoundland and Labrador; Research and Development Council of Newfoundland and Labrador; and the Social Sciences and Humanities Research Council of Canada. And thanks to Roger, who makes it all possible.

# Contents

# Contributors

**Robert J. Anstey**
Department of Archaeology, Memorial University, St. John's,
NL, Canada, A1C 5S7
ranstey@mun.ca

**Trevor Bell**
Department of Geography, Memorial University, St. John's,
NL, Canada, A1C 5S7
tbell@mun.ca

**Stuart C. Brown**
Department of Archaeology, Memorial University, St. John's,
NL, Canada, A1C 5S7
sbrown@mun.ca

**Edward J.H. Eastaugh**
Department of Anthropology, University of Western Ontario,
London, ON, Canada, N6A 5C2
eeastaug@uwo.ca

**John C. Erwin**
Department of Archaeology, Memorial University, St. John's,
NL, Canada, A1C 5S7
jerwin@mun.ca

**Maribeth S. Murray**
Department of Anthropology, University of Alaska Fairbanks,
P.O. Box 757720, Fairbanks, 99708, USA
msmurray@alaska.edu

**M.A.P. Renouf**
Department of Archaeology, Memorial University, St. John's,
NL, Canada, A1C 5S7
mapr@mun.ca

**Karen Ryan**
Department of Anthropology, University of Toronto, 19 Russell Street, Toronto, ON, Canada, M5S 2S2
karen.ryan@utoronto.ca

**Kendra D. Stiwich**
#20, 3208 Gibbins Road, Duncan, BC, Canada, V9L 1G8
kendrastiwich@hotmail.com

**Jeremy Taylor**
School of Archaeology and Ancient History, University of Leicester, Leicester, UK LE1 7RH
jt38@le.ac.uk

**Michael A. Teal**
Archaeological Services, Cultural Resource Management, Ontario Service Centre, Parks Canada, Cornwall, ON, Canada, K6H 6S3
Mike.Teal@pc.gc.ca

**Patricia J. Wells**
Department of Archaeology, Memorial University, St. John's, NL, Canada, A1C 5S7
pwells@mun.ca

# Chapter 1
# Introduction: Archaeology at Port au Choix

**M.A.P. Renouf**

## Introduction

Port au Choix, northwestern Newfoundland, is one of the richest archaeological areas in northeastern North America. Comprised of two linked peninsulas that project out into the ocean where the Strait of Belle Isle widens into the Gulf of St. Lawrence, Port au Choix is at the terminus of a deep marine channel that creates ocean upwelling and high biological productivity (Fig. 1.1). The area is today known for its rich fishery and the large herds of migratory harp seals that appear close to shore at the edge of the late winter ice. Because of its active fishing industry and strategic position mid-way along the west coast of the Northern Peninsula, Port au Choix is an economic growth centre within the region. Port au Choix was also an important place in precontact times and the present volume is about that lengthy and complex occupation.

Port au Choix's precontact cultural history spans almost 5,500 years and includes several Amerindian and Palaeoeskimo populations whose nomenclature and chronology are summarized in Table 1.1. Archaeological data suggest that Port au Choix was a significant locale in the economy and cultural identity of each of these populations. For example, the Maritime Archaic Indians consigned their dead to a sandy terrace at the bottom of Back Arm at the site now known as Port au Choix-3 (Fig. 1.1). There are 117 known and many unknown burials at this site, each richly accoutered with mortuary goods (Tuck 1976). Succeeding populations of Groswater Palaeoeskimos hunted harp seal intensively from Phillip's Garden East (Fig. 1.1) on the Point Riche headland (Renouf 1994). They also used a small terrace nearby from which there was a panoramic view of the ocean and from which they could view the arrival of the winter seals. This site, known as Phillip's Garden West (Fig. 1.1), is associated with an exceptionally well-made and stylistically unique lithic assemblage, the significance of which is not yet fully understood

M.A.P. Renouf (✉)
Department of Archaeology, Memorial University, St. John's, NL, Canada, A1C 5S7
e-mail: mapr@mun.ca

M.A.P. Renouf (ed.), *The Cultural Landscapes of Port au Choix*, Interdisciplinary Contributions to Archaeology, DOI 10.1007/978-1-4419-8324-4_1,

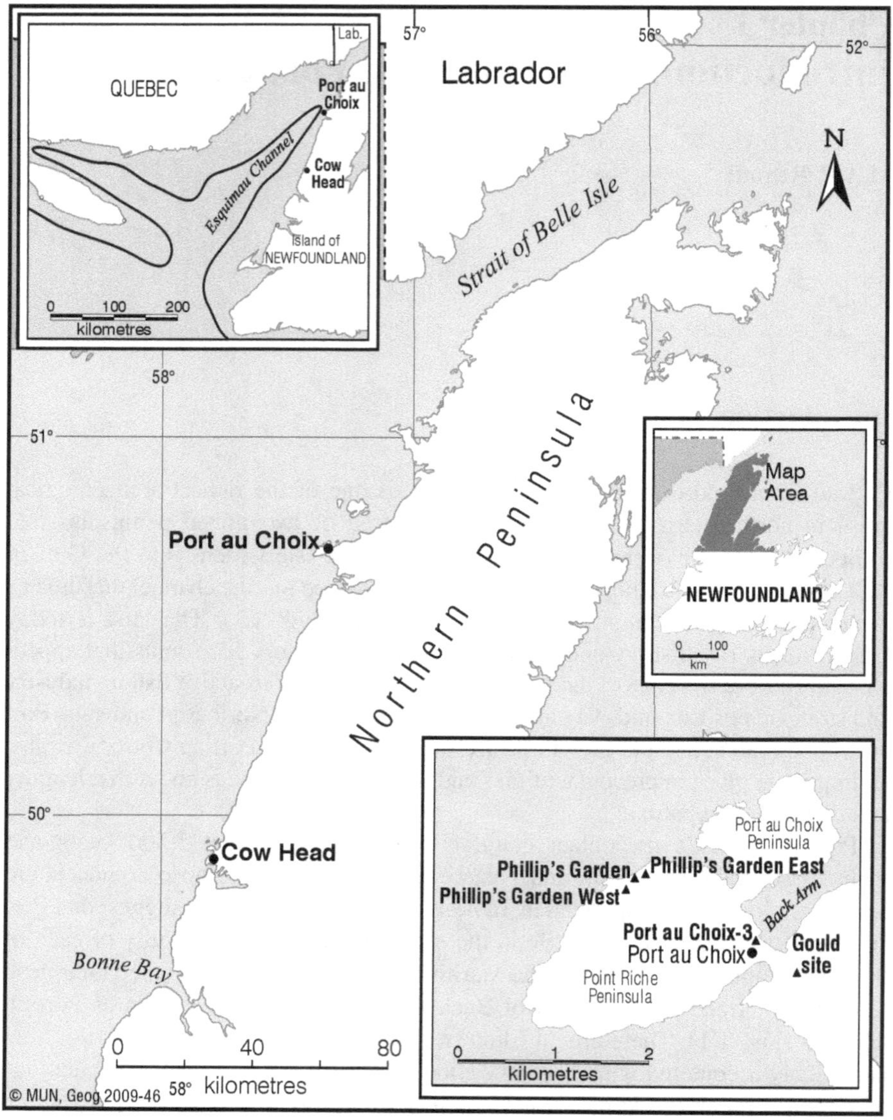


**Fig. 1.1** Map of the Northern Peninsula showing place names mentioned in the text. Also shown is the location of the Esquimau Channel, a deep water channel that creates rich marine conditions off Port au Choix

(Renouf 2005). Situated between these two sites and post-dating them is Phillip's Garden (Fig. 1.1), one of the richest and most intensively occupied Dorset Palaeoeskimo sites within this culture's geographic range. Phillip's Garden was a permanent location that was seasonally occupied for the purpose of hunting harp seals and processing their meat, fat and hides. There are over 60 dwellings

**Table 1.1** Precontact cultures at Port au Choix showing their nomenclature and chronological ranges

| Cultural group | Chronological range in C14 years BP | Chronological range cal BP median age | Chronological range cal BP at 1σ | Chronological range cal BP at 2σ |
|---|---|---|---|---|
| Maritime Archaic Indian | 5440±50 to 3200±100 | 6240–3430 | 6290–3340 | 6310–3160 |
| Groswater Palaeoeskimo | 2760±90 to 1960±80 | 2880–1910 | 2950–1820 | 3140–1720 |
| Dorset Palaeoeskimo | 1970±60 to 1370±90 | 1920–1290 | 1990–1180 | 2110–1070 |
| Recent Indian | 2080±40 to 840±90 | 2050–780 | 2110–680 | 2150–660 |
| Cow Head RI complex | 2080±40 to 1480±70 | 2050–1380 | 2110–1300 | 2150–1290 |
| Beaches RI complex | 1420±70 to 1340±80 | 1330–1250 | 1390–1180 | 1520–1070 |
| Little Passage RI complex | 1020±60 to 840±90 | 940–780 | 1050–680 | 1060–660 |

See Appendix for details of radiocarbon dates. *RI* Recent Indian

at the site, many of which are large multi-family structures, suggesting that Phillip's Garden was a social aggregation site (Renouf 2006). The intensity of human activity at Phillip's Garden changed the landscape, creating a grassy meadow and impacting the ecology of nearby Bass Pond (Bell et al. 2005). Dorset claimed the landscape by burying their dead in three caves, Crow Head Cave, Eastern Point and Gargamelle Rockshelter (Brown 1988). At the same time, Recent Indian populations occupied the sheltered inner shore of Back Arm. The earliest group, the Cow Head complex, lived in a wooded area where we discovered the only known Cow Head complex ceramic assemblage, at the Gould site (Fig. 1.1). Also at the Gould site, we found evidence of direct contact between Dorset Palaeoeskimo and Recent Indian individuals.

A key aspect of Port au Choix's cultural history is its heterogeneity. At between 46.5° and 51.5° north latitude and cooled by the Labrador current, Newfoundland lies at the intersection of Arctic and more temperate regions and, commensurate with this geography, it was a cultural crossroads for Amerindian and Palaeoeskimo populations. The origins of the former were in more temperate areas to the south and west and the origins of the latter were in Arctic areas to the north and west. How populations of these two broad cultural traditions adapted to the Port au Choix environment was filtered through their particular history and world view. The Amerindian populations had a generalized economic pattern based on marine and terrestrial resources and, reflecting this, their sites at Port au Choix were in sheltered coastal locations. In contrast, the Palaeoeskimo populations were marine specialists whose sites at Port au Choix were primarily, although not exclusively, at the exposed outer coast. In addition to adapting to the Port au Choix environment in culturally particular ways, the activities of these populations impacted the landscape differently (Renouf et al. 2009).

Over the past two and a half decades the Port au Choix Archaeology Project has studied aspects of these occupations at a variety of temporal and spatial scales. The project has sought a comparative understanding of how these different cultural groups adapted to the changing physical environment, how they impacted their physical surroundings, how they created cultural landscapes and, where relevant, how they interacted with each other. This work has been carried out by the editor together with colleagues and graduate students. Our research approach has been sharply framed by the concept of place. At the geographic level, our research has focused on the approximately 10 $km^2$ location known today as Port au Choix, a distinct and well-defined landmass characterized by rich marine resources linked to its particular coastal attributes. At the empirical level of settlement, subsistence and social group, our intensive study of this small area has resulted in a high-resolution data set that has been used to address a wide variety of questions across Amerindian and Palaeoeskimo cultures, from individual households, to activity areas within sites, to culturally patterned site location preferences. At the abstract level, we understand Port au Choix as a cultural landscape where, for each culture group, activity, memory and history created layers of meaning through which the place was perceived and acted upon.

This introduction provides context for the succeeding chapters by providing the above cultural historical framework, followed by a summary of early and recent archaeological research at Port au Choix and concluding with a few words about each of the chapters that comprise this book.

## Archaeology at Port au Choix

Archaeological remains from Port au Choix were first published in 1915 by James P. Howley who described artefacts that in 1904 had been collected from a burial cave by Port au Choix resident Frank Hellary (Howley 1915:328; see also Harp 2003:67). Howley's primary interest was in describing the material and intellectual culture of the Beothuk, the indigenous people of Newfoundland at the time of European contact. Howley identified Hellary's collection as Eskimo rather than Indian and included it in his book on the Beothuk culture. This substantiated T.G.B. Lloyd's earlier descriptions of stone-age artefacts from Newfoundland where he proposed an Eskimo presence earlier than the Beothuk (Lloyd 1875, 1876). We now know that the finds described by Howley were from the Dorset Palaeoeskimo cave burial called Eastern Point (Fig. 1.2, Table 1.2), which he referred to as Back Arm.

Howley and Lloyd's reports were followed up in 1929 by William J. Wintemberg, of the National Museums of Canada, who was interested in establishing the range of pre-Beothuk stone-age cultures in Newfoundland. Wintemberg surveyed the west coast of the Northern Peninsula and found several sites. He visited Port au Choix where he test-pitted Phillip's Garden, noting its rich archaeological deposits. He recognized this site as an excellent example of the Cape Dorset culture recently described by his museum colleague, Diamond Jenness (1925). Wintemberg

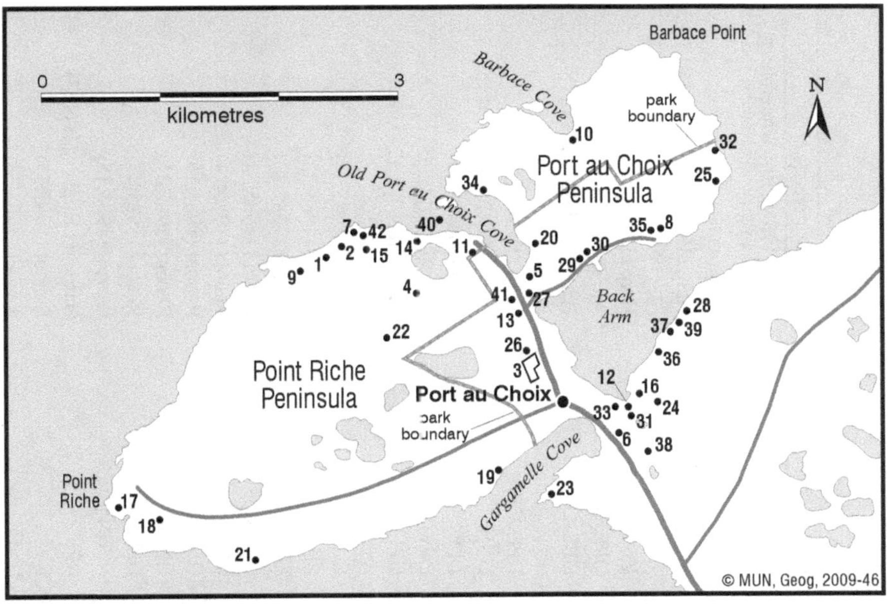


**Fig. 1.2** Map showing the location of registered archaeological sites at Port au Choix; these are keyed to site numbers in Table 1.2. Also shown are the boundary lines of the Port au Choix National Historic Site

reported a smaller Dorset site on the property of James Billard, now known as the Northcott/Rumbolt site (Fig. 1.2, Table 1.2). He described an Amerindian presence in Port au Choix (Wintemberg 1940:313-314) based on a ground stone adz that Alfred Darby found on his property along the main coastal path through the town (Wintemberg 1939:85; see Harp 2003:55) where the Maritime Archaic Indian burial ground was eventually discovered.

Wintemberg's archaeological survey was a beacon for Elmer Harp, Jr., who, as a PhD student from Harvard University, identified an Archaic Amerindian component in Wintemberg's artefact collections (Harp 1964a). Harp's interest lay in studying precontact cultural diversity in the environmentally diverse region of the Strait of Belle Isle which has elements of the Arctic (coastal barrens) and Subarctic (boreal forest) habitats. Harp proposed that the cultural and environmental mix in this region "might have stimulated cultural diffusion and mixture amongst the first peoples who dwelt there" (Harp 2003:53). Port au Choix became the focal point of his research.

In 1949 Harp set out for the first of five field seasons at Port au Choix (Harp 1950, 1951). In his 2003 photographic memoir he described how that first season he and his assistant, Tony Morse, sailed in a bluenose schooner from Woods Hole to southern Labrador, the only way to travel since there were no roads at the time. While Wintemberg hinted at the role Port au Choix residents played in drawing his attention to archaeological material, Harp was more explicit. One of his main

**Table 1.2** Details of registered sites at Port au Choix

| No. on Fig. 1.2 | Site name | Borden # | Parks provenience | Cultural component(s) | Earliest references |
|---|---|---|---|---|---|
| 1 | Phillip's Garden | EeBi-1 | 7A200–381 | DPE, GPE | Harp (1964b:20); Wintemberg (1939:85) |
| 2 | Phillip's Garden East | EeBi-1 | 7A381–399 | GPE | Renouf (1985:45) |
| 3 | Port au Choix-3 | EeBi-2 | n/a | MAI | Harp (1951:216, 1964b:24); Tuck (1976) |
| n/a | Port au Choix-9 | EeBi-3 | n/a | UNPR | Harp (1964b:18) |
| 4 | Crow Head Cave | EeBi-4 | 7A60 | DPE, GPE? | Anderson and Tuck (1974:89) |
| 5 | Northcott/Rumbolt | EeBi-5,-7 | 7A148 | GPE, DPE | Harp (1964a:28); Renouf (1985:24) |
| 6 | Port au Choix-6 | EeBi-6 | n/a | DPE, GPE | Harp (1964b:27) |
| n/a | Port au Choix-8 | EeBi-8 | n/a | UNK | Harp (1964b:30) |
| 7 | Port au Choix-1 | EeBi-9 | n/a | DPE | Harp (1964b:18) |
| 8 | Eastern Point (Back Arm) | EeBi-10 | 7A68 | DPE | Howley (1915:328) |
| 9 | Phillip's Garden West | EeBi-11 | 7A55, 700–800 | GPE | Fitzhugh (1983:121); Renouf (1985:16) |
| 10 | Barbace Cove | EeBi-12 | 7A26 | EU | Renouf (1985:34) |
| 11 | Old Port au Choix West | EeBi-13 | 7A1–25 | EU | Renouf (1985:33) |
| 12 | Offrey Site | EeBi-14 | 7A146 | UNPR | Renouf (1985:32) |
| 13 | Cadet Site | EeBi-15 | 7A51 | UNPR | Renouf (1985:31) |
| 14 | Trike Path | EeBi-16 | 7A52 | GPE | Renouf (1985:14) |
| 15 | Hill East of Phillip's Garden | EeBi-17 | 7A52 | UNPR | Renouf (1985:16) |
| 16 | Lab Site | EeBi-18 | 7A146 | UNPR | Renouf (1985:31) |
| 17 | Lighthouse | EeBi-19 | 7A57 | DPE | Renouf (1985:17) |
| 18 | Point Riche | EeBi-20 | 7A58–59, 500–600 | GPE, DPE | Renouf (1985:18) |
| 19 | Gargamelle Rockshelter | EeBi-21 | 7A65 | DPE | Harp and Hughes (1968:7) |
| 20 | Old Port au Choix East | EeBi-22 | 7A127–129 | EU | Renouf (1985:33) |
| 21 | The First Site | EeBi-23 | 7A145 | UNPR | Renouf (1985:34) |

| | | | | | |
|---|---|---|---|---|---|
| 22 | David's Site | EeBi-24 | 7A151 | GPE | Renouf (1985:13) |
| 23 | Gargamelle Point | EeBi-25 | 7A70 | DPE | Renouf (1991:63) |
| 24 | Joe Offrey | EeBi-26 | 7A75 | UNPR | Renouf (1991:65) |
| 25 | South Side of PAC Peninsula | EeBi-27 | 7A140 | UNPR | Renouf (1985:32) |
| 26 | Find at New RC Church | EeBi-28 | 7A146 | UNPR | Renouf (1985:31) |
| 27 | Cornick Site | EeBi-29 | n/a | GPE | LeBlanc (1996:85) |
| 28 | Party Site | EeBi-30 | 7A71 | GPE, DPE, RI | Renouf (1991:66) |
| 29 | Gaslard's Lane | EeBi-31 | 7A72 | DPE | Renouf (1991:67) |
| 30 | Sid Buckle | EeBi-32 | 7A73 | UNPR | Renouf (1991:67) |
| 31 | Stream Bank | EeBi-33 | 7A74 | PE | Renouf (1991:67) |
| 32 | Dobbin Cave | EeBi-34 | 7A63 | DPE | Renouf (1991:67) |
| 33 | Spence Site | EeBi-36 | 7A76–86 | RI, DPE, GPE | Renouf (1991:69) |
| 34 | Dead Whale | EeBi-37 | 7A99 | UNPR | Renouf (1991:69) |
| 35 | Eastern Point-2 | EeBi-38 | 7A69 | DPE | Renouf (1991:81) |
| 36 | Hamlyn Site | EeBi-39 | n/a | GPE, DPE, RI | Renouf and Bell (1991:27) |
| 37 | Lloyd Site | EeBi-41 | n/a | GPE, DPE, RI | Renouf and Bell (1991:25) |
| 38 | Gould | EeBi-42 | n/a | MAI, RI | Renouf and Bell (1998) |
| 39 | Old Boatyard | EeBi-43 | n/a | MAI, UNPR | Renouf and Bell (1991:15) |
| 40 | Old PAC Hr. Ship Remains | EeBi-44 | n/a | EU | PAO site record form |
| 41 | Old PAC West Extension | EeBi-45 | n/a | EU | PAO site record form |
| 42 | Bird Blind Site | EeBi-46 | n/a | UNK | PAO site record form |

The first column corresponds with site location numbers in Fig. 1.2; n/a refers to sites where the exact location is unknown. Parks provenience refers to the Parks Canada provenience system by which all artefacts are registered within National Historic Sites; the Parks provenience numbers in the fourth column are key to identifying artefacts to site and location within site. In column five, the cultural components are abbreviated as follows: *MAI* Maritime Archaic Indian; *PE* Palaeoeskimo; *GPE* Groswater Palaeoeskimo; *DPE* Dorset Palaeoeskimo; *RI* Recent Indian; *UNPR* unidentified precontact; *UNK* unknown cultural affiliation; *EU* European. In column six PAO refers to Provincial Archaeology Office objectives was to investigate Phillip's Garden. He narrated how it took him and Tony Morse several hours to make their way out to the site because of required socializing at houses en route. At the home of the elderly James and Eugenie Billard, Mrs. Billard recounted how as a child she used to go to Phillip's Garden where she saw stone chips on bare patches in the meadow and large whalebones lined up across the field. In his memoir, Harp wrote about the excitement of his first day at Phillip's Garden when he realized the archaeological potential of this extensive site. That first year, Alfred Darby showed Harp ancient human bones found on his property. Darby's neighbours, Pius and Walter Billard, also showed Harp Archaic artefacts found in their nearby gardens (Harp 2003:61). Harp knew there must be an Archaic Indian burial site in that area but, despite testing, he failed to locate it (Harp 2003:63; Harp and Hughes 1968).

In 1950 Harp returned to Port au Choix, this time with his wife Elaine and their eldest son, Jack. They drove as far as Bonne Bay (Fig. 1.1) and from there travelled north via coastal steamer. Harp returned to Port au Choix in the summers of 1961 to 1963 by which time he could drive all the way, close on the heels of expanding road construction. By this time, he was a faculty member of Dartmouth College and was accompanied by several Dartmouth and other students (Harp 2003:xii–xiii). The 1949, 1950 and 1961 field seasons included archaeological survey in White Bay, on the Northern Peninsula, and in southern Labrador where Harp found many Amerindian and Palaeoeskimo sites. His description of material from these sites (Harp 1964a) was the basis for all subsequent archaeological work in that region.

Over the course of the 1961–1963 field seasons Harp, his students and a few local youths excavated seven and extensively tested 13 dwelling structures at Phillip's Garden. Material from Phillip's Garden and the other Dorset sites Harp had discovered at Port au Choix and elsewhere on the Northern Peninsula formed the basis of his published PhD thesis characterizing the Dorset culture in Newfoundland and examining its archaeological relationship to the older Boreal Archaic occupations of that area (Harp 1964b).

Harp's subsequent discussions of Phillip's Garden focused on households and site evolution (Harp 1976). He reconstructed house architecture and related it to social group and season of occupation. He also reconstructed changing population size. In addition, he reported two Dorset burials, one an infant burial found in House 12 at Phillip's Garden and the second a collection of human remains and artefacts found at Gargamelle Rockshelter in 1953 by Port au Choix resident Finton Gould (Harp and Hughes 1968). Harp described carved amulets from these burials and from Phillip's Garden (Harp 1970) noting that, compared to Dorset amulets elsewhere, they were distinctively abstract and two-dimensional.

In 1967, a few years after Harp completed his Port au Choix fieldwork, Theodore Farwell accidentally disinterred at least eight human skeletons along with ground stone tools on the same sandy terrace where Alfred Darby and Pius and Walter Billard had found artefacts and human remains almost 20 years before (Tuck 1976:8). Memorial University's first archaeologist, James A. Tuck, was called to Port au Choix to examine Mr. Farwell's finds and was quick to understand their significance. Tuck returned to Port au Choix in 1968 to excavate the initial find area and to test the sandy terrace for other burial locations; he returned for a short period in 1969. He found two more concentrations of burials which he excavated and reported in full (Tuck 1970, 1971, 1976). Based on the location of the burial ground close to the shore and the inclusion of many items relating to fishing and marine mammals, Tuck characterized these Archaic people as marine-oriented and on this basis he defined the Maritime Archaic Tradition (Tuck 1976:98). Tuck's work highlighted the material and spiritual richness of these Maritime Archaic Indians. Physical anthropological work established the robust health of the buried population (Anderson 1976; Kennedy 1980, 1981; see Jerkic 1993:216).

The presence of two important archaeological sites, the Maritime Archaic Indian burial ground and Phillip's Garden, led in 1984 to the designation of 8.3 $km^2$ on the Point Riche and Port au Choix Peninsulas as a National Historic Site. That same year Parks Canada funded an archaeological survey to establish an archaeological site inventory of the area. As a new faculty member of Memorial University, I directed that survey which became the basis for the Port au Choix Archaeology Project. In 1984 we identified known sites and discovered several new ones (Renouf 1985). Notable among them was a second large Dorset site, Point Riche; although extensive, it was only one-third the size of Phillip's Garden. We located two Groswater Palaeoeskimo sites, Phillip's Garden East and Phillip's Garden West; the latter had been discovered by William Fitzhugh of the Smithsonian Institution during a short visit to Port au Choix in 1982 (Fitzhugh 1983). We also identified a number of eighteenth-to-nineteenth century French fishing sites. We eventually shifted our survey focus to Amerindian sites, in 1990 finding the Recent Indian Spence site and in 1997 finding the Maritime Archaic Gould site. In total, we registered 45 sites at Port au Choix, both within and beyond the boundaries of the National Historic Site. These are shown in Fig. 1.1 and listed in Table 1.2.

In 1997 Trevor Bell, a Memorial University geographer, joined the project to provide a crucial palaeoenvironmental context for Port au Choix and its archaeological sites, building upon earlier work by his Memorial University colleague, Joyce Macpherson (1995). His reconstruction of the 5,000 year-old shoreline of the area was the basis for the discovery of the Gould site (Renouf and Bell 2000), which became a major focus of our project from 1998 to 2001. Bell and a number of his colleagues from Canadian and U.S. universities worked on various projects associated with the site, the results of which were published in a theme issue of the journal *Newfoundland and Labrador Studies* (Bell and Renouf 2005). Part of this interdisciplinary group, Memorial University archaeologist, Michael Deal, collected and analyzed seeds from the Gould site's soil deposits (Deal 2005).

The archaeology of Port au Choix was interpreted in a small museum near the Maritime Archaic burial ground in the heart of town. This museum was initially a local initiative and in 1984 it was taken over by Parks Canada. In 1997 a new, larger museum was opened on the southern bluffs of the Point Riche Peninsula, directly above Gargamelle Rockshelter. The exhibition was expanded to include the story of Port au Choix from the Maritime Archaic occupation up to the recent past, based on the work of Harp, Tuck and the Port au Choix Archaeology Project. Elmer and Elaine Harp were present at the opening ceremonies.

It has been over 60 years since Elmer Harp, Jr., began his archaeological investigations at Port au Choix, over 40 years since James A. Tuck excavated the Maritime Archaic Indian burial ground and over 25 years since the Port au Choix Archaeology Project initiated its first survey. During this time, many students situated their honours and graduate research projects on the archaeology of Port au Choix. A number of honours (Evans 1991; Marshall 1990; Reader 1990) and Master's (Kennedy 1980, 1981) theses and one PhD thesis (Jelsma 2000, 2006) focused on the biological anthropology of the skeletons found at Port au Choix-3; D'Entremont (1978) examined the significance of the grave goods. In her Master's thesis, Reid (2007) discussed the earliest Maritime Archaic Indian component of the Gould site. Other honours (Lavers 2006; Ryan 1997) and Master's (Kennett 1990; LeBlanc 1996, 2000a; Wells 2002, 2005; Wheatley 2004) projects focused on aspects of Groswater Palaeoeskimo sites. Several honours students (Anstey 2008; Bates 1991; Culleton 1991; Gracie 2004; Howse 2002; Linehan 1990) and Master's students (Cogswell 2006; Colligan 2006; Eastaugh 2002, 2003; Erwin 1995; Knapp 2008; Murray 1992; Renouf and Murray 1999) situated their research on the Dorset occupation of Port au Choix. In her PhD thesis, LeBlanc (2000b, 2008) included a sample of lithic artefacts from Phillip's Garden. In her Master's, Lyons (1983) included Dorset amulets from Port au Choix in her comparative sample; this art was also discussed in Wells (2009). Bambrick (2009) examined pond sediment data relating to the Dorset occupation of Phillip's Garden and Wells (1988) compared Groswater and Dorset cut marks on seal bones. Hiseler (1997) examined harp seal canine dentine incremental structures to determine the seasonality of Phillip's Garden and Phillip's Garden East. O'Driscoll (2003) included a sample of soapstone artefacts from Phillip's Garden in her geochemical analysis of Dorset soapstone artefacts from sites in western Newfoundland. Hodgetts (2005a, 2005b; see also Hodgetts et al. 2003) focused a postdoctoral project on Phillip's Garden faunal assemblages. Teal's (2001) Master's thesis explored the Recent Indian occupation of the Gould site and Lavers (2010) included lithic samples from the Recent Indian Gould and Spence sites in her geochemical analysis of Recent Indian lithic sources and stone tools on the Northern Peninsula. The results of some of these projects are presented as chapters in this volume.

## The Chapters

The chapters in this book progress thematically and chronologically, starting with the present chapter which provides historical context and followed by Chap. 2 which provides environmental context. The chapters then progress through a sequence of topics relating to the Maritime Archaic Indian, Groswater Palaeoeskimo, Dorset Palaeoeskimo and Recent Indian occupations of Port au Choix. A final chapter draws together these aspects of Port au Choix's prehistory.

In Chap. 2 Trevor Bell and I synthesize the physical environment of Port au Choix, summarizing the physical and cultural setting, followed by a discussion of data relating to landscape history, vegetation history and climate change. We relate this to human history, noting that there is correspondence between warm conditions and Maritime Archaic occupations and cold conditions and Groswater Palaeoeskimo occupations. Dorset Palaeoeskimo and Recent Indian occupations occurred during generally warmer but fluctuating conditions. We discuss how human responses to these climate changes were culturally particular. We also note the importance of understanding how changing climate affected sea ice conditions which are directly connected to the distribution of harp seal herds, of prime importance for all precontact populations who lived at Port au Choix.

In Chap. 3 Trevor Bell and I describe and discuss the Maritime Archaic Indian levels of the Gould site. Our interpretation of the site as an intermittently used short-stay camp associated with the nearby burial ground changes how we look at Maritime Archaic Indians at Port au Choix and elsewhere, shifting the focus from sites to paths of movement through the landscape. We argue that the location of the burial ground and the Gould site on either side of a narrow marine channel is key to understanding both sites.

Groswater Palaeoeskimos represent the so-called transitional period between the Pre-Dorset and Dorset Palaeoeskimo occupation of Labrador; there is no Pre-Dorset material known from Newfoundland aside from Tuck's (1978) suggestion of a small amount of Pre-Dorset material at the Spearbank site in Cow Head (Fig. 1.1). Groswater is the term applied to Newfoundland and Labrador transitional period material and is similar to Independence II assemblages from Greenland, but not closely similar to transitional period material from the Canadian Arctic. The Groswater occupation of Newfoundland was anything but transitional in nature; rather it was a vigorous cultural presence that lasted for over a 1,000 years. One of the best-preserved Groswater sites in Newfoundland is Phillip's Garden East and its lithic and bone assemblage is representative of Groswater material culture elsewhere in Newfoundland and in Labrador. Phillip's Garden West is a small Groswater site nearby with, as mentioned, a distinct and finely made lithic assemblage. Quantification of metric and non-metric attributes of Phillip's Garden East and Phillip's Garden West lithics was the basis for the definition of the Phillip's Garden West lithic variant (Renouf 2005). Of the 22 calibrated radiocarbon dates from both sites, 20 overlap at the one-sigma probability range.

Chap. 4 by Patricia Wells is a detailed examination of four faunal assemblages from these two sites. Three of the four assemblages are from the period of site overlap and the fourth post-dates the overlap. All assemblages are dominated by phocid remains, which point to similar subsistence and seasonality. However, Wells demonstrates that there are significant differences in the relative frequency of phocid skeletal elements between the two sites, in particular the relative proportion of cranial elements. She suggests that the sites are contemporaneous and functionally connected for the greater part of their occupations. She further suggests that this function might be linked to the ritual dimension of the harp seal hunt, an aspect of the ethnographically well-documented relationship between hunter and prey animal. This study is a reminder that faunal assemblages are not the result only of a series of pragmatic actions related to killing and butchering, followed by cooking, eating and discarding, but they also result from the ritual dimension of all stages of the hunt. Like other studies of Phillip's Garden West (Renouf 2005; Ryan 1997) and similar material from elsewhere in Newfoundland (Melnik 2007), Wells' chapter raises as many questions as it answers about this curious site.

Chap. 5 by Karen Ryan focuses on diachronic variation within the Phillip's Garden West lithic variant as expressed in the endblades. Based on radiocarbon dates and quantitative and qualitative endblade attributes, she argues that the extraordinarily well-made endblades on the upper terrace of the site, compared to the less well-made examples from the hillside midden deposits, represent the end of site occupation and, by extension, the end of the Groswater occupation of Newfoundland. Importantly, she identifies Phillip's Garden West-type endblades in Groswater, Dorset and Recent Indian contexts in sites throughout Newfoundland. Ryan notes that outside the west coast of the Northern Peninsula, only the late form of Phillip's Garden West endblades occurs and that they are isolated specimens within a larger site component. Ryan explores possible reasons for this geographically broad distribution of Phillip's Garden West-type material and concludes that it represents the distribution and curation of isolated specimens rather than movement of actual Phillip's Garden West populations.

Ryan also discusses the possibility that Groswater populations briefly interacted with Dorset and/or Recent Indian populations, especially at Port au Choix where sites of all three cultures occur and which seems to be the geographic focus of the Phillip's Garden West variant. This is a logical possibility, based on overlapping radiocarbon date ranges, and it is an intriguing idea. However, since the chronological overlap is minimal and based on few dates, it remains to be seen if it is an anomaly or a pattern.

Chap. 6 by Kendra Stiwich shifts the focus away from Groswater seal hunting sites on the Point Riche headland to smaller and more ephemeral Groswater occupations in the sheltered shore of Back Arm. These sites are either heavily disturbed by road and house construction or they are hidden within dense spruce forest and beneath 1–2 m of peat. Wheatley presents data from one of these sites, the Party site. She reports on two excavation areas and argues that they represent temporally and functionally different warm-weather occupations. This study of an inner bay and warm-weather Groswater site provides important details about this lesser-known aspect of Groswater occupation of Port au Choix.

The next five chapters focus on the Dorset Palaeoeskimo occupation of Phillip's Garden, a 2 ha site occupied for 800 years. The site has excellent organic preservation and therefore we have tens of thousands of animal bones and many bone, ivory and antler artefacts. The site is relatively undisturbed and therefore the imprint of dwelling architecture is discernable. An excavated sample of 24 dwellings and a small number of middens have yielded an artefact collection of over 35,000 items. The wealth and range of artefactual and structural data make this site ideal for high-resolution spatial and temporal studies.

In Chap. 7 I review the highlights of our research at Phillip's Garden since 1984. I synthesize data relating to the importance of harp seal hunting, seal hide processing, large multi-family dwellings and diversification of a range of other activities, in particular obtaining and using whalebone. I tie these threads together in an argument for labour intensive and coordinated economic activity comparable to precontact Thule whaling in the eastern Arctic and salmon fishing and processing among Pacific coast hunter-fishers. This scale of economic and social activity is unknown for Dorset outside Newfoundland and is linked to the predictable abundance of harp seals which were briefly available directly offshore of Phillip's Garden.

In Chap. 8 John Erwin looks at the important issue of house contemporaneity at Phillip's Garden. Erwin calculates the number of likely contemporaneous houses based on radiocarbon dates and three possibilities for house longevity: 25, 20 and 75 years. His results show a battleship curve of initial low population followed by population growth, peak and eventual decline. This independently substantiates Harp's (1976) and our (Renouf and Bell 2009) calculations based on the number of chronologically overlapping houses, but where we did not take house longevity into account. Erwin also looks at how site use changed over time, arguing for a shift from an initial short-term seasonal camp to a larger and more permanent occupation, returning to a short-term seasonal occupation. Erwin reminds us of the importance of examining Phillip's Garden diachronically to understand how various aspects must have changed over its long occupation. Finally, Erwin looks at spatial patterning of Phillip's Garden dwellings within his chronological framework, which shows that adjacent dwellings are more likely to be contemporaneous than distant dwellings; in other words, he traces what were likely social clusters.

In Chap. 9 Edward Eastaugh and Jeremy Taylor apply the geophysical technique of magnetometry to record sub-surface anomalies in a 1,400 $m^2$ test area of Phillip's Garden. Magnetometry proved surprisingly sensitive to the subtle anomalies created by Dorset activities. Eastaugh and Taylor found four previously unknown dwelling depressions that were hidden beneath midden deposits. Interestingly, although the hidden houses were picked up by the magnetometer, those visible on the surface of the ground were not. Eastaugh and Taylor suggest that this is because, except where dwellings were filled in with organically rich midden material, there is little difference between the limestone-constructed dwellings and the surrounding limestone shingle substrate. Extrapolating from their test area to the entire site, Eastaugh and Taylor suggest a total of 20 hidden dwellings. Their magnetometry results also show a pattern of small features 5–10 m from any dwelling. These data suggest that it would be useful to excavate beyond the perimeter of dwellings, which have been the focus of almost all Phillip's Garden excavations until very recently (Renouf 2009). These exciting results show that there is a great deal of potential for magnetometry at this and possibly other hunter-gatherer sites.

In Chap. 10 Robert Anstey and I look at the relative proportions of Ramah chert in the lithic assemblage of a sample of six Phillip's Garden dwellings that collectively span site occupation; the only source of Ramah chert is in northern Labrador. Ramah chert is present in small amounts in all Phillip's Garden dwellings sampled, indicating minor but sustained communication between Newfoundland and Labrador. The relative amount of Ramah chert is greatest in the youngest dwelling. This dates to the end of site occupation which corresponds to the end of the Dorset occupation of Newfoundland. Anstey and I argue that this increase in relative frequency represents increased communication with Labrador just before the abandonment of Phillip's Garden. We argue that this increase was a way of mitigating the risk inherent in the processes that were undermining the Phillip's Garden occupation at this crucial time.

In Chap. 11 Maribeth Murray addresses the issue of cultural choice in the exploitation of harp seals, in particular with regard to animal size, condition of pelt, and ease of exploitation. She introduces harp seal biology as a context for discussing how Newfoundland small-boat sealers categorize seals at seven stages of their life based on the changing condition of their pelage and their migratory behaviour. Since the pelage changes with age, these seven categories correspond to stage of life, or age, categories. She uses this framework to examine the faunal collection from one of the earliest dwelling features from Phillip's Garden, Feature 1. Using harp seal stage of life categories established for modern harp seals by Storå (2000) based on epiphyseal fusion sequence, Murray classifies the faunal material from Feature 1 into four age classes. She concludes that the Dorset hunters who occupied Feature 1 focused on sexually immature seals 1–4 years old (bedlamers) as well as young (turners, harps) and old adult (harps) seals. In contrast, they did not target whitecoats (seals under two weeks old) or yearlings (ragged jackets and beaters). Since ragged jackets in particular have poor quality coats, this suggests a hunting strategy focused on larger seals with better quality pelage. This paper reminds us that hides as well as meat and fat were an important consideration in the all-important seal hunt. It also reminds us that, like Inuit who have many words to describe the different properties of snow, Dorset seal hunters likely had a complex view of the many properties of seals and that hunting decisions were made on that multi-dimensional view.

In Chap. 12 Stuart Brown summarizes the results of his salvage excavations of two Dorset cave burials at Port au Choix and uses these as a springboard for synthesizing the small amount of published material on Dorset burials in the Arctic, much of it dating to the 1960s and 1970s. Where appropriate, he uses Arctic ethnography as a framework for interpreting archaeological patterns within an explicitly deductive approach. On these bases, he is able to suggest several patterns in Dorset burial behaviour. One of the most interesting is the special treatment accorded to crania and mandibles, which is consistent with Wells' conclusions about seal crania in Groswater faunal assemblages, discussed in Chap. 4.

In Chap. 13 the focus shifts to the Recent Indian occupation of Port au Choix. There is a small number of Recent Indian find spots along the eastern shore of Back Arm (Fig. 1.2). The Recent Indian occupation of Newfoundland comprises three cultural complexes: Cow Head, Beaches and Little Passage (Table 1.1). In Chap. 13 I present, together with Michael Teal and Trevor Bell, the Cow Head complex component of the Gould site. This occupation was found by chance as we tested an ancient shoreline in search of a Maritime Archaic Indian site. We found the Gould site which unexpectedly had a Recent Indian component in the peat that overlay the Maritime Archaic levels on and directly above the ancient beach. At the time of use the Recent Indian occupation would have been in a wooded area, near a small stream and set back about 350 m from the shoreline. This was the first

Recent Indian site found in a near-coastal location, which we call the coastal margin, and perhaps unsurprisingly it was unique in many respects. We argue that the Cow Head complex component of the Gould site is a residential base in contrast to the other Cow Head complex sites on the Northern Peninsula, which are at the shore and have a significant lithic quarry and workshop function. The discovery of a residential site in the woods set back a short distance from the coast suggests that we should include the coastal margin as a focus of our future archaeological surveys.

In Chap. 14, the concluding chapter, I synthesize the Amerindian and Palaeoeskimo occupations of Port au Choix from a landscape perspective. I discuss how the different cultures adapted to Port au Choix via subsistence, mobility and site location preferences. I then discuss the life history of three Port au Choix landscapes, examining how each occupation created cultural landscapes through impacting the natural surroundings, creating a built environment, and investing locations with cultural meaning. I discuss how these cognized landscapes structured subsequent use and how populations were aware of prior landscapes.

The Appendix presents all the radiocarbon dates from the Port au Choix Archaeology Project as of 2010, including calibrations and contextual information.

In conclusion, the papers in this volume collectively summarize the Amerindian and Palaeoeskimo occupations of Port au Choix that spanned 5,500 years. The multiple perspectives and methods employed provide a detailed reconstruction and understanding of the long-term history of Port au Choix. This in turn sets the stage for a comparative understanding of how these different populations adapted to and created the landscape that today is Port au Choix. In doing so, this volume demonstrates the value of an intensive archaeological focus on a single region, or place.

**Acknowledgements** The Port au Choix Archaeology Project has been generously funded by many organizations over the past 25 years. Parks Canada provided significant funding for the 1984–1986 and 1990–1992 field seasons. They continued to support our work from 1997 to 2002 when we were excavating outside the boundaries of the Port au Choix National Historic Site and provided supplementary but significant funding when we returned to work within the Site, from 2003 until present. Since 1991 the Social Sciences and Humanities Research Council of Canada has given sustained support for our research in Port au Choix and elsewhere on the Northern Peninsula. Several funding sources at Memorial University have regularly supported our work: the Institute of Social and Economic Research, the J.R. Smallwood Foundation for Newfoundland and Labrador Studies and the Dean of Arts Office; we received a substantial Memorial University Special New Initiatives Grant for our Maritime Archaic research. Since 2001 we have been generously assisted by the Canada Research Chairs Program. In 2001 and 2009 the Canadian Foundation for Innovation provided funds to update our equipment and transform our university labs; one of the most important achievements to date is completing the re-examination and re-cataloguing of the entire Phillip's Garden artefact collection. In 2009 the Industrial Research and Innovation Fund of the Research Council of Newfoundland and Labrador also funded our research infrastructure. The Provincial Archaeology Office, Government of Newfoundland and Labrador, contributed funds to several of our field seasons. The town of Port au Choix was a funding partner during 1997–2001 when we worked on the Gould site which is within the town boundaries.

Parks Canada Atlantic Region and Parks Canada Western Newfoundland and Labrador gave us sustained logistical support. I would like to mention Chip Bird, Geoff Hancock, Charles Lindsay, Millie Spence and Trudy Taylor-Walsh. The staff of the Port au Choix National Historic

Site helped us in any way they could. Similar assistance was provided by the Town of Port au Choix. Port au Choix residents have maintained a keen interest in our research. They shared in the joint celebration hosted by the Port au Choix National Historic Site in July 2009 in honour of the 25th anniversary of the Port au Choix Archaeology Project and the 60th anniversary of Elmer Harp's first visit to the area.

Many students, non-students and colleagues have worked on the Port au Choix Archaeology Project and I thank them all. They are: Claire Allum, Robert Anstey, Juanita Atkins, Terry Aylward, Jennifer Baird, Darlene Balkwill, Jillian Bambrick, Patty Barefoot, George Barron, Alison Bates, Greg Beaton, Trevor Bell, Bradea Billard, Gail Billard, Pauline Billiard, Helena Biggin, Mary Biggin, Paul Biggin, Scott Biggin, Paula Broaders, Jeanette Bromley, Preston Bromley, Stuart Brown, Don Butler, Aaron Butt, James Cahill, Aiden Caines, Ainslie Cogswell, Dave Culleton, Jenneth Curtis, Julia Daly, Michael Deal, Tessa de Roo, Michael Devine, Tammy Dobbin, Trish Dunphy, Ed Eastaugh, Jarvis Eastman, Tracy Ellsworth, John Erwin, Clifford Evans, Helen Evans, Cara Finn, Lisa Fogt, J.P. Foster, Joan Fuller, Todd Garlie, Alisa Gaslard, Brenda Gaslard, Trevor Gaslard, Vincent Genneax, Morley Gil, Barb Gould, Evelyn Gould, Ian Gould, Jolene Gould, Lois Gould, Rose Gould, Rosemary Gould, Sofia Gould, Suzanne Gould, Troy Gould, Wade Greeley, Shane Greene, Tana Green, Vaughan Grimes, Trudy Harris, Caitlyn Hill, Peter Hiscock, Lisa Hodgetts, Sam Horwood, Lesley Howse, Brendan Hughes, Hazel Hynes, Kyla Hynes, Johan Jelsma, Sonja Jerkic, Sue Kearsey, Alice Kelley, Vernon Kelly, Rebecca Knapp, Janet Kreda, Carol Krol, Rob Lackowicz, Brian Lavers, Dominique Lavers, Steve Lenser, Donna LeShano, Chris Lewis, Deea Linehan, Ryan Lloyd, Gerald Lowe, Shaunda Lowe, Earl Luffman, Nichola MacIllfhinnein, Jeannette Macey, Joyce Macpherson, Aiden Mahar, Bryant Mahar, Darcy Mahar, Danny Major, Kevin McAleese, Elizabeth McManus, Mike Meadows, Jerry Melbye, Mary Melnik, Maribeth Murray, Meghan Negrijn, Deirdre Ni Bhuacalla, Doug Nixon, Gary Nobles, Verdun Noel, Cindy O'Driscoll, Emmett O'Keefe, Rita Offrey, Ralph Pastore, Mark Penney, Eric Phaneuf, Roger Pickavance, Rhonda Pickett, Mindy Pitre, Sarah Player, Kim Ploughman, Nadine Ploughman, Krista Plowman, Peter Pope, Andrea Powers, David Reader, Heather Reid, Griffith Roberts, Pamela Rogers, Cavel Rose, Nicholas Rose, Dolores Rumbolt, Karen Ryan, Jeanette Ryan, Katherine Scott, Adrienne Selby, John Shea, David Simpson, Mike Spence, Marianne Stopp, Juanita Sutton, Mike Teal, Jeremy Taylor, Sheila Vardy, Jim Walsh, Patrick Warner, Patty Wells, Kendra Wheatley, Andrea Wilson and Brent Wimmer.

## References

Anderson, J.E. 1976. Appendix A: The human skeletons. In *Ancient people of Port au Choix*, ed. J.A. Tuck, 124–131. Newfoundland Social and Economic Studies 17, Institute of Social and Economic Research, St. John's.

Anderson, J.E., and J.A. Tuck. 1974. Osteology of the Dorset people. *Man in the Northeast* 8: 89–97.

Anstey, R.J. 2008. Ramah Chert Use at Phillip's Garden (EeBi-1), Port au Choix, Newfoundland. Honour's dissertation, Department of Anthropology and Archaeology, Memorial University of Newfoundland, St. John's.

Bambrick, J. 2009. Palaeoeskimo Sealskin Processing in Port au Choix, Northwestern Newfoundland: A Paleoenvironmental Analysis. Master's thesis, Department of Archaeology, Memorial University of Newfoundland, St. John's.

Bates, A. 1991. A Use Wear Analysis of Dorset Palaeo-Eskimo Microblades from Port au Choix, Newfoundland. Honour's dissertation, Department of Anthropology, Memorial University of Newfoundland, St. John's.

Bell, T., and M.A.P. Renouf (ed.). 2005. *Newfoundland and Labrador Studies* 20(1).

Bell, T., J.B. Macpherson, and M.A.P. Renouf. 2005. Late prehistoric impact on Bass Pond, Port au Choix. *Newfoundland and Labrador Studies* 20(1): 107–129.

Brown, S.C. 1988. Archaeological Investigations at Crow Head Cave and the Gargamelle Rockshelter in the Port au Choix National Historic Park, Newfoundland: Report of 1986 Field Activities. On file, Parks Canada, Archaeology, Atlantic Region, Halifax.

Cogswell, A.E. 2006. House 18 and the Middle Phase Occupation at Phillip's Garden (EeBi-1). Master's thesis, Department of Anthropology, Memorial University of Newfoundland, St. John's.

Colligan, E.M. 2006. Newfoundland Dorset Endblades: Measuring Regional Variability. Master's thesis, Department of Anthropology, Hunter College, The City University of New York, New York.

Culleton, D. 1991. Scratching the Surface of Usewear Studies: A Micrwear Analysis of Middle Dorset Endscrapers from Phillip's Garden, Port au Choix, Northwest Newfoundland. Honour's dissertation, Department of Anthropology, Memorial University of Newfoundland, St. John's.

Deal, M. 2005. Palaeoethnobotanical research at Port au Choix. *Newfoundland and Labrador Studies* 20(1): 131–156.

d'Entremont, S.A. 1978. The Determination of Status at Port au Choix. Master's thesis, Department of Anthropology, Memorial University of Newfoundland, St. John's.

Eastaugh, E.J.E. 2002. The Dorset Palaeoeskimo Site at Point Riche, Newfoundland: An Intra-Site Analysis Master's thesis, Department of Anthropology, Memorial University of Newfoundland, St. John's.

Eastaugh, E.J.E. 2003. A Middle Dorset semi-subterranean dwelling at Point Riche, Port au Choix. *Études/Inuit/Studies* 27(1–2): 451–471.

Erwin, J.C. 1995. An Intra site Analysis of Phillip's Garden: A Middle Dorset Palaeo-Eskimo Site at Port au Choix, Newfoundland. Master's thesis, Department of Anthropology, Memorial University of Newfoundland, St. John's.

Evans, H. 1991. Stress Indicators of an Archaic Indian Population. Honour's dissertation, Department of Anthropology, Memorial University of Newfoundland, St. John's.

Fitzhugh, W.W. 1983. Archaeological Surveys in the Strait of Belle Isle. In *Archaeology in Newfoundland and Labrador 1982*, ed. J.S. Thomson and C. Thomson, 118–132. Historic Resources Division, Government of Newfoundland and Labrador, St. John's.

Gracie, N. 2004. Function and Manufacture of Tabular Ground Slate Tools from Phillip's Garden, Newfoundland. Honour's dissertation, Department of Anthropology, Memorial University of Newfoundland, St. John's.

Harp, E. 1950. A detective story from the far north. *Dartmouth Alumni Magazine*, November: 17–19.

Harp, E. 1951. An archaeological survey in the Strait of Belle Isle area. *American Antiquity* 16(3): 203–220.

Harp, E. 1964a. Evidence of Boreal Archaic culture in southern Labrador and Newfoundland. In *Contributions to Anthropology 1961-62: Part I*, pp. 184–260. National Museums of Canada Bulletin 193, Ottawa.

Harp, E. 1964b. *The cultural affinities of the Newfoundland Dorset Eskimo*. National Museums of Canada Bulletin 200, Ottawa.

Harp, E. 1970. Late Dorset Eskimo art from Newfoundland. *Folk* 12: 109–124.

Harp, E. 1976. Dorset settlement patterns in Newfoundland and Southeastern Hudson Bay. In *Eastern Arctic prehistory: Paleoeskimo problems*, ed. M.S. Maxwell, 119–138. Washington, D.C.: Society for American Archaeology.

Harp, E. 2003. *Lives and landscapes: A photographic memoir of outport Newfoundland and Labrador, 1949–1963*. Montreal: McGill-Queens University Press.

Harp, E., and D.R. Hughes. 1968. Five prehistoric burials from Port au Choix. *Polar Notes* 8: 1–47.

Hiseler, G. 1997. Harp Seal (Phoca groenlandica) Canine Dentine Incremental Annuli as Indicators of Age and/or Season of Death. Master's thesis, Department of Anthropology, Memorial University of Newfoundland, St. John's.

Hodgetts, L.M. 2005a. Using bone measurements to determine the season of harp seal hunting at Phillip's Garden. *Newfoundland and Labrador Studies* 20(1): 91–106.

Hodgetts, L.M. 2005b. Dorset Palaeoeskimo harp seal exploitation at Phillip's Garden (EeBi-1), northwestern Newfoundland. In *The exploitation and cultural importance of sea mammals*, Proceedings of the 9th Conference of the International Council of Archaeozoology, Durham, August 2002, ed. G. Monks, 62–76. Oxford: Oxbow Books.

Hodgetts, L.M., M.A.P. Renouf, M.S. Murray, D. McCuaig-Balkwill, and L. Howse. 2003. Changing subsistence practices at the Dorset Paleoeskimo site of Phillip's Garden, Newfoundland. *Arctic Anthropology* 40(1): 106–120.

Howley, J.P. 1915. *The Beothucks or Red Indians: The aboriginal inhabitants of Newfoundland.* Cambridge: Cambridge University Press.

Howse, L. 2002. A Comparison of Two Faunal Assemblages from Phillip's Garden (EeBi-1), Port au Choix, Newfoundland. Honour's dissertation, Department of Anthropology, Memorial University of Newfoundland, St. John's.

Jelsma, J. 2000. A Bed of Ochre: Mortuary Practices and Social Structure of a Maritime Archaic Indian Society at Port au Choix, Newfoundland. PhD dissertation, University of Groningen, Netherlands.

Jelsma, J. 2006. Three social status groups at Port au Choix: Maritime Archaic mortuary practices and social structure. In *The Archaic of the far northeast*, ed. D. Sanger and M.A.P. Renouf, 83–103. Orono: University of Maine Press.

Jenness, D. 1925. A new Eskimo culture in Hudson Bay. *The Geographical Review* 15: 428–437.

Jerkic, S. 1993. Burials and bones: A summary of burial patterns and human skeletal research in Newfoundland and Labrador. *Newfoundland Studies* 9(2): 213–234.

Kennedy, B.V. 1980. Marriage Patterns in an Archaic Population: A Study of Skeletal Remains from Port au Choix, Newfoundland. Master's thesis, Department of Anthropology, Memorial University of Newfoundland, St. John's.

Kennedy, B.V. 1981. *Marriage patterns in an Archaic population: A study of skeletal remains from Port au Choix, Newfoundland.* Mercury Series Archaeology Paper 34, National Museums of Canada, Ottawa.

Kennett, B.L. 1990. Phillip's Garden East: An Examination of the Groswater Palaeo-Eskimo Phase. Master's thesis, Department of Anthropology, Memorial University of Newfoundland, St. John's.

Knapp, R.E. 2008. An Analysis of Tabular Slate Tools from Phillip's Garden (EeBi-1), A Dorset Palaeoeskimo Site in Northwestern Newfoundland. Master's thesis, Department of Anthropology, Memorial University of Newfoundland, St. John's.

Lavers, D. 2006. The Groswater Palaeoeskimo Component at Phillip's Garden (EeBi-01), Port au Choix, Newfoundland. Honour's dissertation, Department of Anthropology and Archaeology, Memorial University of Newfoundland, St. John's.

Lavers, D. 2010. The Recent Indian Cow Head Complex Occupation of the Northern Peninsula, Newfoundland: A Geochemical Approach to Investigating Cow Head Chert Acquisition. Master's thesis, Department of Archaeology, Memorial University of Newfoundland, St. John's.

LeBlanc, S. 1996. A Place with a View: Groswater Subsistence-Settlement Patterns in the Gulf of St. Lawrence. Master's thesis, Department of Anthropology, Memorial University of Newfoundland, St. John's.

LeBlanc, S. 2000a. Groswater technological organization: A decision-making approach. *Arctic Anthropology* 37(2): 23–37.

LeBlanc, S. 2000b. Middle Dorset (1900-1100 BP) regional variability on the island of Newfoundland and in Saint-Pierre et Miquelon. In *Identities and cultural contacts in the Arctic*, ed. M. Appelt, J. Berglund, and H.C. Gulløv, 97–105. Copenhagen: The Danish National Museum and Danish Polar Center.

LeBlanc, S. 2008. Middle Dorset Variability and Regional Culture Traditions: A Case Study from Newfoundland and Saint-Pierre and Miquelon. PhD thesis, Department of Anthropology, University of Alberta, Edmonton.

Linehan, D. 1990. The Analysis and Discussion of Four Midden Squares at Phillip's Garden, Port au Choix, Newfoundland. Honour's dissertation, Department of Anthropology, Memorial University of Newfoundland, St. John's.

Lloyd, T.G.B. 1875. On the "Beothucs," a tribe of Red Indians, supposed to be extinct, which formerly inhabited Newfoundland. *Journal of the Royal Anthropological Institute of Great Britain and Ireland* 4: 21–39.

Lloyd, T.G.B. 1876. A further account of the Beothucs of Newfoundland. *Journal of the Royal Anthropological Institute of Great Britain and Ireland* 5: 220–248.

Lyons, D. 1983. Regionalism of Dorset Art Style: A Comparative Analysis of Stylistic Variability in Five Dorset Art Samples. Master's thesis, Department of Anthropology, University of Calgary, Calgary.

Macpherson, J. B. 1995. Palaeoenvironments. In *West coast of Newfoundland field trip guide*, ed. D. Liverman and M. Batterson, 55–61. CANQUA-CGRG meeting, St. John's.

Marshall, K. 1990. Patterns of Activity in a Maritime Archaic Population in Newfoundland Derived from Observations of Activity-Induced Degenerative Joint Disease. Honour's dissertation, Department of Anthropology, Memorial University of Newfoundland, St. John's

Melnik, M.E. 2007. Salmon Net (EfAx-25): An Investigation of Groswater Material Culture Stylistic Variability in Newfoundland. Master's thesis, Department of Anthropology, Memorial University of Newfoundland, St. John's.

Murray, M.S. 1992. Beyond the Laundry List: The Analysis of Faunal Remains from a Dorset Dwelling at Phillip's Garden (EeBi 1), Port au Choix, Newfound-land. Master's thesis, Department of Anthropology, Memorial University of Newfoundland, St. John's.

O'Driscoll, C. 2003. The Application of Trace Element Geochemistry to Determine the Provenance of Soapstone Vessels from Dorset Palaeoeskimo Sites in Western Newfoundland. Master's thesis, Department of Anthropology, Memorial University of Newfoundland, St. John's.

Reader, D. 1990. An Examination of Molar Attrition, Mandibular Torus, and Temporo-mandibular Joint Arthritis: Their Relations and Cultural and Genetic Affiliations in a Maritime Archaic and a Thule Skeletal Group. Honour's dissertation, Department of Anthropology, Memorial University of Newfoundland, St. John's.

Reid, H. 2007. A Study of Southern Variant Maritime Archaic Sites from the Northern Peninsula and Strait of Belle Isle. Master's thesis, Department of Anthropology, Memorial University of Newfoundland, St. John's.

Renouf, M.A.P. 1985. Archaeology of the Port au Choix National Historic Park: Report of 1984 Field Activities. On file, Parks Canada, Archaeology, Atlantic Region, Halifax.

Renouf, M.A.P. 1991. Archaeological Investigations at the Port au Choix National Historic Park: Report of the 1990 Field Season. On file, Parks Canada, Archaeology, Atlantic Region, Halifax.

Renouf, M.A.P. 1992. The 1991 Archaeological Field Season, Port au Choix National Historic Park: Report of Archaeological Excavations On file, Parks Canada, Archaeology, Atlantic Region, Halifax.

Renouf, M.A.P. 1994. Two transitional sites at Port au Choix, northwestern Newfoundland. In *Threads of Arctic prehistory: Papers in honour of William E. Taylor, Jr.*, ed. D. Morrison and J.-L. Pilon, 165–196. Mercury Series Archaeology Paper 149, Canadian Museum of Civilization, Ottawa.

Renouf, M.A.P. 2005. Phillip's Garden West: A Newfoundland Groswater variant. In *Contributions to the study of the Dorset Palaeo-Eskimos*, ed. P.D. Sutherland, 57–80. Mercury Series Archaeology Paper 167, Canadian Museum of Civilization, Gatineau.

Renouf, M.A.P. 2006. Re-investigating a middle phase Dorset dwelling at Phillip's Garden, Port au Choix, Newfoundland. In *Dynamics of northern societies*, Proceedings of the SILA/NABO Conference on Arctic and North Atlantic Archaeology, Copenhagen, May 10th–14th, 2004, ed. J. Arneborg and B. Grønnow, 119–128. Publications from the National Museum Studies in Archaeology and History Vol. 10, Copenhagen.

Renouf, M.A.P. 2009. 2008 Field season at Phillip's Garden, Port au Choix National Historic Site. *Provincial Archaeology Office Archaeology Review* 7: 133–135.

Renouf, M.A.P., and M. Murray. 1999. Two winter dwellings at Phillip's Garden, a Dorset site in northwestern Newfoundland. *Arctic Anthropology* 36(1–2): 118–132.

Renouf, M.A.P., and T. Bell. 1998. Searching for the Maritime Archaic Indian Habitation Site at Port au Choix, Newfoundland: An Integrated Approach using Archaeology, Geomorphology and Sea Level History. On file, Provincial Archaeology Office, Government of Newfoundland and Labrador, St. John's.

Renouf, M.A.P., and T. Bell. 2000. Integrating sea level history and geomorphology in targeted archaeological site survey: The Gould site (EeBi-42), Port au Choix, Newfoundland. *Northeast Anthropology* 59: 47–64.

Renouf, M.A.P., and T. Bell. 2009. Contraction and expansion in Newfoundland prehistory, AD 900-1500. In *The northern world AD 900–1400*, ed. H. Maschner, O. Mason, and R.J. McGhee, 263–278. Salt Lake City: University of Utah Press.

Renouf, M.A.P., T. Bell, and J. Macpherson. 2009. Hunter-gatherer impact on Subarctic vegetation: Amerindian and Palaeoeskimo occupations of Port au Choix, northwestern Newfoundland. *Arctic Anthropology* 46(1–2): 176–190.

Ryan, K. 1997. Groswater Palaeo-Eskimo Tool-makers: Phillip's Garden West and Beyond. Honour's dissertation, Department of Anthropology, Memorial University of Newfoundland, St. John's.

Storå, J. 2000. Skeletal development in the grey seal *Halichoerus grypus*, the ringed seal *phoca hispida botnica*, the harbour seal *Phoca vitulina vitulina*, and the harp seal *Phoca groenlandica*: Epiphyseal fusion and life history. *Archaeozoologia* 11: 199–222.

Teal, M.A. 2001. An Archaeological Investigation of the Gould Site (EeBi-42) in Port au Choix, Northwestern Newfoundland: New Insight into the Recent Indian Cow Head Complex. Master's thesis, Department of Anthropology, Memorial University of Newfoundland, St. John's.

Tuck, J.A. 1970. An Archaic Indian cemetery in Newfoundland. *Scientific American* 226: 112–121.

Tuck, J.A. 1971. An Archaic cemetery at Port au Choix, Newfoundland. *American Antiquity* 36: 343–358.

Tuck, J.A. 1976. *Ancient people of Port au Choix*. Newfoundland Social and Economic Studies 17, Institute of Social and Economic Research, Memorial University of Newfoundland, St. John's.

Tuck, J.A. 1978. Excavations at Cow Head, Newfoundland: An interim report. *Études/Inuit/Studies* 2(1): 138–141.

Wells, P.J. 1988. An Analysis of the Butchery Practices of Two Palaeo-Eskimo Groups. Honour's dissertation, Department of Anthropology, Memorial University of Newfoundland, St. John's.

Wells, P.J. 2002. An Analysis of Faunal Remains from Two Groswater Palaeoeskimo Sites at Port au Choix, Northwestern Newfoundland: Phillip's Garden West (EeBi-11) and Phillip's Garden East (EeBi-1). Master's thesis, Department of Anthropology, Memorial University of Newfoundland, St. John's.

Wells, P.J. 2005. Animal exploitation and season of occupation at the Groswater Palaeoeskimo site of Phillip's Garden West. *Newfoundland and Labrador Studies* 20(1): 75–90.

Wells, P.J. 2009. Constructed seascapes of Newfoundland inshore fishers: a model for precontact small scale societies. In *On the track of the Thule culture, from Bering Strait to east Greenland*, Proceedings of the SILA Conference "The Thule Culture – New Perspectives in Inuit Prehistory," ed. B. Grønnow, 105–118. Publications from the National Museum Studies in Archaeology and History Vol. 15, Copenhagen.

Wheatley, K.D. 2004. The Party Site (EeBi-30) and Beyond: An Interpretation of Groswater Mobility and Landscape from Port au Choix, Newfoundland and Labrador. Master's thesis, Department of Anthropology, Memorial University of Newfoundland, St. John's.

Wintemberg, W.J. 1939. Eskimo sites of the Dorset culture in Newfoundland, Part I. *American Antiquity* 5(1): 83–102.

Wintemberg, W.J. 1940. Eskimo sites of the Dorset culture in Newfoundland, Part II. *American Antiquity* 5(4): 309–333.

# Chapter 2
# By Land and Sea: Landscape and Marine Environmental Perspectives on Port au Choix Archaeology

**Trevor Bell and M.A.P. Renouf**

## Introduction

This chapter provides a palaeoenvironmental context for prehistoric human settlement at Port au Choix. It describes the physical setting of the Port au Choix region and documents the changes in landscape, climate and vegetation that have occurred over the period of human occupation, that is, over the last 6,000 years. Using terrestrial and marine proxy climate records, chronologies of climate change are compared with the settlement history of six different cultural groups.

## Physical Setting

Port au Choix is a coastal community on the west side of the Northern Peninsula of northwestern Newfoundland. It occupies sheltered parts of the Point Riche and Port au Choix peninsulas, which extend 1–2 km into the Gulf of St. Lawrence, and straddles the isthmus linking these bedrock promontories with the Northern Peninsula coast (Fig. 2.1).

The local landscape typifies the physiography of the West Newfoundland Coastal Lowland, which generally lies below 50–70 m elevation, has subdued local relief and may extend up to several tens of kilometres inland (Sanford and Grant 1976) (Fig. 2.2). The lowland is underlain by Ordovician-age (490–440 million years old) dolomite and limestone deposited in shallow, tropical shelf seas. Gently inclined rock strata produce low, parallel forested ridges separated by lakes and bogs (Knight 1991a; Knight and Boyce 1984). Near the coast, limestone terrain produces distinctive coastal barrens that are largely devoid of trees, support a high biodiversity of vascular plants and exhibit karst weathering features (Damman 1983) (Fig. 2.3).

T. Bell (✉)
Department of Geography, Memorial University, St. John's, NL, Canada, A1C 5S7
e-mail: tbell@mun.ca

M.A.P. Renouf (ed.), *The Cultural Landscapes of Port au Choix*, Interdisciplinary Contributions to Archaeology, DOI 10.1007/978-1-4419-8324-4_2,

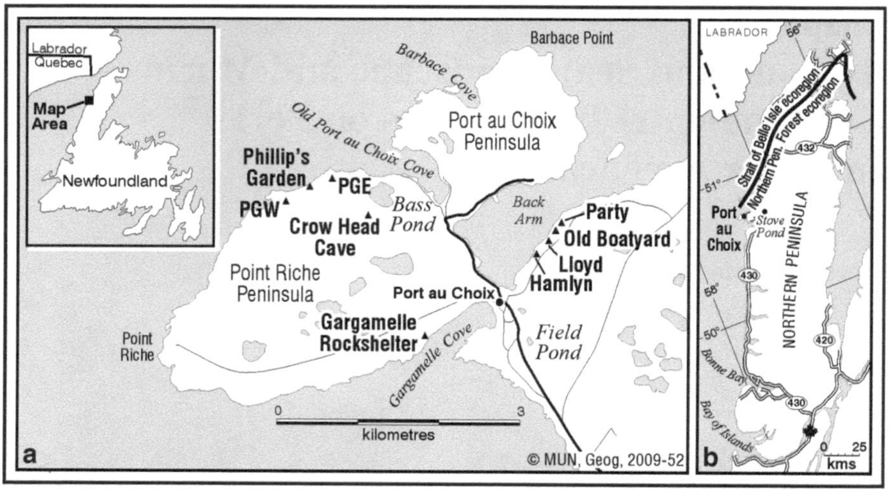


**Fig. 2.1** (**a**) Map of Port au Choix with location of place names mentioned in the text. (**b**) Map of the Northern Peninsula showing ecoregion boundary between Strait of Belle Isle and the Northern Peninsula Forest zones. *PGE* Phillip's Garden East; *PGW* Phillip's Garden West

The lowlands abut the Great Northern Highlands physiographic region, which comprises the Long Range Mountains and the Highlands of St. John northeast of Port au Choix (Sanford and Grant 1976) (Fig. 2.2). The Long Range Mountains form a distinct topographic block along the entire west coast of the island, reaching elevations between 300 and 500 m above sea level (asl) on the Northern Peninsula (Fig. 2.1). They are composed mainly of Precambrian granitic gneisses (up to 1.4 billion years old) that form the southeastern edge of the Canadian Shield (Bostock et al. 1983). The Long Range Mountains show abundant evidence of glacial scouring in the form of ice-moulded bedrock and numerous lake basins (Grant 1994). The Highlands of St. John are composed of Cambrian-age quartzite and limestone (~500 million years old) and produce a particularly imposing landscape as one drives north along the coastal highway (Route 430) because of their sheer western margin, largely unvegetated summits and proximity to the shore of St. John Bay (Figs. 2.1 and 2.2).

## Geology

The Port au Choix region is underlain by Lower and Middle Ordovician platformal rocks of the St. George and Table Head Groups (Knight and Boyce 1984). They consist of limestone and dolostone (carbonate rock that contains a high percentage of the mineral dolomite) deposited in shelf and shallow shoreline settings during several transgression-regression cycles of the tropical platform. Four rock formations

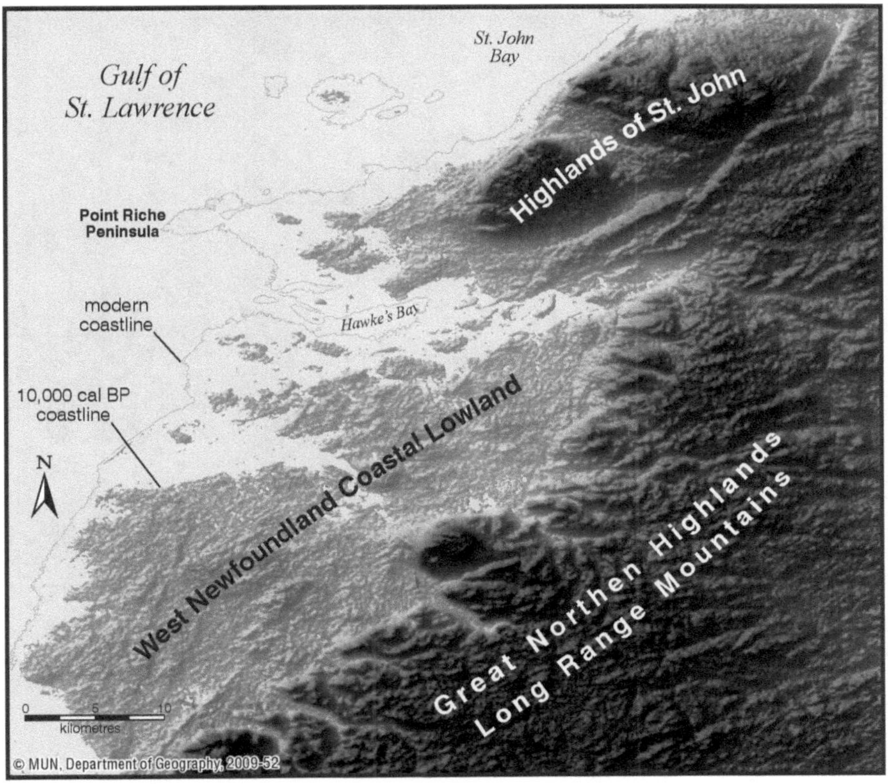


**Fig. 2.2** Digital elevation model of the west-central portion of the Northern Peninsula. The main physiographic regions are clearly demarcated by elevation: the West Newfoundland Coastal Lowland and the Great Northern Highlands. The model ends at 40 m above present sea level, the coastline reconstructed for 10000 cal BP. The modern shoreline is also shown. Source: ASTER (Advanced Spaceborne Thermal Emission and Reflection Radiometer) DGEM (Global Digital Elevation Model) which is a product of Ministry of Economy, Trade and Industry of Japan (METI) and United States National Aeronautics and Space Administration (NASA)

outcrop across the study area; from oldest to youngest, they are: Boat Harbour Formation, Catoche Formation, Aguathuna Formation and Table Point Formation (Knight 1991b) (Fig. 2.4a). The formations are dominated by dolomitic limestone, in places bioturbated and fossiliferous, and dolostone with minor shale, chert pebble conglomerate beds and dolomitic-argillaceous seams (Figs. 2.4a and 2.5). A fault running through Back Arm and Gargamelle Cove offsets bedrock formations to the northeast (Fig. 2.4a).

Wave action and subaerial weathering have exploited weaker lithologies in the bedrock to produce tidal platforms and sea cliffs both around the modern coast and inland, where they are related to higher postglacial sea levels over the past 15,000 years (Fig. 2.5). These coastal terraces were preferentially selected by prehistoric inhabitants either as cliff-top sites with wide seascape views, for example, Phillip's Garden West (EeBi-11) (see Chaps. 4 and 5) (Fig. 2.6), or cliff-bottom sites with

**Fig. 2.3** Photograph of the coastal limestone barrens on the Northern Peninsula. At this site, the barrens consist of frost-shattered limestone bedrock and wave-washed till. Low vegetation is typically confined to mudboils or isolated areas of fine-grained sediment. Isolated shrub thickets of spruce and fir in centre background are locally called tuckamore

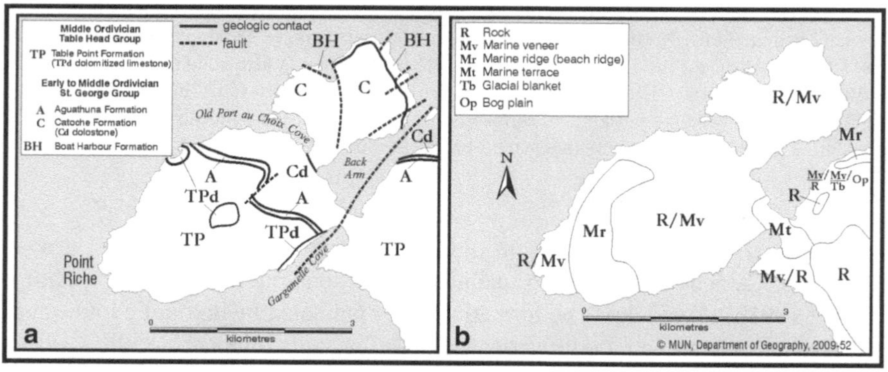


**Fig. 2.4** (**a**) Bedrock geology map of the Port au Choix region, also showing selected faults (after Knight 1991a and 1991b). See text for bedrock description. (**b**) Surficial geology map of the Port au Choix region showing the distribution of deposits by depositional environment and surface morphology. Each category is listed in order of dominance and is separated from the other categories by a slash (e.g. R/Mv). Categories separated by a horizontal dash indicate the relative stratigraphic position of each (e.g. Mv/R represents Marine veneer underlain by Rock). See Proudfoot and St. Croix (2001) for further details

**Fig. 2.5** Oblique photograph of the northern end of the Point Riche Peninsula looking southeast towards Port au Choix (background). The rocky coastline exposes well-bedded dolostone of the Catoche Formation and Aguathuna Formation (*right-hand side* of photograph only). Phillip's Garden (**a**) consists of a series of gravelly beach ridges capped with thin peat and supports a grassy meadow. The remains of at least 68 Dorset dwellings are distributed across the beach ridges. To either side of Phillip's Garden are the Groswater Palaeoeskimo sites of Phillip's Garden West (**b**) and East (**c**). Bass Pond (**d**) is located about 500 m from Phillip's Garden East and was the focus of palaeoenvironmental investigations related to occupation of the Phillip's Garden sites

substantial wind shelter, for example, Hamlyn (EeBi-39), Lloyd (EeBi-41), Old Boatyard (EeBi-43) and Party (EeBi-30) (see Chap. 6) sites in Back Arm (Fig. 2.1). Weathered or wave-eroded limestone caves on the Point Riche Peninsula were used as burial sites by Dorset Palaeoeskimo, for example, Crow Head Cave (EeBi-4) and Gargamelle Rockshelter (EeBi-21) (see Brown 1988, Chap. 12).

The Quaternary geology of the Port au Choix region is largely dominated by postglacial marine sediments and organic deposits (peat) that directly overlie carbonate bedrock (Proudfoot and St. Croix 2001) (Fig. 2.4b). Marine sediments drape much of the landscape and in places occur in greater thickness as ridges and terraces. Raised beach deposits vary widely from sand to pebble gravel to boulders, depending on sediment source and amount of wave action (Fig. 2.7). The Maritime Archaic Indian burial ground was located on a sandy spit and beach ridge complex on the sheltered landward side of the Point Riche Peninsula, a rare extensive sand deposit along the Strait of Belle Isle coast (Fig. 2.8). The centre of the modern town of Port au Choix is built on a sandy marine terrace that emerged from the sea over the past 2,000 years (Bell et al. 2005a) (Fig. 2.8).

**Fig. 2.6** View of Phillip's Garden West from the water showing its geomorphic setting on a bedrock bench (*midground*), backed by vegetated rock cliffs. The site takes advantage of a wide ocean view with local shelter at the rear. The modern beach is backed by an eroding terrace composed of raised marine sediments

**Fig. 2.7** Raised gravel beach sediments at the head of Barbace Cove on the Port au Choix Peninsula

## Climate

The northern two thirds of the Northern Peninsula fall within a single climatic zone named the Northern Peninsula by Banfield (1983). The region has short, cool summers with high average cloudiness and occasional warm days of 25°C and long, cold winters with continuous snow cover duration averaging up to three months (Banfield 1983). Extreme temperature can reach as high as 30°C in the summer and

**Fig. 2.8** Oblique aerial view of the town of Port au Choix straddling the isthmus between the mainland coast of the Northern Peninsula (*foreground*) and Port au Choix and Point Riche peninsulas (*background*). The town is built on raised marine sediments overlying limestone bedrock. The Maritime Archaic Indian burial ground (**a**) is located in sandy beach sediments, whereas the associated habitation site (**b**, Gould site, showing two areas) has a broad areal extent across sandy gravel and weathered limestone. Field Pond (**c**) was the focus of palaeoenvironmental investigations related to occupation of the Gould site (Renouf et al. 2009)

as low as −37.5°C in the winter. The average frost-free season is 120 days with the last spring frost occurring about June 10 (Banfield 1983). Annual precipitation is 900–950 mm near the coast, except 760–900 mm near the Strait of Belle Isle. About 30% of the annual total precipitation falls as snow and the coastal lowlands experience ca. 300 cm of mean annual snow accumulation.

Sea ice is present on average from 60 to 80 days/year off the west-central coast of the Northern Peninsula (e.g. Bay of Islands) (Fig. 2.1) to >120 days/year in the Strait of Belle Isle, first appearing at the beginning of January and disappearing by early May along the Port au Choix coast (Drinkwater et al. 1999). Winter (February) sea surface temperatures (SST) are below 0°C all over the Gulf of St. Lawrence, while summer (August) temperatures are more variable, ranging, for example, from 12–14°C off Port au Choix to 14–16°C off Bay of Islands (Petrie et al. 1996).

## *Vegetation*

The Port au Choix region lies at the coastal boundary of two distinct ecoregions: the Strait of Belle Isle ecoregion and the Northern Peninsula Forest ecoregion (Damman 1983) (Fig. 2.1). To the north, the Strait of Belle Isle region is characterized by rocky coastal barrens largely without forest cover. Isolated shrub thickets of spruce (*Picea glaucea* and *Picea mariana*) and fir (*Abies balsamea*) are common (Fig. 2.3). Peatlands are small, shallow and minerotrophic. The growing season length at 110 days is the shortest on the island.

The Northern Peninsula Forest ecoregion includes the forested lowlands along the west coast of the Northern Peninsula, inland of the coastal barrens. Forest fires are very infrequent (Meades and Moores 1989) and consequently balsam fir (*Abies balsamea*) is the predominant forest tree. Towards the coast, open woodland and heathland (including *Empetrum*, juniper, willows and sedges) are more common. A strong climate-controlled change in the flora of the Northern Peninsula occurs near the southern boundary of the ecoregion where well over 100 species reach their northern limit, including tree species such as white pine (*Pinus strobus*), yellow birch (*Betula alleghaniensis*) and aspen (*Populus tremuloides*) (Damman 1983). Ombrotrophic bogs are common on the extensive marine deposits of the coastal lowlands.

## Cultural Setting

Port au Choix was occupied by a series of Palaeoeskimo and Amerindian populations. There were two Palaeoeskimo cultures, Groswater (dating at 2950–1820 cal BP[1] at Port au Choix) and Dorset (1990–1180 cal BP). Groswater and Dorset Palaeoeskimo sites at Port au Choix are the largest and richest found anywhere in Newfoundland, and indeed, throughout their greater Arctic range (Chaps. 4, 5 and 7). Two Amerindian culture groups were present, Maritime Archaic Indian (dated to 6290–3340 cal BP in Port au Choix) and Recent Indian (dated to 2100–680 cal BP in Port au Choix). An extensive and rich Maritime Archaic burial ground is situated in Port au Choix, indicating the importance of this region to this culture (Chap. 3). Recent Indians comprised three archaeological complexes, the Cow Head complex (dating from 2110 to 1300 cal BP in Port au Choix), the Beaches complex (1390–1180 cal BP in Port au Choix) and the Little Passage complex (1050–680 cal BP in Port au Choix), the latter identified as ancestral to Newfoundland's historic period Beothuk people. The only known Recent Indian ceramic assemblage in Newfoundland is found at Port au Choix, associated with one of only two identified Cow Head complex dwellings to date (Chap. 13; Lavers 2010). The prehistoric cultural heterogeneity at Port au Choix and the substantial nature of many of the sites reflect the region's rich resource base and show how important this area was to these different cultures.

The single most important aspect of the resource base was the large harp seal herds that migrated past Port au Choix twice a year, in December and again in March–April (Chap. 11). All groups exploited those herds to a greater or lesser degree, depending on their particular historical background and culturally determined world view.

---

[1] Except where indicated, all calendar dates in this chapter were calibrated using Calib 6.0html (Stuiver and Reimer 1993) and are represented by the one sigma probability range.

The Palaeoeskimo groups were more specialized seal hunters compared to the Amerindians who folded seal hunting into a more generalized marine-terrestrial subsistence pattern. Since the distribution of harp seal herds was directly tied to sea ice conditions, changes in climate, in particular as they affected sea surface temperature and sea ice distribution, would have directly affected the economic base of all Palaeoeskimo and Amerindian populations living in Port au Choix. Since the economic base was culturally defined, changes in climate would have played out differently in different cultural contexts (Renouf and Bell 2009).

## Palaeoenvironmental Change

### *Landscape History*

Large scale elements of the coastal landscape around Port au Choix are geological in origin and have been in place for tens of millions of years. Modification of that landscape over the last several million years has been primarily by the effects of glaciation. The last Ice Age was largely responsible for creating the extensive, discontinuous, glacially scoured limestone pavement along the west coast of the Northern Peninsula and the distribution of glacial sediment that blankets interior regions and underlies many of the lowland bogs (Grant 1994). Distinctive landforms associated with the final retreat of glacier ice in the region include moraine ridges and eskers. The first people to visit the Strait of Belle Isle region were not directly impeded by the retreating ice, but they would have been aware of its waning presence on uplands in Labrador and the Long Range Mountains.

An important legacy of the last glaciation was the glacio-isostatic adjustment that occurred when ice melted and the earth's crust rebounded. By the time of the first known human presence at Port au Choix shortly before 6,000 years ago, the glacier ice from the last Ice Age had melted and oceans had largely returned to preglacial volumes. In contrast, the earth's crust, and especially its underlying upper mantle, was recovering at a much slower pace, causing local rebound to lag behind ice retreat by thousands of years (Bell et al. 2005a). Postglacial relative sea-level (RSL) adjustments, which combine the effects of both vertical land movements and ocean volume changes, are preserved on the Port au Choix landscape in the form of raised beaches and marine sediments, the latter representing the former seafloor (Figs. 2.4b, 2.5, 2.7 and 2.8).

Radiocarbon-dating of the remains of marine fauna that lived in these higher seas provides a chronology of RSL fall and shoreline migration seaward. In the early postglacial period, RSL fall was relatively rapid – about 2.1 m/century – but declined to ~0.4 m/century between 6200 and 5200 cal BP (Bell et al. 2005a). The impact of RSL fall on coastal configuration and change is largely dependent on the landscape gradient – greatest where the slope is gentle and least where it is steep. For example, the reconstructed coastline at ~10000 cal BP around Hawke's Bay shows significant changes from its modern position and configuration (Fig. 2.2);

the bay extended 7 km farther inland, was surrounded to the north and east by highlands, and had numerous islands within and at the mouth of the bay. In contrast, there are relatively minor changes along the steep coastal topography below the western slopes of the Highlands of St. John (Fig. 2.2).

Two important coastal physiographic changes occurred in the Port au Choix region over the past 6,000 years: the islands that today constitute the Point Riche Peninsula and Port au Choix Peninsula were joined about 3,000 years ago and eventually connected to the mainland around 2,000 years ago (Fig. 2.9). Extensive areas of former seabed emerged from the sea to form coastal lowlands, including those between Barbace Cove and Old Port au Choix Cove, at the western end of Point Riche Peninsula, around Phillip's Garden, and along much of the modern town site of Port au Choix (Fig. 2.9).

The RSL history and evolution of the coastline of Port au Choix are important factors in interpreting the spatial pattern of archaeological sites in the region, although factors other than proximity to shore (e.g. shelter and view) were also important (Bell et al. 2005a). For example, the Maritime Archaic Indian occupation of Port au Choix (6290–3340 cal BP) was largely confined to the Gould site, which at the time was located on the mainland coast across a protected channel from what was then an island (Fig. 2.9). The Gould site afforded open water access to the

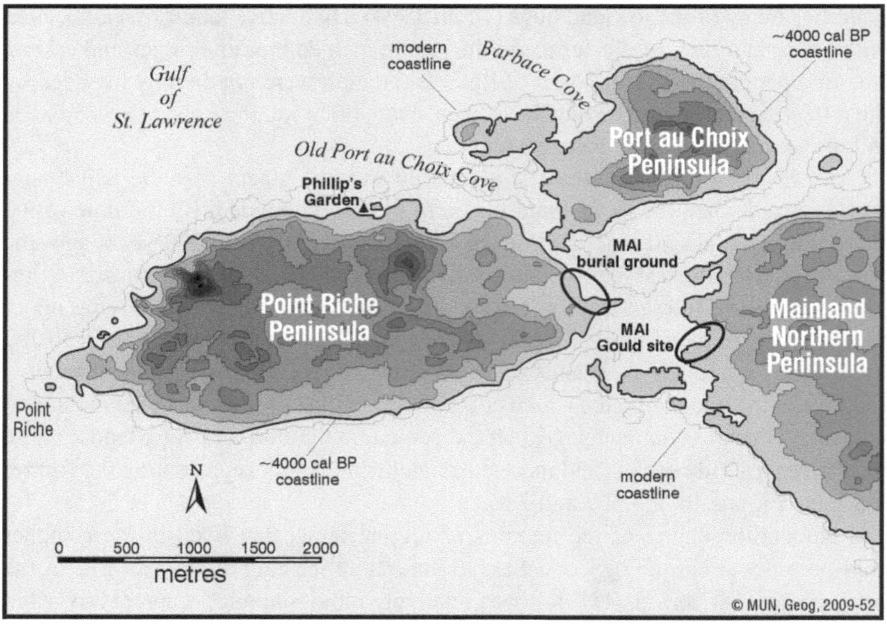


**Fig. 2.9** Palaeogeography of the Port au Choix area at ~4000 cal BP. Note the marked changes in coastal palaeogeography between Barbace Cove and Phillip's Garden, near Point Riche and around the modern town site of Port au Choix. The locations of the Maritime Archaic Indian (MAI) burial ground and Gould site are roughly shown (Source: ASTER GDEM which is a product of METI and NASA)

north, west and south while providing protection from both sea and land-based storms in all directions (Fig. 2.9). Of particular note is the appearance and expansion of a series of islands and eventually a headland that sheltered the interior waterways of Port au Choix from south westerly winds, which have a long fetch over the Gulf of St. Lawrence and would have led to large and choppy seas. The Gould site had a view of the Maritime Archaic Indian burial ground across the water (Fig. 2.9), an important factor in site selection (see Chap. 3).

## *Climate History*

The past 6,000 years of climate history in the Port au Choix region are largely inferred from proxy data sources found in the natural record. For example, local pond water temperature is inferred from fossil insect (*Chironomidae* or non-biting midges) evidence preserved in the sediment that slowly accumulated in the bottom of freshwater ponds (Rosenberg et al. 2005). Similarly, sea surface conditions are reconstructed from preserved phytoplankton (*Dinoflagellata*) remains in nearshore marine sediments (Levac 2003). In both examples, there is a close statistical relationship between the modern distribution of these species and associated environmental conditions, which provides the basis for the use of transfer functions or inference models to reconstruct century-scale changes in past conditions (Rochon et al. 1999; Walker et al. 1997). A similar approach was used by McCarthy et al. (1995) to calculate palaeo-temperatures and palaeo-precipitation from pollen records preserved in two ponds located 400 km south of Port au Choix along the west coast of Newfoundland.

The fossil midge record from Bass Pond on the Point Riche Peninsula (Figs. 2.1 and 2.5) shows fluctuations in midge assemblages, based on the percentage abundances of individual midge species. Major changes in midge community composition were statistically classified into zones that were interpreted to represent significant changes in palaeoclimate, palaeosalinity or both (Rosenberg et al. 2005) (Fig. 2.10). The application of a midge-temperature inference model to the Bass Pond data generated a record of maximum summer lake surface temperatures (LSTs) that are typically 4°C higher than mean July air temperatures (Livingstone et al. 1999). The oldest midge zone overlapping with the earliest occupation of Port au Choix (Zone BPC-3; 6500–4900 cal BP) shows an increase in warm-adapted midge taxa and a sustained warm-water period of between 22 and 23°C, or 4–5°C warmer than present (mean July air temperature for Daniel's Harbour, 50 km to the south, was 14°C during the Canadian Climate Normal period 1971–2000) (Environment Canada 2010). Zone BPC-4, dated between 4900 and 2400 cal BP, shows a decline in warm-adapted species and an inferred cooling of 6°C over the 2,500-year period, with one distinct warming interval centred on 3300 cal BP. The youngest zone (Zone BPC-5; 2400 cal BP to present) begins several centuries before the summer LST minimum at 2100 cal BP (3°C cooler than present) and generally represents a resurgence of temperate midge taxa. The warming trend is

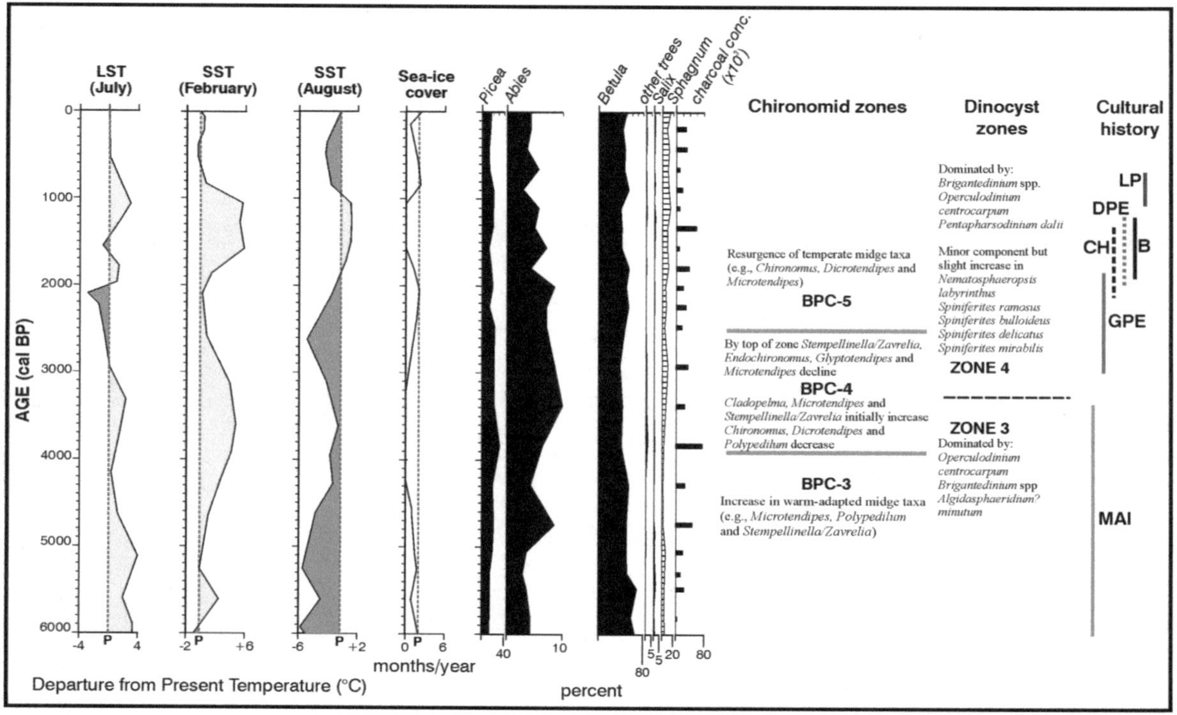


**Fig. 2.10** Inferred lake surface temperature (LST) record for Bass Pond and associated chironomid assemblage zones (Rosenberg et al. 2005), inferred sea surface temperature (SST) and sea-ice cover records for Humber Arm and associated dinoflagellate cyst assemblage zones (Levac 2003), selected pollen and spore taxa percentages and charcoal concentrations from Bass Pond (Bell et al. 2005b) and the chronology of culture occupation of Port au Choix (*MAI* Maritime Archaic Indian; *GPE* Groswater Palaeoeskimo; *DPE* Dorset Palaeoeskimo; *CH* Cow Head complex Recent Indian; *B* Beaches complex Recent Indian; *LP* Little Passage complex Recent Indian). *Vertical dashed line* on Bass Pond and Humber Arm records indicate present conditions (P) and associated inferred temperature and sea ice changes relative to present conditions. See text for discussion

initially abrupt and overall is irregular, attaining a maximum of 21°C at 1100 cal BP (3°C warmer than present), later decreasing to 18°C over the last 500 years. The low rate of accumulation of the top 10 cm of the Bass Pond sediment core precludes finer temporal resolution of this recent part of the record (Bell et al. 2005b).

Reconstructed marine climate for the west coast of the Northern Peninsula is inferred from century-scale, fossil dinoflagellate cyst assemblages preserved in the sediments of the Bay of Islands, a large bay located about 200 km south of Port au Choix (Levac 2003) (Fig. 2.1). The assemblages were divided into four zones, based on the proportions of two dominant species and fluctuations in accompanying species (see Levac 2003) (Fig. 2.10). Only the youngest two zones coincide with human occupation of Port au Choix and the transition between them occurs at 3100 cal BP (Fig. 2.10). Transfer functions based on the distribution of dinoflagellate cysts and sea surface conditions in the modern North Atlantic (Rochon et al. 1999) were used to reconstruct August and February SST and the duration of sea ice cover (SIC) on more than 50% of the sea surface in the Bay of Islands (Fig. 2.10). Sea surface conditions were largely similar to those of the present day between 7000 and 5000 cal BP (August SST 17°C, February SST −0.5°C, 2 months SIC), except for August temperature, which was lower than today by about 3°C. Between 5000 and 1000 cal BP, SST was periodically much higher than today's average, up to 5°C higher in February and 2°C higher in August. SIC was probably absent at 4200–3200 and 1500–1000 cal BP. Winter SST cooled and SIC increased about 1,000 years ago, while August SST showed an increase of 2°C in the last 400 years. The sampling interval (20–30 cm) of the uppermost part of the core does not permit finer analysis of recent climate trends (Levac 2003).

In summary, with the exception of the 5000–4000 cal BP period, there appears to be reasonably close correspondence between palaeoclimate trends derived from proxy lake evidence in Port au Choix and marine fossil evidence in Bay of Islands (Fig. 2.10). Between 6000 and 3200 cal BP, there is progressive winter and summer SST warming, with peak seasonal temperatures reached at about 3600 cal BP and a reduction in SIC to zero. In contrast, there is a steady decline in LST of about 4°C between 5000 and 4000 cal BP, although LST maxima were consistently warmer than present during this period. Between 3200 and 2100 cal BP, there is a synchronous cooling in freshwater and marine systems and a corresponding increase in SIC. Winter conditions declined largely to those of present day, while peak summer temperatures were significantly cooler than present (by 3–4°C). The terrestrial minimum was reached at 2100 cal BP, about 500 years after the summer SST minimum. The period from 2100 to ~1000 cal BP is generally characterized by warming of both marine and freshwater systems with SST maxima maintained between 1500 and 1000 cal BP and a LST maximum at 1100 cal BP. SIC was more or less absent between 1500 and 1000 cal BP. The period 1000 to 500 cal BP was characterized by cooling to near-modern conditions, except for summer SST which was cooler than today over the last 1,000 years. For comparison, McCarthy et al. (1995) resolved a decrease in July air temperature of 1–2°C and an increase in effective precipitation over the past 6,000 years from lake pollen data from southwest Newfoundland.

### *Vegetation History*

The vegetation encountered by the first humans on the coastal lowlands of the Northern Peninsula was largely the same as today, with perhaps more widespread forest and less raised bog. Our knowledge of this past vegetation of the region is derived from pollen grains, spores and macrofossils preserved in lake sediments and bog peat (Bell et al. 2005b; Macpherson 1997). These records are carefully assembled and interpreted to reveal shifts in regional vegetation composition, local pond or bog plants, and forest fire occurrence due to climate change, natural forest disturbances (e.g. fire, insect outbreaks) and paludification (bog development). The vegetation history of the Port au Choix region is based on pollen, spore and charcoal records from Stove Pond, a small 4 ha lake located between moss-dominated bog and forested moraine on the coastal lowlands about 11 km east of Port au Choix (Bell et al. 2005b) (Fig. 2.1). Although other palaeo-vegetation sites have been documented closer to Port au Choix, their records have revealed human disturbance of the natural vegetation (Renouf et al. 2009) and are not discussed here.

Bog inception and expansion coincided with the first documented appearance of Maritime Archaic Indians in the Port au Choix region about 6000 cal BP. Raised groundwater levels facilitated the growth of mosses and the accumulation of peat in former forested depressions. Given that this was around the time of the local postglacial climatic optimum, elevated water tables must have been a response to increased precipitation rather than reduced evaporation. Retreat of the forest cover to well-drained slopes was also accompanied by a gradual shift in composition in response to warmer growing seasons, though a lag in vegetation response is apparent. For instance, balsam fir peaked between 4800 and 2000 cal BP, the maximum of spruce occurred at 3800 cal BP, and the pollen signature of temperate trees, such as Mountain ash, peaked at 4300 cal BP (Fig. 2.10), suggesting an expansion of its northern range from Bonne Bay which is its northern limit today. The decline in sphagnum and other spores of wetland or moist habitat plants between 5000 and 3000 cal BP indicates drier conditions (and probable wetland contraction) and coincides with local peaks in charcoal concentration which signify more abundant forest fire activity and greater summer warmth in the region (Fig. 2.10). After 3000 cal BP, wetland shrubs and herbs rebound, coinciding with cooler summer LST and SST. A decline in tree birch pollen between 1600 and 1100 cal BP is likely a response to the expansion of fir and spruce as summer temperatures rise (Fig. 2.10). A second notable peak in charcoal concentration occurred within this period.

## Environmental Change and Cultural Occupations

In the 1980s, researchers broadly correlated Maritime Archaic Indian and Palaeoeskimo occupations of Port au Choix with warmer and cooler intervals than present, respectively (e.g. Fitzhugh and Lamb 1985; Macpherson 1981).

In the 1990s, Macpherson (1997) attempted to document the palaeoenvironmental evidence to support these correlations. She used pollen composition and marl and shell isotope geochemistry to tentatively identify warm and cold intervals between 5500 and 4500 and 3800 and 2600 cal BP, respectively. Although the chronology of Maritime Archaic Indian and Palaeoeskimo occupations of Port au Choix, as it was known at that time, only narrowly overlapped with Macpherson's proxy temperature record, it stimulated further research on local palaeoenvironmental reconstructions (e.g. Bell and Renouf 1998; Bell et al. 2005b; Renouf and Bell 2000, 2009).

The data used in this current correlation (Fig. 2.10) differ from Macpherson's in several aspects: (i) they are presented in a calibrated radiocarbon timescale; (ii) an expanded radiocarbon and archaeological site database has refined the chronology of prehistoric culture periods at Port au Choix with the most notable modifications being longer occupation of each culture and temporally overlapping cultures; (iii) the proxy palaeoenvironmental records are more varied and avoid those that may be directly impacted by human activities (e.g. Bass Pond) (Bell et al. 2005b); and (iv) they include estimates of SST and SIC from a coastal marine embayment farther south in Bay of Islands with the underlying assumption that marine conditions were broadly similar along this part of the Northern Peninsula coast. Moreover, our growing understanding of the settlement, subsistence and economy of different culture groups in Port au Choix has illustrated the multi-faceted nature of human responses to physical and social environmental change (e.g. Renouf and Bell 2009). Initially, such climate terms as warmer and cooler were primarily interpreted in terms of the extent and duration of pack ice in the Strait of Belle Isle and the opportunities for spring hunting of the Gulf of St. Lawrence harp seal herd as it moved with the northward retreating ice (Macpherson 1997; Renouf 1993; see Chap. 7). We now recognize that climate changes can affect different cultures differently and that changes in one culture can indirectly impact another (Renouf and Bell 2009).

## *Maritime Archaic Indians*

Maritime Archaic Indian occupation of Port au Choix (6290–3340 cal BP) began during the postglacial climatic optimum in the region when summer LST was as much as 2–4°C higher than present and winter SST was close to or above present. SIC duration was 1–2 months. The last 1,000–1,500 years of Maritime Archaic Indian occupancy were characterized by increasingly warmer SST, both winter and summer, and the likely absence of winter sea ice. The summer LST was less consistent during this period with both warmer and cooler episodes, whereas the terrestrial pollen and spore records suggest a period of strong summer warmth, increased forest fire activity and drier wetlands. All of the proxy climate records suggest that the onset of climate cooling began one to two centuries before 3000 cal BP, when Maritime Archaic Indians are no longer recorded in the archaeological record of Port au Choix, or anywhere else in Newfoundland and Labrador.

The Maritime Archaic Indian presence in Port au Choix consists of a burial ground, the Gould occupation site and the component at the Old Boatyard site. Although the burial ground is extensive, containing at least 117 individuals and intermittently used over 1,200 years, the occupation site is interpreted as a field camp solely occupied for mortuary activities associated with the burial ground (Chap. 3). The two sites straddle the inner coastal travel route along the west coast of the Northern Peninsula, and as such, their usage reflects continuous Maritime Archaic mobility and travel along this coast. Any correlation with climate must therefore be interpreted in terms of regional settlement and mobility patterns beyond Port au Choix and perhaps extending to Labrador.

## *Groswater Palaeoeskimos*

Groswater Palaeoeskimo occupation of Port au Choix (2950–1820 cal BP) coincides very closely with (i) a period of cooler summer and winter SST, (ii) the coolest summer LST of at least the past 7,000 years, (iii) SIC conditions similar to present, (iv) a relative increase in birch and a decrease in spruce and fir in the boreal forest, and (v) an expansion of wetlands (Fig. 2.10). Essentially, Groswater arrived in Port au Choix when climate was colder and left when it began to warm up.

At least eight Groswater sites or site components are identified at Port au Choix (Table 1.1, Chap. 1). The two largest are situated on the Point Riche headland where in March–April the retreating edge of the pack ice in the Strait of Belle Isle occurs close to the Port au Choix shore. Harp seals regularly occur in large numbers both on the ice and in the water at this ice edge and are accessible by small boat technology (Chap. 7). Faunal material and hunting equipment confirm that seal hunting was the primary focus of the Groswater sites at Point Riche, although a wide range of bird and small mammals was also exploited in lesser amounts (Renouf 1994; Wells 2005, Chap. 4).

Occupation of Port au Choix by Groswater Palaeoeskimos coincided with cooler climate conditions, specifically lower winter SST and inter-annual winter SIC similar to today. Groswater people likely first arrived in Port au Choix by crossing the Strait of Belle Isle from southern Labrador where their earliest sites date from about 2900 cal BP (Stopp 1997). The first appearance in over 1,000 years of winter SIC in the Bay of Islands was at ~3000 cal BP (Fig. 2.10). These sea ice conditions, and more importantly, the late winter/early spring ice lead may have induced Groswater Palaeoeskimos to extend their range across the Strait in pursuit of harp seal hunting opportunities in Port au Choix and elsewhere along the coast. Their disappearance from the archaeological record of Port au Choix and elsewhere in Newfoundland about 1,000 years later may be interpreted in terms of poorer winter sea ice conditions – increased winter SST and decreased SIC – that affected the duration of the seal herd availability and, of great importance, the predictability of the harp seal hunt.

## *Dorset Palaeoeskimos*

Dorset Palaeoeskimos first appeared in Port au Choix at Phillip's Garden around 2000 cal BP. The site was occupied for approximately 800 years and the remains of at least 68 dwellings have been documented (Chap. 7). Site occupation is divided into three temporal phases: an initial low-to-medium population between 1950 and 1550 cal BP; a period of maximum occupation between 1550 and 1350 cal BP; and a return to a medium occupation level prior to abandonment at 1180 cal BP (Chaps. 7 and 8). The proxy climate records suggest that the initial occupation phase occurred during conditions similar to today, with 1–2 months SIC, but the trend was one of warming (Fig. 2.10). The short period of maximum site occupancy coincides with the warmest episode since Maritime Archaic Indian occupation. Winter SST was up to 5°C warmer than present, SIC was more or less absent and summer LST and SST were above present and increasing (Fig. 2.10). Tree and shrub pollen proportions also indicate warmer growing seasons with increased spruce, fir, alder and shrub birch and decreased tree birch between 1600 and 1100 cal BP (Bell et al. 2005b) (Fig. 2.10). Climate conditions during the final occupancy phase were similarly warm with a continuation of sea ice free winters (Fig. 2.10).

Dorset Palaeoeskimos were specialized seal hunters. Faunal assemblages from the site (Hodgetts et al. 2003; Chap. 13) show that their initial arrival and subsequent establishment at Phillip's Garden were related to access to the late winter/early spring harp seal herds. LeBlanc (1996) and Renouf (Chap. 7) link Phillip's Garden as a prime harp seal hunting locale to the persistence of an ice lead that opens up a short distance offshore which concentrates the harp seals and makes them easily accessible and exploitable in large numbers, as was the case with the Groswater Palaeoeskimos.

Dorset expanded and intensified their use of the Phillip's Garden location during a notable warming trend. On the one hand, this may have affected the ice in such a way that its timing, extent and duration were not as predictable as during the earlier Groswater period when colder winter sea conditions likely resulted in more predictable ice conditions. On the other hand, warmer conditions might have been conducive to maintaining the ice lead that makes this an ideal harp seal hunting location. We have yet to fully understand the impact of changing temperatures on the sea ice conditions and harp seal herds at Port au Choix.

Phillip's Garden continued to be occupied for another couple of centuries. Faunal remains suggest that hunting strategies diversified to include birds and fish, which Hodgetts et al. (2003) argue reflects diminishing intensity of the harp seal hunt. By 1180 cal BP, Phillip's Garden was abandoned. This may be linked to increasingly warm conditions, indicated by a peak at about 1100 cal BP in the Bass Pond chironomid record (Rosenberg et al. 2005) (Fig. 2.10). Renouf and Bell (2009) suggested that this warmth undermined the sea ice conditions and attendant harp seal availability.

## *Recent Indians*

The Cow Head complex and Beaches complex populations experienced the same climate as Dorset Palaeoeskimo populations during their shared occupation of Port au Choix. Cow Head complex people first arrived just before the peak cool period at 2100 cal BP in LST and winter SST and remained until half-way through the subsequent warming of marine conditions and a series of minor temperature fluctuations in summer lake water. The 700-year Beaches complex occupation was about 200 years later than the Cow Head complex occupation and the Beaches people experienced a climate more or less the same or slightly warmer than today's. The Little Passage complex occupation postdates the Beaches complex occupation and the Little Passage people were exposed to a relatively rapid cooling trend in both terrestrial and marine climate, though conditions were generally warmer than today. Their arrival coincides with the return of a winter SIC in Bay of Islands. By 700 cal BP, when the Little Passage complex transitioned into the historic Beothuk period, the modern climate was well established with perhaps the one exception of a slightly cooler summer SST.

Both Cow Head complex and Beaches complex occupations of Port au Choix arrive when marine conditions were close to present, including SIC. Their record ends during the peak in late Holocene SST and when sea ice was largely absent. In contrast, the Little Passage complex populations arrive in Port au Choix as winter SIC re-appears in Bay of Islands and summer climate conditions are cooler than those of today. These Recent Indian occupations therefore overlap with a wide range of climate conditions (the largest range in Port au Choix prehistory), which suggests they may not have been dependent on any one resource linked to a particular climate state. The archaeological evidence supports this view of a generalized marine-terrestrial subsistence pattern for Recent Indian cultural complexes (Chap. 13).

# Conclusions

For the past 6,000 years, prehistoric people strategically located themselves on the landscape of Port au Choix for a variety of reasons: shelter, access to resources, proximity to their ancestral burial grounds, view of migrating seal herds, or position along a regional coastal route. As the landscape changed in response to coastal emergence, new locations were exploited and old ones attracted new settlers. In some cases, the nature of the land was an important added attraction, for example, the suitability of a sandy spit for burial grounds next to a strategic coastal route, as was the case for the Maritime Archaic Indian burial ground.

During its prehistory, Port au Choix experienced climate fluctuations that coincided with major shifts in cultural occupation. The Arctic-adapted Palaeoeskimos settled during cooler periods and the more boreal and temperate-adapted Amerindians during warmer periods. Although this basic correlation is for the most part accurate, the proxy climate and culture history records are more complicated than this broad generalization and we are only beginning to understand past human–environment interactions at Port au Choix.

**Acknowledgements** Multi-year funding has been provided by the Natural Sciences and Engineering Research Council of Canada, the Social Sciences and Humanities Research Council of Canada, Parks Canada, and Memorial University. Of particular note was funding by the New Initiatives Research program of the Office of the Vice-President (Research), Memorial University that initially supported our ongoing interdisciplinary research on the prehistoric landscape and archaeology of the Northern Peninsula. Diagrams were drafted by Charles Conway, Memorial University Geography Department. We thank Alice Kelley and Peta Mudie for their comments on an earlier draft of this chapter.

## References

Banfield, C.E. 1983. Climate. In *Biogeography and ecology of the island of Newfoundland*, ed. G.R. South, 37–106. The Hague: W. Junk.

Bell, T., and M.A.P. Renouf. 1998. Integrating archaeology and sea level history at Port au Choix, northwestern Newfoundland. In *Assessing the state of the environment of Gros Morne National Park*, ed. D.W. Anions and A.R. Berger, 109–113. Ecosystem Science Review Report 011. Parks Canada, Western Newfoundland and Labrador Division, Rocky Harbour.

Bell, T., I.R. Smith, and M.A.P. Renouf. 2005a. Postglacial sea-level history and coastline change at Port au Choix, Great Northern Peninsula, Newfoundland. *Newfoundland and Labrador Studies* 20(1): 9–31.

Bell, T., J.B. Macpherson, and M.A.P. Renouf. 2005b. Late prehistoric impact on Bass Pond, Port au Choix. *Newfoundland and Labrador Studies* 20(1): 107–129.

Bostock, H.H., L.M. Cummings, H. Williams, and W.R. Smyth. 1983. *Geology of the Strait of Belle Isle area, northwestern insular Newfoundland, southern Labrador and adjacent Québec.* Geological Survey of Canada Memoir 400, Ottawa.

Brown, S. 1988. Archaeological Investigations at Crow Head Cave and the Gargamelle Rockshelter in the Port au Choix National Historic Park, Newfoundland: Report of 1986 Field Activities. On file, Parks Canada, Archaeology, Atlantic Region, Halifax.

Damman, A.W.H. 1983. An ecological subdivision of the island of Newfoundland. In *Biogeography and ecology of the island of Newfoundland*, ed. G.R. South, 163–206. The Hague: W. Junk.

Drinkwater, K.F., R.G. Pettipas, G.L. Bugden, and P. Langille. 1999. Climatic data for the northwest Atlantic: A sea ice database for the Gulf of St. Lawrence and the Scotian Shelf. *Canadian Technical Report of Hydrography and Ocean Sciences* 199: 1–134.

Environment Canada. 2010. Canadian climate normals or averages 1971-2000. Electronic document, www.climate.weatheroffice.gc.ca/climate_normals/index_e.html, accessed 27 Jan 2010. National Climate Data and Information Archive, Environment Canada.

Fitzhugh, W.W., and H.F. Lamb. 1985. Vegetation history and culture change in Labrador prehistory. *Arctic and Alpine Research* 17(4): 357–370.

Grant, D.R. 1994. *Quaternary Geology of Port Saunders Map Area, Newfoundland.* Geological Survey of Canada Paper 91-20, Ottawa.

Hodgetts, L.M., M.A.P. Renouf, M.S. Murray, D. McCuaig-Balkwill, and L. Howse. 2003. Changing subsistence practices at the Dorset Paleoeskimo site of Phillip's Garden, Newfoundland. *Arctic Anthropology* 40(1): 106–120.

Knight, I. 1991a. *Geology of Cambro-Ordovician Rocks in the Port Saunders [NTS 12I/11], Castors River [NTS 12I/15], St. John Island [NTS 12I/14] and Torrent River [NTS 12I/10]*

*Map Areas*. Government of Newfoundland and Labrador, Department of Mines and Energy, Geological Survey Branch Report 91-04, St. John's.

Knight, I. 1991b. *Geological Map for the Port Saunders/Torrent River areas, Newfoundland.* Government of Newfoundland and Labrador, Department of Mines and Energy, Geological Survey Map 91-174, St. John's.

Knight, I., and W.D. Boyce. 1984. Geological mapping of the Port Saunders (12I/11), St. John Island (12I/14) and parts of the Torrent River (12I/10) and Bellburns (12I/6) map sheets, northwestern Newfoundland. In *Current research: Report 84-1*, ed. M.J. Murray, J.G. Whelan, and R.V. Gibbons, 114–124. Newfoundland Department of Mines and Energy, Mineral Development Division, St. John's.

Lavers, D. 2010. The Recent Indian Cow Head Complex Occupation of the Northern Peninsula, Newfoundland: A Geochemical Approach to Investigating Cow Head Chert Acquisition. Master's thesis, Department of Archaeology, Memorial University of Newfoundland, St. John's.

LeBlanc, S. 1996. A Place with a View: Groswater Subsistence-Settlement Patterns in the Gulf of St. Lawrence. Master's thesis, Department of Anthropology, Memorial University of Newfoundland, St. John's.

Levac, E. 2003. Palynological records from Bay of Islands, Newfoundland: direct correlation of Holocene paleoceanographic and climate changes. *Palynology* 27: 135–154.

Livingstone, D.M., A.F. Lotter, and I.R. Walker. 1999. The decrease in summer surface water temperature with altitude in Swiss alpine lakes: a comparison with air temperature lapse rates. *Arctic, Antarctic and Alpine Research* 31: 341–352.

Macpherson, J.B. 1981. The development of the vegetation of Newfoundland and climatic change during the Holocene. In *The Natural environment of Newfoundland, past and present*, ed. A.G. Macpherson and J.B. Macpherson, 189–217. Memorial University of Newfoundland, St. John's.

Macpherson, J.B. 1997. Palaeoenvironments. In *CAG '97 field guide: Great Northern Peninsula and southern Labrador*, ed. T. Bell, D. Liverman, S. Evans, and M. Batterson, 46–52. Department of Geography, Memorial University.

McCarthy, F.M.G., E.S. Collins, J.H. McAndrews, H.A. Kerr, D.B. Scott, and F.S. Medioli. 1995. A comparison of postglacial Arcellacean ("Thecamoebian") and pollen succession in Atlantic Canada, illustrating the potential of Arcellaceans for paleoclimatic reconstruction. *Journal of Paleontology* 69(5): 980–993.

Meades, W.J., and L. Moores. 1989. *Forest site classification manual: A field guide to the Damman forest types of Newfoundland.* Canadian Forest Service, Newfoundland and Labrador Region, St John's.

Petrie, B., K. Drinkwater, A. Sandstrom, R. Pettipas, D. Gregory, D. Gilbert, and P. Sekhon. 1996. Temperature, salinity and sigma-t atlas for the Gulf of St. Lawrence. *Canadian Technical Report of Hydrography and Ocean Sciences* 178: 1–256.

Proudfoot, D.N., and L. St. Croix. 2001. *Landforms and Surficial Geology of the Port Saunders Map Sheet (NTS 12I/11).* Newfoundland Department of Mines and Energy, Geological Survey Map 2001-35, St. John's.

Renouf, M.A.P. 1993. Palaeoeskimo seal hunters at Port au Choix, northwestern Newfoundland. *Newfoundland Studies* 9(2): 185–212.

Renouf, M.A.P. 1994. Two transitional sites at Port au Choix, northwestern Newfoundland. In *Threads of Arctic prehistory: Papers in honour of William E. Taylor, Jr.*, ed. D. Morrison and J.-L. Pilon, 165–196. Mercury Series Archaeology Paper 149, Canadian Museum of Civilization, Ottawa.

Renouf, M.A.P., and T. Bell. 2000. Integrating sea level history and geomorphology in targeted archaeological site survey: The Gould site (EeBi-42), Port au Choix, Newfoundland. *Northeast Anthropology* 59: 47–64.

Renouf, M.A.P., and T. Bell. 2009. Population contraction and expansion in Newfoundland prehistory, AD 900-1500. In *The northern world AD 1100-1350*, ed. H. Maschner, O. Mason and R.J. McGhee, 263–278. Salt Lake City: University of Utah Press.

Renouf, M.A.P., T. Bell, and J.B. Macpherson. 2009. Hunter-gatherer impact on Subarctic vegetation: Amerindian and Palaeoeskimo occupations of Port au Choix, northwestern Newfoundland. *Arctic Anthropology* 46(1–2): 176–190.

Rochon, A., A. de Vernal, J.L. Turon, J. Matthiessen, and M.J. Head. 1999. Distribution of recent dinoflagellate cysts in surface sediments from the North Atlantic Ocean and adjacent seas in relation to sea-surface parameters. *American Association of Stratigraphic Palynologists, AASP Contribution Series* 35: 1–152.

Rosenberg, S.M., I.R. Walker, and J.B. Macpherson. 2005. Environmental changes at Port au Choix as reconstructed by faunal midges. *Newfoundland and Labrador Studies* 20(1): 57–73.

Sanford, B.V., and G.M. Grant. 1976. *Physiography, Eastern Canada and Adjacent Areas*. Geological Survey of Canada Map 1399A, 1:2,500,000. Geological Survey of Canada, Ottawa.

Stopp, M.P. 1997. Long-term coastal occupancy between Cape Charles and Trunmore Bay, Labrador. *Arctic* 50(2): 119–137.

Stuiver, M., and P.J. Reimer. 1993. Extended 14C database and revised CALIB radiocarbon calibration program. *Radiocarbon* 35(215): 230.

Walker, I.R., A.J. Levesque, L.C. Cwynar, and A.F. Lotter. 1997. An expanded surface-water palaeotemperature inference model for use with fossil midges from eastern Canada. *Journal of Paleolimnology* 18: 165–178.

Wells, P. 2005. Animal exploitation and season of occupation at the Groswater Palaeoeskimo site of Phillip's Garden West. *Newfoundland and Labrador Studies* 20(1): 75–90.

# Chapter 3
# Across the Tickle: The Gould Site, Port au Choix-3 and the Maritime Archaic Indian Mortuary Landscape

**M.A.P. Renouf and Trevor Bell**

## Introduction

This chapter links two Maritime Archaic Indian (MAI) sites in Port au Choix, the extensive mortuary site of Port au Choix-3 (EeBi-2) and the younger Maritime Archaic component of the Gould site (EeBi-42) (Fig. 3.1). This is the first detailed description of the latter site and we argue that it was an intermittently used field camp connected directly to Port au Choix-3. Understanding the 5,000 year-old coastal topography is fundamental to establishing this connection. We argue that the importance of Port au Choix-3 lay in its key position on the coastal travel route along the Northern Peninsula and that it was an important landmark for Maritime Archaic not just at Port au Choix, but throughout the larger region.

## Context

### *Changing Coastal Topography*

At the time of Maritime Archaic occupation, the Northern Peninsula landscape was much different than it is today (see Chap. 2). On the Northern Peninsula, post-glacial relative sea level (RSL) has been continuously falling (Liverman 1994) so that all sites that were coastal at the time of occupation are today above water, with older coastal sites at higher elevations than younger (Bell et al. 2005a). We estimate that at 5,000 years ago the Port au Choix shoreline was approximately 3–4 m higher than present, submerging the isthmus that today connects both the Port au Choix

M.A.P. Renouf (✉)
Department of Archaeology, Memorial University, St. John's, NL, Canada, A1C 5S7
e-mail: mapr@mun.ca

M.A.P. Renouf (ed.), *The Cultural Landscapes of Port au Choix*, Interdisciplinary Contributions to Archaeology, DOI 10.1007/978-1-4419-8324-4_3,

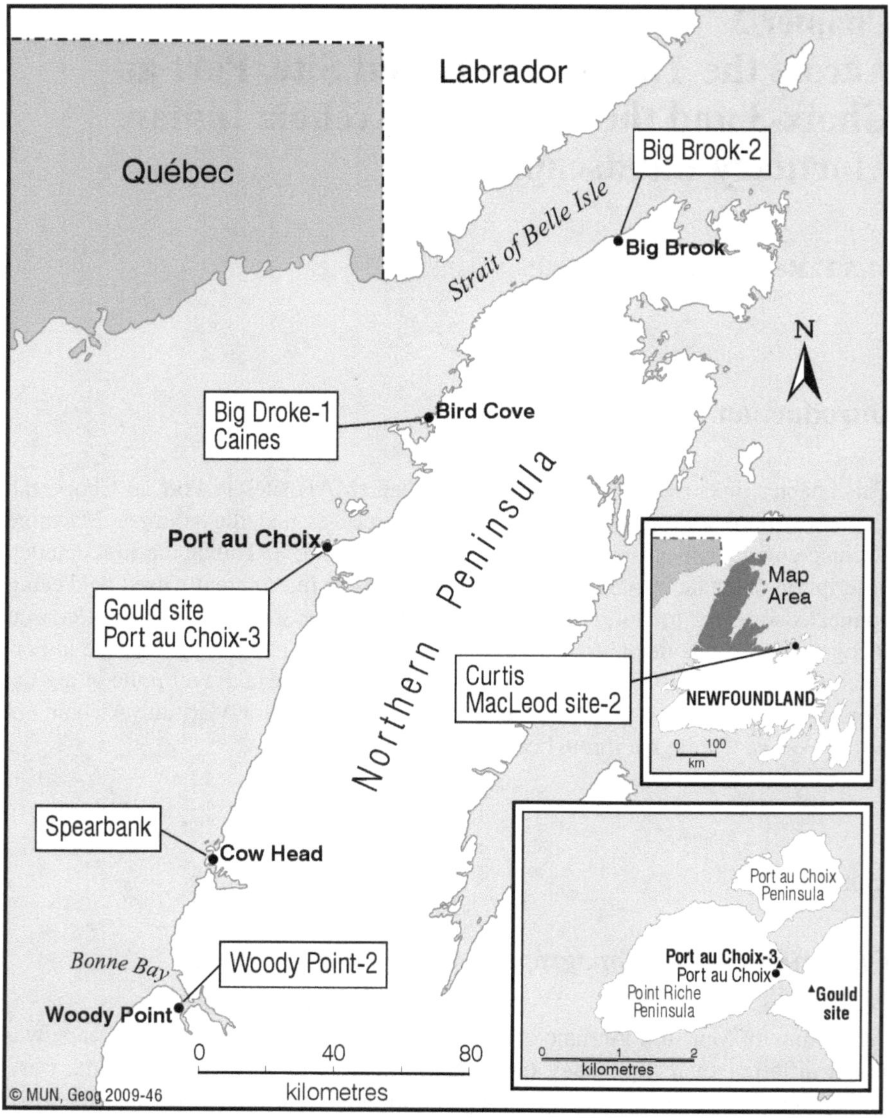


**Fig. 3.1** Place names and site names mentioned in the text

and Point Riche peninsulas to the mainland Northern Peninsula (Renouf and Bell 1999) (Fig. 3.2). This created a single island that was separated from the mainland by a narrow waterway which in Newfoundland and Labrador terminology is called a tickle (DNE 1990:566). The Gould site was on the mainland side of the tickle and Port au Choix-3 was on the island directly across.

**Fig. 3.2** Port au Choix shoreline reconstructed at 5000 BP, showing the Gould site and the extent of Port au Choix-3 as based on local reports

## *Port au Choix-3: A Maritime Archaic Indian Burial Ground*[1]

Much of the significance of the younger Maritime Archaic occupation of the Gould site lies in its potential association with Port au Choix-3, a site with at least three burial loci, each comprising multiple burials. Locus 2 was the largest with 93 individuals in three burial clusters (Jelsma 2000, 2006; Tuck 1976). This site was well published by its excavator (Tuck 1970, 1971, 1976) and is one of only three known MAI burial grounds in Newfoundland. The others are the Curtis site (DjAq-1) and the MacLeod Site-2 (DjAq-7) on North Twillingate Island, on the northeast coast (MacLeod 1967a, 1967b, 1968; Temple 2007; Thibaudeau 1993; Wells and Renouf 2008) (Fig. 3.1). Of the three burial grounds, only Port au Choix-3 had bone preservation. Port au Choix-3 burial loci were on a 4–6 m above sea level (asl) sandy

[1]Although "cemetery" is in common usage when referring to hunter-gatherer mortuary sites, we prefer to use the term "burial ground" which is without assumptions about size, delineation or formalization and which instead focuses attention on location and landscape.

terrace that parallels the modern shoreline in the centre of the town of Port au Choix. Residents tell us that other burials in this area were disturbed by episodes of house construction. We mapped the location of these reports and, if correct, they indicate that burials extended the full length of the terrace, including up to the edge of the 5,000-year-old shoreline closest to the Gould site (Fig. 3.2).

Locus 2 was the main focus of archaeological, osteological, and biochemical analyses. Osteological data indicated that the burial population was healthy and robust (Anderson 1976) and that marriage practices were exogamous (Kennedy 1981). DNA and MtDNA data supported Kennedy's conclusions and isotopic analysis indicated that marine mammals were a significant part of the diet (Jelsma 2000, 2006). Artefactual data suggested that the population was technologically sophisticated and had a world view consistent with ethnographically known hunter-gatherers of the circumpolar north (Tuck 1976). Original (Tuck 1976) and subsequent (d'Entremont 1978; Jelsma 2000; Rothchild 1983) analyses of grave goods and biological data indicated that status was ascribed by age and sex as would be expected of an egalitarian hunter-gatherer population.

An associated habitation site was assumed to exist (Tuck 1976:85), although in several site surveys none was found (Renouf 1985, 1991; Tuck 1976). In 1997 we developed a non-mortuary Maritime Archaic site location model incorporating elements of palaeogeography, geomorphology and the landscape setting of known Maritime Archaic sites in Newfoundland (Renouf and Bell 1999, 2000a). A key factor was the island location of Port au Choix-3. For many hunter-gatherers, islands were favoured burial locations and, in addition, burial grounds were commonly separate from habitation areas (Littleton and Allen 2007; Robinson 2006; Walthall 1999). Therefore, we focused our attention away from what had once been an island and onto what had once been the nearby mainland. Since in other Archaic contexts a view of a burial ground was important (Charles and Buikstra 2002), we further focused on mainland locations from which Port au Choix-3 could clearly be seen. One such area had a number of geographic attributes frequently associated with Newfoundland Maritime Archaic sites: level and well-drained ground, shelter from prevailing winds, and proximity to a stream. This area was a partially wooded terrace, set back from the current shoreline, with up to 2 m of peat overlying limestone gravel substrate. At 5,000 years ago, this would have been a partially vegetated terrace close to the shore (Renouf and Bell 1998). In 1997 we intensively tested this wooded area and beneath the peat found the Gould site (Renouf and Bell 2000a), which we excavated between 1998 and 2000 (Renouf and Bell 1999, 2000b, 2001).

## *Northern Peninsula Maritime Archaic Sites*

There are at least 15 Maritime Archaic sites or site components identified on the Northern Peninsula (Renouf and Bell 2006:20), few of which have been excavated (Fig. 3.1). In addition to the Gould site and Port au Choix-3 these are: Big Droke-1 (EgBf-11) and the Caines site (EgBf-15) in the community of Bird Cove; Big

Brook-2 (EjBa-2) in the community of the same name; the Spearbank site (DlBk-1) in the community of Cow Head; and Woody Point-2 (DjBl-4) in the community of that name. The Caines site, Big Brook-2 and Spearbank are lithic workshop sites and Woody Point-2 has a workshop component. Big Droke-1 is the only site identified as a non-workshop habitation. Big Droke-1, Caines, Spearbank and Big Brook-2 have unformalized and extensive hearth features characterized by fire-coloured soil and some fire-cracked rock.

Big Droke-1 and the Caines site are within 70 m of each other on a heavily forested terrace at 8–10 m asl (Reader 1999). At the time of Maritime Archaic occupation, this was the sheltered coast of a small offshore island (Renouf and Bell 2006:37). Big Droke-1 is about 500 $m^2$ and eight charcoal-based radiocarbon dates range from 4530±60 BP (Beta 108599) to 3470±50 BP (Beta 120398) (Reid 2008:35). Although Reader originally reported 434 artefacts from Big Droke-1, Reid's (2008) re-examination of the assemblage reduced the total to 78. This discrepancy can in part be attributed to initial field misidentifications and Reid's exclusion of 227 utilized flakes and 21 whetstones which were part of Reader's (1999:11) original tally. There were 12 hearths defined on the basis of charcoal over pink, burned subsoil and the presence of small amounts of fire-cracked rock (Reader 1999:7). There was a 9 $m^2$ midden which included a small amount of burned bone, some of which was tentatively identified as bird (Reader 1999:12). Reader (1999:12) interpreted Big Droke-1 as a palimpsest of short-term domestic occupations.

The Caines site is 40 $m^2$ and was characterized by extensive burning defined by charcoal and discoloured soil; two hearths showed evidence of heat-treated lithics. Two dates from these hearths are 3600±60 BP (Beta 108562) and 3490±80 BP (Beta 108562) (Reader 1998:11). In addition to seven hearths, there was a flake concentration and two caches, one of which consisted of biface blanks and preforms (Reader 1998:12). Reid's (2008) re-examination of the assemblage totalled 100 artefacts, close to Reader's original count. Based on the extensive hearths, the wide range of lithic manufacturing debris and evidence of heat-treating, Reader interpreted this site as a lithic workshop and speculated that it might have been connected to Big Droke-1 (Reader 1999:2).

Big Brook-2 is situated on a heavily wooded 6–8 m asl terrace set back about 255 m from the current beach and <100 m from Big Brook itself (Beaton 2004). At the time of occupation, the site was situated on a long narrow point that separated the seashore from the river. The site was originally found in a small clearing (Carignan 1975; Renouf and Bell 2002), which became the focus of subsequent excavation (Beaton 2004). In addition to finding flakes and cores in test pits throughout the clearing, debitage was found in several test pits in the woods, suggesting that the site is quite large. Throughout the 66 $m^2$ excavation area, there were many flakes and cores that together represent the early stages of tool manufacture (Beaton 2004:81). There was extensive burning characterized by pink, black and grey discoloured soil and scattered fire-cracked rock. Beaton defined five hearths and three activity areas. Two dates are within Maritime Archaic temporal range, 4090±40 (Beta 177106) and 3820±40 BP (Beta 171715), and a third date of 2830±40 BP

(Beta 171714) may represent a later Intermediate Indian site component (Beaton 2004:118). A total of 545 artefacts was found, but there were few formalized tools. Cores, hammerstones, preforms, linear flakes and pieces of raw material comprised 89% of the total (Beaton 2004:90). There were >27,000 flakes, most primary and secondary (Beaton 2004:81). Beaton (2004:78, 102) interpreted Big Brook-2 as a lithic workshop and suggested that the source of the predominant raw material was at the nearby coast and the source of the secondary raw material was upriver. He acknowledged that his interpretation did not necessarily characterize the unexcavated remainder of what appeared to be a very large site.

Spearbank is a stratified multi-component site (Tuck 1978). The Maritime Archaic component was a small area on the upper terrace, at approximately 8–10 m asl. There were four unformalized hearths and a workshop area defined on the basis of many core fragments, flakes and preforms (Reid 2008). One hearth was dated at 4130 ± 150 BP (DAL 326) (Reid 2008:66). There were few finished tools among the 160 artefacts (Reid 2008:68). Tuck (1978) and Reid (2008) interpreted the Maritime Archaic component of Spearbank as a workshop and quarry occupation. Raw material from this component was visually similar to chert cobbles on the beach.

Woody Point-2 is on a 2–4 m asl terrace paralleling the shoreline of Woody Point (Provincial Archaeology Office Site Database 2009). Its relatively low elevation is consistent with the RSL of this area where submergence followed initial post-glacial emergence (Liverman 1994). This resulted in Maritime Archaic sites that are today at elevations very close to their original coastal position during occupation (Bell and Renouf 2003). Woody Point-2 had been significantly disturbed by gardening and modern construction activities. Schwarz and Skanes (2005) excavated 27 $m^2$ as part of an archaeological assessment prior to proposed construction of an access road and parking area; they estimated that originally the site had been quite large. They found thousands of pieces of debitage and at least 80 artefacts including cores, bifaces, ground slate axes, bayonets and lances as well as slate blanks and preforms. They concluded that slate tool manufacturing took place at the site; because water-worn slate was naturally present on the beach, they inferred that the source was local (Schwarz 2009, personal communication).

## The Gould Site

The Gould site is on a broad, level 8–10 m asl terrace. A stream flows south from Field Pond (Fig. 3.3) to the ocean about 350 m away. At the time of Maritime Archaic occupation, the Gould site was close to the shoreline and had a clear view across the tickle to the island beyond.

The site has two cultural and four temporal components. The younger two components are Recent Indian Cow Head occupations (see Chap. 13) and the older components are Maritime Archaic occupations. The focus of this chapter is the younger of the two Maritime Archaic components which are situated primarily on the north side of the stream (Fig. 3.3). Eight associated radiocarbon dates range from 4060 ± 80 BP

**Fig. 3.3** Aerial view of the Gould site showing Field Pond and the four main activity areas described in the text; view is looking southwest

to 3200±100 BP (Table 3.1). The older Maritime Archaic component is on the south side of the stream and three associated radiocarbon dates give a radiocarbon age at or close to 5430±50 BP (Beta 148518). This component is described in Renouf and Bell (2000b) and Reid (2008). In all site areas, soil stratigraphy consists of up to 2 m of peat over limestone gravel. Maritime Archaic components are on and slightly above the basal stratum, which suggests that during Maritime Archaic occupation the area was a partially vegetated limestone terrace (Renouf and Bell 1998).

The younger Maritime Archaic component extended over approximately 0.7 ha and comprised four concentrations of cultural material with some additional cultural material scattered in between.

Area 1 (Fig. 3.3) comprised three Maritime Archaic soil levels. From upper to lower, Level 3c was the compact bottom of the peat overburden; cultural material in this level was found at the interface with Level 4. Level 4 was a dark clay-rich soil layer immediately above the limestone substrate; cultural material in this level was found within it and at the interface with Level 5. Level 5 was the weathered limestone substrate, consisting of angular beach gravel in a matrix of yellow-brown clay; cultural material was found on top of and within this matrix.

There was little cultural material in Level 3c aside from a few flakes, charcoal, a few fire-cracked rocks and a small piece of preserved wood.

In Level 4, there was some fire-cracked rock, two small hearths, scattered charcoal, five charcoal concentrations and a cut and partially burned spruce log which was dated to 3450±70 BP (Table 3.1, Fig. 3.4a); four smaller pieces of wood were

**Table 3.1** Radiocarbon dates from the younger MAI occupation of the Gould site and Port au Choix-3. Calibrations using Calib 6.0html (Stuiver and Reimer 1993)

| Context | C14 years BP | Lab No. | Technique | Material | Median age cal BP | Cal BP 1σ | Cal BP 2σ |
|---|---|---|---|---|---|---|---|
| MAI younger component | 3200±100 | Beta 132364 | Radiometric | Charcoal | 3430 | 3560–3340 | 3690–3160 |
| MAI younger component | 3270±50 | Beta 108099 | AMS | Charcoal | 3500 | 3560–3450 | 3620–3390 |
| MAI younger component | 3420±60 | Beta 134153 | Radiometric | Charcoal | 3680 | 3820–3580 | 3840–3490 |
| MAI younger component | 3450±70 | Beta 120795 | Radiometric | Wood | 3720 | 3830–3640 | 3900–3490 |
| MAI younger component | 3460±50 | Beta 134148 | AMS | Charcoal | 3730 | 3830–3640 | 3850–3590 |
| MAI younger component | 3740±50 | Beta 107795 | Radiometric | Charcoal | 4100 | 4210–3990 | 4240–3930 |
| MAI younger component | 3850±100 | Beta 121295 | Radiometric | Charcoal | 4260 | 4420–4100 | 4520–3980 |
| MAI younger component | 4060±80 | Beta 146081 | Radiometric | Charcoal | 4570 | 4800–4430 | 4830–4300 |
| Port au Choix-3, Locus 4 | 3230±220 | I 4380 | Radiometric | Bark | 3460 | 3820–3160 | 3980–2870 |
| Port au Choix-3, Locus 2 | 3370±80 | Y 2608 | Radiometric | Charcoal[a] | 3620 | 3700–3480 | 3830–3440 |
| Port au Choix-3, Locus 2 | 3690±90 | I 4682 | Radiometric | Charcoal[a] | 4040 | 4150–3900 | 4380–3730 |
| Port au Choix-3, Locus 2 | 3777±44 | AA 33919 | AMS | Beaver incisor | 4150 | 4230–4090 | 4340–3990 |
| Port au Choix-3, Locus 2 | 3825±60 | AA 29482 | AMS | Caribou bone | 4230 | 4380–4100 | 4420–4110 |
| Port au Choix-3, Locus 2 | 3826±38 | AA 33920 | AMS | Caribou bone | 4230 | 4290–4150 | 4410–4090 |
| Port au Choix-3, Locus 2 | 4290±110 | I 3788 | Radiometric | Charcoal[a] | 4870 | 5040–4630 | 5280–4530 |

Sources: Beta # (Renouf and Bell 2001, 2006); I #, Y # (Tuck 1976; CAA Radiocarbon Database); AA # (Robinson 2006)

[a] All samples run from same charcoal sample (Tuck 1976:162)

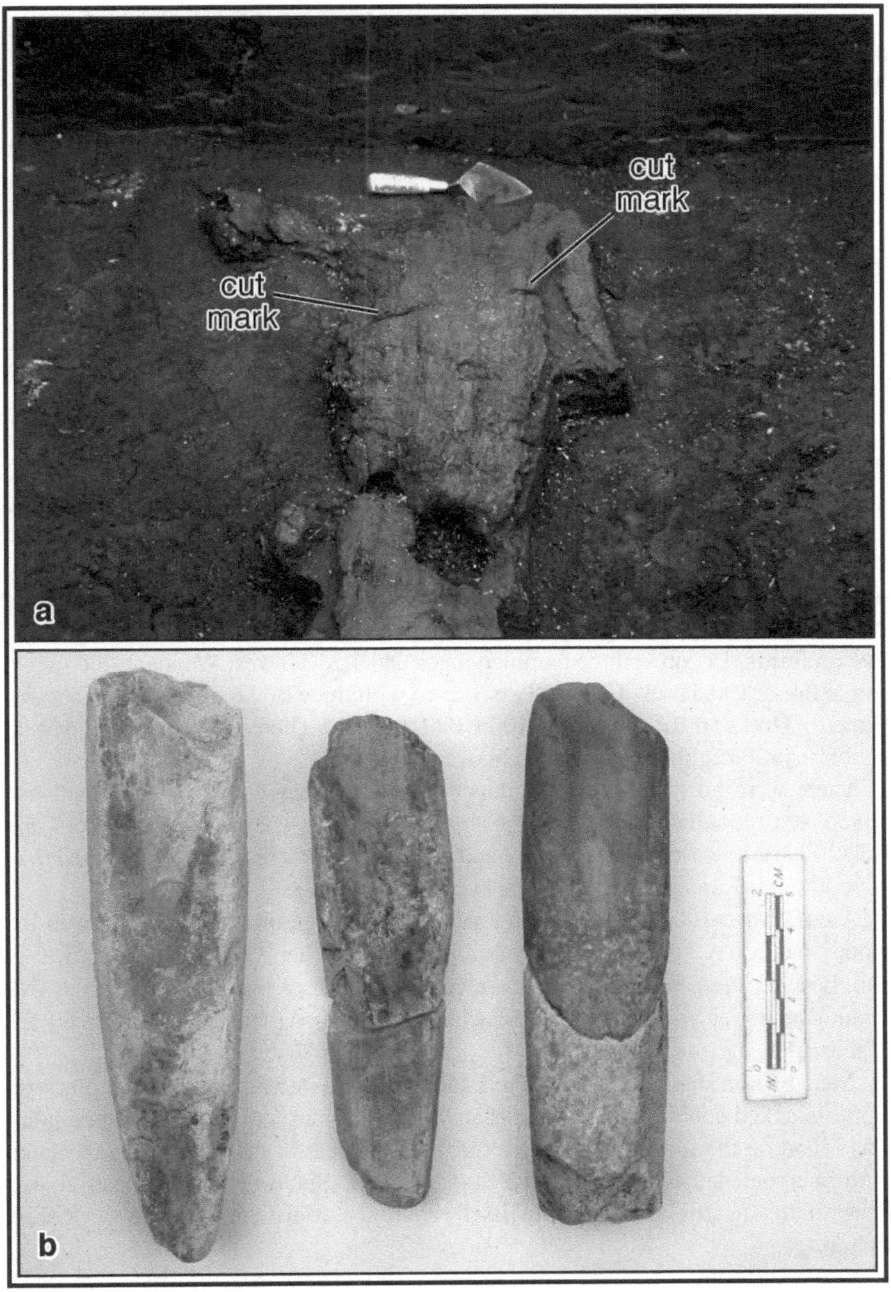


**Fig. 3.4** (**a**) Cut and partially burned spruce log from the first activity area; cut marks are indicated. (**b**) Heavily worn gouges found near the cut log

likely associated. A well-worn gouge was found near the log and two similar gouges were found a few metres away (Fig. 3.4b). Nearby, there was a small refuse dump, Feature 1, dated to 3270±50 BP (Table 3.1). This feature included a small amount of highly fragmented bones, some of which we identified as bird and fish. There were some flakes, a nodule of red ochre, at least two highly fragmented, small, barbed bone points (Fig. 3.5a) and a concentration of 56 pea-sized white quartz pebbles (Fig. 3.5b).

Within Level 5, there were three small hearths, one of which was dated to 3850±100 BP (Table 3.1).

Other cultural material in Levels 4 and 5 included 195 flakes, three bifaces, three cores, three hammerstones, one scraper and two chipped stone tool fragments. There were four small, smooth, round or oval pebbles that were white, grey or brown. Since these kinds of pebbles do not occur naturally within the gravel substrate, we interpret them as manuports, that is, natural objects that were moved from their original context (Fig. 3.6a).

Area 2 was about 55 m to the southwest, near the stream bank (Fig. 3.3). Here the stratigraphy differed from Area 1: there was a brown sandy layer, Level 6, found above a substrate of rounded limestone gravel, Level 7. Almost all cultural material occurred in Level 6. This included fragments of four projectile points found within a few cm of each other (Fig. 3.6b), in addition to eight bifaces and biface fragments, five preforms, 14 cores, three hammerstones and 1,240 flakes. We identified only a single fire-cracked rock. Charcoal associated with three of the projectile points was dated to 3460±50 BP and 3200±100 BP (Table 3.1). There were six round or oval pebbles similar to those mentioned above.

There were no features in this area other than dozens of large and small pits which we originally thought were post-holes based on size and shape (Renouf and Bell 2000b); however, their ubiquity and the absence of a distinct fill indicated they were natural (Renouf and Bell 2001:11).

Area 3 was >100 m northeast of Area 2, at the edge of a small wet area in the woods that today becomes a water-pan in rainy weather. Overlooking this water-pan, beneath the peat, was a well-formed oval hearth of fire-cracked cobbles. No cultural material was found and associated charcoal was dated to 3420±60 BP (Table 3.1).

Area 4 was a small hearth close to Field Pond. This was a small, round platform of fire-cracked cobbles beneath the peat. The base of a small, slate, stemmed point was found on the limestone substrate beneath the hearth; the point was associated with charcoal dated to 3740±50 BP (Table 3.1). Since the dated material was beneath the hearth, the structure itself could be Maritime Archaic or Recent Indian.

Small amounts of cultural material were scattered in between these four areas, including a handful of flakes, a biface, three cores and a hematite nodule. In one location not far from the stream, a charcoal concentration associated with a few flakes was dated to 4060±80 BP (Table 3.1).

Other material found in Maritime Archaic levels throughout the site included 21 pieces of unburned wood. Deal (2005) floted 61 litres of soil from Maritime Archaic

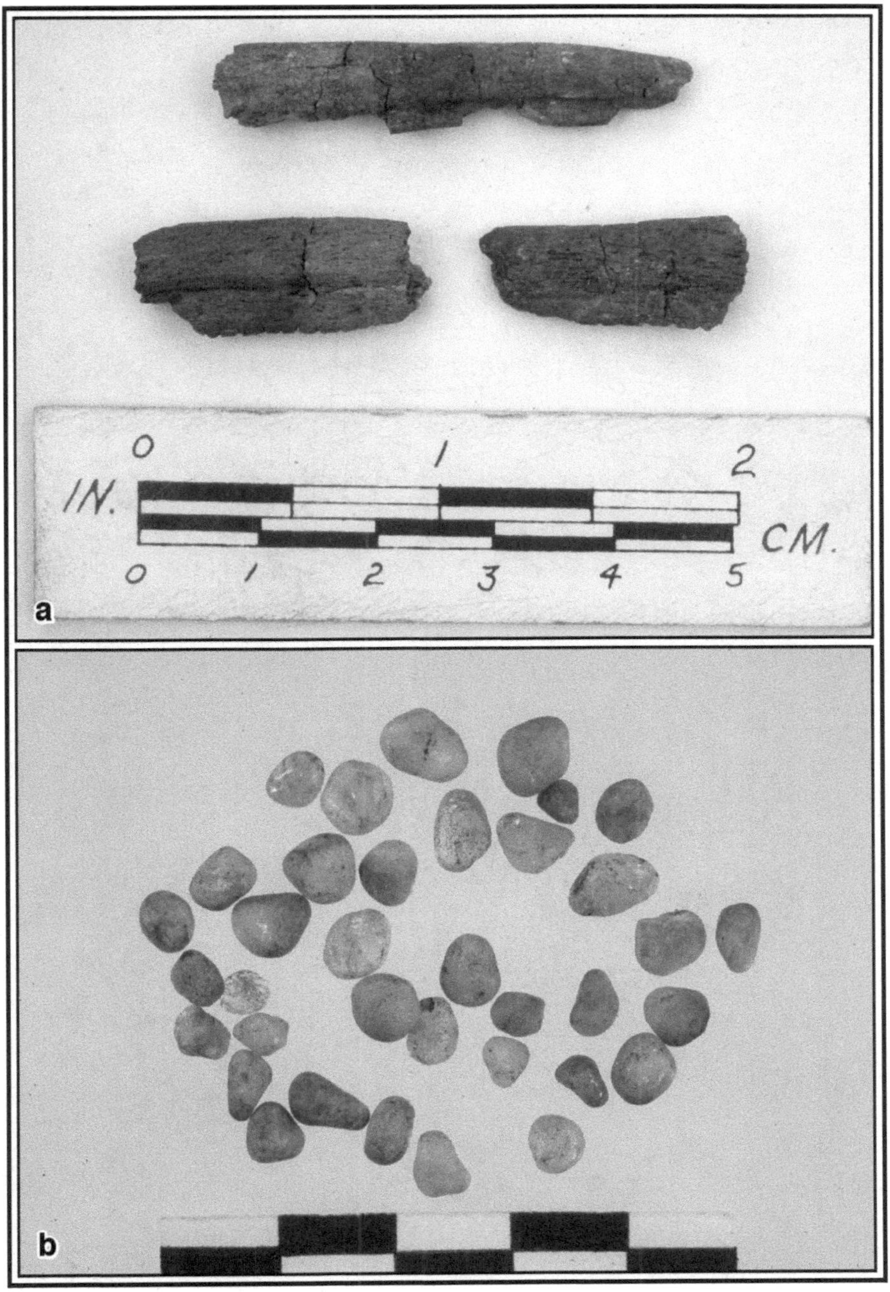


**Fig. 3.5** (**a**) Fragments of two small barbed bone points from Feature 1. (**b**) Pea-sized white quartz pebbles from Feature 1

**Fig. 3.6** (**a**) Smooth, round or oval pebbles from Area 1 and Area 2. Since these do not occur naturally in the substrate, we conclude they are manuports. (**b**) Projectile points found near each other in Area 2. The two outer examples are broken just above the side notches

levels which yielded 3,568 charred spruce or fir needles. A small number of charred seeds from edible berries were present in Maritime Archaic levels: blueberry ($n=1$), raspberry ($n=1$), elderberry ($n=1$) and pin cherry ($n=7$). There were 28 uncharred raspberry seeds which were either a modern intrusion or ancient; the preservation of uncharred wood in Maritime Archaic levels suggested the latter (Deal 2005:145). There were four unburned cinquefoil seeds from Feature 1.

Analysis of Field Pond sediments showed a sustained period of vegetation disturbance temporally overlapping with the younger Maritime Archaic occupation (Bell et al. 2005b; Renouf et al. 2009). The disturbance included a charcoal spike, a decline of tree pollen, an increase in shrub pollen, a decrease in mosses and an increase in aquatic algae. We argued that these disturbances were anthropogenic, caused by local fires, tree-cutting, trampling and increased nutrient input from human activities at or near Field Pond.

In addition to evidence of local fires in these sediments, there was evidence of burning at the site itself: 20 charcoal concentrations, scattered fire-cracked rock and the burned and cut log mentioned above.

## Interpretation

Data from the younger Maritime Archaic component of the Gould site suggest a palimpsest of brief occupations. Few artefacts and features are concentrated in four areas across a large area, with cultural material scattered in between. Seventy artefacts including 11 pebble manuports were recovered in 350 m$^2$ (Table 3.2), which is a density of only 0.2 artefacts/m$^2$ of excavated area. Table 3.2 compares this to artefact densities at four of the five Maritime Archaic sites reviewed above with a range of 1.0/m$^2$ for Big Droke-1 to 6.3/m$^2$ for the Caines site. In comparison, the Gould site artefact density is very low. This could in part be connected to the fact that four of the five sites in the comparison are either lithic workshop sites or have a workshop component, and these kinds of sites tend to have large assemblages that include many cores and preforms. However, the Gould site density is also low compared to Big Droke-1, which is interpreted as an exclusively domestic occupation.

This meagre artefact accumulation occurred over several centuries. Radiocarbon dates (Table 3.1) bracket occupations over a maximum of 1,040 and minimum of 860 radiocarbon years. This suggests that Gould occupations were intermittent and each was brief, perhaps only overnight. This, in turn, suggests that the Gould site was a field camp in the sense of Binford (1980) who described a field camp as a short-term, task-specific site connected to a residential base. We argue that the Gould site was connected, not to a residential base, but to the sacred burial ground that is Port au Choix-3. We base this on the chronological overlap of the sites, the presence of a few similar items at each and their complementary landscape settings.

The radiocarbon dates from the younger Maritime Archaic component of the Gould site and Port au Choix-3 fully overlap (Fig. 3.7). Of the 19 radiocarbon dates from Port au Choix-3 (Jelsma 2000; Robinson 2006; Tuck 1976), we include only

**Table 3.2** Comparison of artefact frequencies and density/m$^2$ from Northern Peninsula MAI sites summarized in the text. Numerous cores were present at Spearbank, but have yet to be counted and catalogued. Data are not available for Woody Point-2

| Artefacts | Gould | Big Droke-1 | Caines | Big Brook-2 | Spearbank |
|---|---|---|---|---|---|
| Core | 22 | 12 | 10 | 110 | |
| Biface | 7 | 21 | 17 | 3 | 2 |
| Projectile point | 4 | 1 | 1 | 2 | |
| Preform | 5 | | 9 | 91 | 87 |
| Bipoint | | | 1 | | 10 |
| Tool blank | | | | | 26 |
| Microblade | 4 | | 4 | | |
| Blade-like flake | | 27 | 2 | | |
| Retouched flake | | | 14 | 8 | 2 |
| Gouge | 3 | 1 | 1 | 1 | |
| Celt | | | | | 1 |
| Unidentified ground stone tool | 2 | 12 | 7 | 10 | 7 |
| Scraper | 2 | | 2 | 3 | 1 |
| Uniface | 2 | | | | |
| Barbed bone point | 2 | | | | |
| Unidentified tool fragment | 2 | | 1 | 11 | |
| Blade | 1 | | | | |
| Hammerstone | 1 | | | 19 | 15 |
| Red ochre | 1 | | | | |
| Abrader | | | | 20 | |
| Plummet | | | | | 1 |
| 56 Pea-sized pebbles | 1 | | | | |
| Pebble manuport | 11 | | | | |
| Total | 70 | 74 | 69 | 278 | 152 |
| Excavated area (m$^2$) | 350 | 73 | 11 | 66 | 100 |
| Artefact density/m$^2$ | 0.2 | 1 | 6.3 | 4.2 | 1.5 |

seven (Table 3.1), three on wood charcoal, one on carbonized bark (Tuck 1976:162),[2] two on caribou bone and one on beaver tooth (Robinson 2006:363). We exclude 12 AMS dates on human bone (Jelsma 2000:191), which would be affected by the marine reservoir effect linked to the marine diet suggested by Tuck (1976) and established by Jelsma (2000). However, Robinson's AMS dates on caribou bone and beaver tooth suggest that the marine reservoir effect is about 300 years (Robinson 2009, personal communication).

The younger Maritime Archaic Gould component has a number of artefacts that could be mortuary-related. Most striking are the 56 pea-sized white quartz pebbles (Fig. 3.5b) found in a concentration in Feature 1 which, following Tuck (1976:71), are not local to Port au Choix and must have been intentionally brought to the area. These tiny pebbles are identical to those found in 21 of the Port au Choix-3 burials where they

[2]In Tuck (1976:162) the radiocarbon dates Y-2608 and I-4380 are misreported, the first as 3,770 ± 80 BP instead of 3,370 ± 80 BP and the second as 4,230 ± 220 BP instead of 3,230 ± 220 BP; we use the dates reported in the CAA radiocarbon website which are based on Wilmeth (1978).

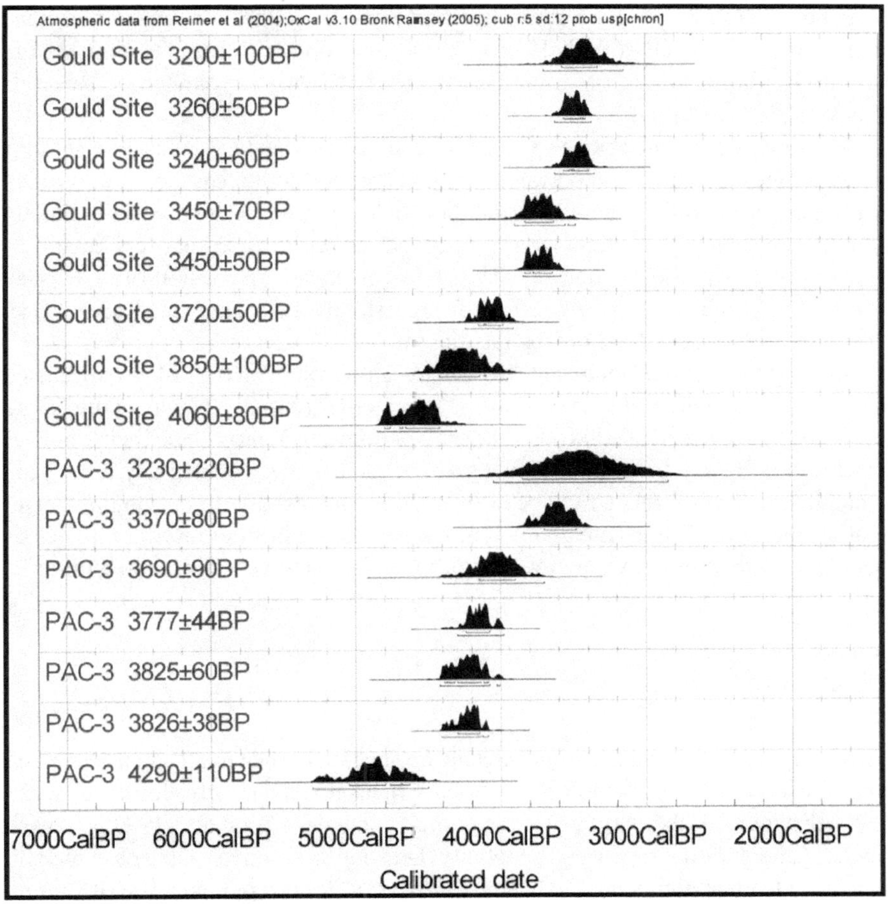


**Fig. 3.7** Overlapping calibrated radiocarbon dates from the younger MAI component of the Gould site and Port au Choix-3. Calibrations by Oxcal 5.2

commonly occurred as concentrations. The concentration of 56 such pebbles in Feature 1 suggests that they were originally in a bag or pouch, ready for use at Port au Choix-3. To speculate further, we wonder if the four cinquefoil seeds found in Feature 1, but not in any of the other 61 flot samples, were meant as a burial offering.

In the Gould site assemblage there are 11 small, smooth, round or oval pebbles (Fig. 3.6a) similar to those found in 13 Port au Choix-3 burials. Tuck suggested those from the burials were likely magic, religious, or decorative items (Tuck 1976:55). He also noted their resemblance to birds' eggs (Tuck 1976:71), which is consistent with the wide array of bird-related items found in most burials, such as merganser heads carved on bone pins, natural or modified limestone concretions resembling birds, swan ulna whistles, great auk beaks, wing skeletal elements, and bones from a variety of 30 different bird species.

There was a nodule of red ochre in Feature 1, a pigment abundantly used in all the Port au Choix-3 burials. Also in Feature 1 were fragments of at least two small bone points (Fig. 3.5a), one with square barbs similar to the small and larger bone points from the burials (Tuck 1976:Plate 30) and one with tiny decorative edge serrations similar to that on a number of Port au Choix-3 ground slate bayonets (Tuck 1976:Plate 21:4–5).

The Gould site and Port au Choix-3 have complementary landscape settings which suggest a connection between them. As previously noted, the burial ground was on an island at the time of use and the Gould site was on the mainland directly across. The sites were simultaneously near and far, <1 km apart but separated by a tickle. There was a clear line of vision between them. The visibility of Port au Choix-3 was enhanced by low piles of large stones that demarcated each burial (Tuck 1971:345) and stood out against the sand.

We propose that the Gould site was a field camp from which a task group carried out mortuary activities at Port au Choix-3. Each task group that occupied the Gould site would have occasioned little in the way of domestic debris since their stay was short and narrowly focused. Activities at the Gould site would likely have been intangible in nature, related to social interactions involved with burying and celebrating the dead. These interactions may have taken place around beach fires as people looked across to where their family members and ancestors were buried.

## Discussion

This interpretation of the Maritime Archaic occupation of the Gould site has implications for how we think about Port au Choix-3. In his review of early Holocene mortuary behaviour in the North American mid-continent, Walthall (1999:3) linked hunter-gatherer burial practices to mobility. Drawing from hunter-gatherer ethnography, he noted that highly mobile hunter-gatherers, or foragers in the sense of Binford (1980), generally had expedient disposal of the dead, such as burial in a midden or a shallow grave at the place of death (Walthall 1999:3). He contrasted this with less mobile, logistically organized, hunter-gatherers, or collectors (Binford 1980), who disposed of their dead in more designated places at a remove from habitations. Logistical organization is the norm for temperate and northern regions which are strongly seasonal with predictable temporal and spatial patterning of resource availability. Walthall (1999:23) argued that the regular seasonal movements of these logistically organized collectors took place within delineated territories which gave them strong connections to places, including burial grounds, to which they regularly returned. He noted that burial grounds were often situated on prominent landforms from which they functioned as visible symbols of ancestral territorial claims. These locations were often in areas of high resource potential such as rivers and coasts which were also significant routes of communication (Walthall 1999:19). According to Walthall (1999:23), burial grounds commonly functioned as a ritual node drawing together an otherwise dispersed population. These population aggregations were made possible by the locality's productive and predictable resources (Walthall 1999:23).

This is consistent with Robinson's (2006) analysis of Late Archaic burial grounds in the Gulf of Maine where he noted that the earliest of these were situated on rivers at prime spots for anadromous fish. They were also on prominent locations, for example, above major waterfalls or major portage routes. He interpreted these burial grounds as signposts at geographical obstructions overlooking major communication routes between distinct interior and coastal Archaic populations (Robinson 2006:356). Robinson (2006:255) considered these burial grounds as proxies for social aggregations.

Littleton and Allen (2007) took a somewhat different approach in their analysis of pre-contact aboriginal burial grounds along the Murray River in Australia. Following Schlanger (1992), they preferred the more fluid concept of persistent place to the commonly used and more static category "cemetery" with implicit notions of territorial ownership. Like Robinson, they interpreted burial grounds as signposts on the landscape, marking symbolically and socially important places along tracks and paths of movement. This is consistent with Zedeňo et al. (1997) who described cultural landscapes not as places and resources, but as networks of relations and connections among nodes of activity which they called landmarks. Some landmarks, such as burial grounds, were more significant than others and exerted a greater pull on that network. Rather than linking burial grounds to social aggregations, Littleton and Allen (2007:295) observed that burial grounds could be created in persistent places by the accumulation of one or a few burials over time, in this way creating a mortuary landmark that structured subsequent use.

Goldstein (2002) and Charles and Buikstra (2002) argued that visibility was an important aspect of a mortuary landmark. Charles and Buikstra (2002:18) observed that Middle and Late Archaic mortuary sites in Illinois were often situated on bluff-top knolls and terminal ridges from which they could overlook the landscape and be viewed by the living. This is similar to Robinson's (2006) example of Late Archaic burial grounds situated above essential travel routes from which they could see and be seen. Goldstein (2002:203) suggested that the visibility of burial grounds was in itself an important link among ancestors, the landscape and the living.

Port au Choix-3 is consistent with some but not all of these observations. It is a locality of high marine productivity, which could support social aggregations associated with a burial ground. On the basis of available resources, Tuck (1976) suggested that the Maritime Archaic who created Port au Choix-3 lived in the area throughout the warmer months, either at a central site in conjunction with satellite camps or else cycling through a series of smaller base camps. When we discovered the extensive Gould site, we thought it was the central site suggested by Tuck and, further, that it was a good candidate for a social aggregation site (Renouf and Bell 2000a). However, unless there is an as yet undiscovered Maritime Archaic habitation site at or near Port au Choix, it appears that Port au Choix-3 was connected to a briefly and intermittently occupied field camp. While this is anomalous in the context of our original expectations, it is consistent with our subsequent study of Maritime Archaic site location patterns in Newfoundland that suggested the Maritime Archaic were more mobile than previously appreciated and highlighted Maritime Archaic movement across the landscape – along the arms of bays and inlets and through the Newfoundland interior along rivers and lakes (Renouf and Bell 2006). In this light, a small transient camp associated with a burial ground begins to make sense.

**Fig. 3.8** Port au Choix shoreline reconstructed at 5000 BP showing Port au Choix-3 in relation to the inside passage which is represented by the *dashed line*

Following Littleton and Allen (2007), Port au Choix-3 may have functioned not as a focus of social aggregation but as a signpost on the landscape along an important travel path. At 4,000 years ago, Port au Choix-3, on a sandy spit, overlooked the narrowest point of what was the inside passage of the coastal travel route along the west coast of the Northern Peninsula (Fig. 3.8). Port au Choix-3 would have looked upon every traveller moving north and south along this route. As the ancestors observed all travellers passing through the tickle, so too the travellers looked upon the ancestors in their special place. These would have included individuals travelling to and from Big Brook-2, Big Droke-1, the Caines site, Spearbank, Woody Point-2, and other Maritime Archaic sites as yet undiscovered.

The position of Port au Choix-3 may or may not have included staking ancestral or territorial claims. The lack of identified Maritime Archaic habitation sites at Port au Choix, other than the Gould site, suggests that if any claims were made on the basis of the burial ground, they were not specific to that locality but rather were more general to the larger region that was traversed by mobile and wide-ranging Maritime Archaic family groups.

## Summary and Conclusions

We interpret the Gould site as an intermittently used field camp associated with the Port au Choix-3 burial ground a short distance away, directly across a narrow channel of water. This is contrary to the general expectation that hunter-gatherer burial grounds are associated with social aggregations and to the particular expectation that a social aggregation site would be associated with Port au Choix-3. That Port au Choix-3 is instead associated with a short-term field camp is consistent with suggestions that Maritime Archaic were more mobile than previously thought. This shifts the focus away from individual sites to movement across the landscape. We suggest that the significance of Port au Choix-3 lies in its position overlooking a narrow channel that was a constriction on an important coastal travel route along the west coast of the Northern Peninsula. The burial grounds monitored travellers as they moved up and down the coast, who could in return look upon ancestral burials to complete the linkage of past, present and place.

**Acknowledgements** Gould site excavations were initially and generously funded by a Memorial University New Initiatives Research Grant from the Office of the Vice-President (Research). Subsequent funding was provided by: the Social Sciences and Humanities Research Council of Canada; the Institute of Social and Economic Research, Memorial University; the J.R. Smallwood Foundation for Newfoundland and Labrador Studies, Memorial University; Parks Canada; the Provincial Archaeology Office of the Government of Newfoundland and Labrador; and the Town of Port au Choix. Facilities for analysis were funded by the Canada Foundation for Innovation, the Canada Research Chairs Program and Memorial University. We would like to thank: Peter Scott for tree species identification, Larry Nolan for developing our GIS software and assisting in its application, Charles Conway for drafting the figures, Chris Hammond, Dominique Lavers and Patty Wells for artefact photographs, and the many crew members who spent several summers working on our Gould site excavations. I would like to thank Fred Schwarz for providing unpublished data on the Woody Point-2 site, Elaine Anton of The Rooms Provincial Museum for granting me access to the Port au Choix-3 collection, and the Provincial Archaeology Office for answering endless queries about Maritime Archaic sites. Thanks also to David Sanger and Brian Robinson for their helpful comments on an earlier draft of this paper.

## References

Anderson, J.E. 1976. Appendix A: The human skeletons. In *Ancient people of Port au Choix*, ed. J.A. Tuck, 124–131. Newfoundland Social and Economic Studies 17, Institute of Social and Economic Research, St. John's.

Beaton, G. 2004. A Chip off the Old Block: Investigations of a Maritime Archaic Lithic Workshop/Quarry Site in Big Brook (EjBa-2), Northwestern Newfoundland. Master's thesis, Department of Anthropology, Memorial University of Newfoundland, St. John's.

Bell, T., and M.A.P. Renouf. 2003. Prehistoric cultures, reconstructed coasts: Maritime Archaic Indian site distribution in Newfoundland. *World Archaeology* 35(3): 350–370.

Bell, T., J.B. Macpherson, and M.A.P. Renouf. 2005a. Late prehistoric impact on Bass Pond, Port au Choix. *Newfoundland and Labrador Studies* 20(1): 107–129.

Bell, T., I.R. Smith, and M.A.P. Renouf. 2005b. Postglacial sea-level history and coastline change at Port au Choix, Great Northern Peninsula, Newfoundland. *Newfoundland and Labrador Studies* 20(1): 9–31.

Binford, L.R. 1980. Willow smoke and dogs' tails: Hunter-gatherer settlement systems and archaeological site formation. *American Antiquity* 45(1): 4–20.

Carignan, P. 1975. Archaeological survey 1975. Report for the Newfoundland Museum. On file, Provincial Archaeology Office, Government of Newfoundland and Labrador, St. John's.

Charles, D.K., and J.E. Buikstra. 2002. Siting, sighting, and citing the dead. In *The space and place of death*, ed. H. Silverman and D.B. Small, 13–25. Archaeological Papers of the American Anthropological Association 11. American Anthropological Association, Arlington, VA.

Deal, M. 2005. Palaeoethnobotanical research at Port au Choix. *Newfoundland and Labrador Studies* 20(1): 131–156.

d'Entremont, S.A. 1978. The Determination of Status at Port au Choix. Master's thesis, Department of Anthropology, Memorial University of Newfoundland, St. John's.

DNE. 1990. *Dictionary of Newfoundland English*, ed. G.M. Story, W.J. Kirwin, and J.D.A. Widdowson. Toronto: University of Toronto Press.

Goldstein, L. 2002. Visible death: Mortuary site and mortuary landscape in diachronic perspective. In *The space and place of death*, ed. by H. Silverman and D.B. Small, 201–205. Archaeological Papers of the American Anthropological Association 11. American Anthropological Association, Arlington, VA.

Jelsma, J. 2000. A Bed of Ochre: Mortuary Practices and Social Structure of a Maritime Archaic Indian Society at Port au Choix, Newfoundland. PhD dissertation, University of Groningen, Groningen.

Jelsma, J. 2006. Three social status groups at Port au Choix: Maritime Archaic mortuary practices and social structure. In *The Archaic of the far northeast*, ed. D. Sanger and M.A.P. Renouf, 83–103. Orono: University of Maine Press.

Kennedy, B.V. 1981. *Marriage patterns in an Archaic population: A study of skeletal remains from Port au Choix, Newfoundland.* Mercury Series Archaeology Paper 104, National Museums of Canada, Ottawa.

Littleton, J., and H. Allen. 2007. Hunter-gatherer burials and the creation of persistent places in southeastern Australia. *Journal of Anthropological Archaeology* 26(2): 282–298.

Liverman, D.G.E. 1994. Relative sea-level history and isostatic rebound in Newfoundland, Canada. *Boreas* 23: 217–230.

MacLeod, D. 1967a. Field Season Report. Manuscript #254. On file, Canadian Museum of Civilization, Hull, QC.

MacLeod, D. 1967b. A Red Paint Burial Site in Northeastern Newfoundland. Paper presented to the Society for American Archaeology 32nd annual meeting, Ann Arbor.

MacLeod, D. 1968. Field Season Report. Manuscript #255. On file, Canadian Museum of Civilization, Hull, QC.

Provincial Archaeology Office. 2009. Archaeological Site Database. Department of Tourism, Culture and Recreation, Government of Newfoundland and Labrador, St. John's.

Reader, D. 1998. The 1997 Archaeological Survey of the Bird Cove Area, Northwestern Northern Peninsula, Newfoundland. On file, Provincial Archaeology Office, Government of Newfoundland and Labrador, St. John's.

Reader, D. 1999. Report of the 1998 Bird Cove Archaeological Project. On file, Provincial Archaeology Office, Government of Newfoundland and Labrador, St. John's.

Reid, H.S. 2008. A Study of Southern Variant, Maritime Archaic Sites from the Northern Peninsula and Strait of Belle Isle, Newfoundland and Labrador. Master's thesis, Department of Anthropology, Memorial University of Newfoundland, St. John's.

Renouf, M.A.P. 1985. Archaeology of the Port au Choix National Historic Park: Report of 1984 Field Activities. On file, Parks Canada, Archaeology, Atlantic Region, Halifax.

Renouf, M.A.P. 1991. Archaeological Investigations at the Port au Choix National Historic Park: Report of the 1990 Field Season. On file, Parks Canada, Archaeology, Atlantic Region, Halifax.

Renouf, M.A.P., and T. Bell. 1998. Searching for the Maritime Archaic Indian Habitation Site at Port au Choix, Newfoundland: An Integrated Approach using Archaeology, Geomorphology and Sea Level History. On file, Provincial Archaeology Office, Government of Newfoundland and Labrador, St. John's.

Renouf, M.A.P., and T. Bell. 1999. Report of the Port au Choix Archaeology and Sea Level History Project: Preliminary Report of the 1998 Field Season. On file, Provincial Archaeology Office, Government of Newfoundland and Labrador, St. John's.

Renouf, M.A.P., and T. Bell. 2000a. Integrating sea level history and geomorphology in targeted archaeological site survey: The Gould site (EeBi-42), Port au Choix, Newfoundland. *Northeast Anthropology* 59: 47–64.

Renouf, M.A.P., and T. Bell. 2000b. Gould Site, Port au Choix, 1999 Report of Field Activities. On file, Provincial Archaeology Office, Government of Newfoundland and Labrador, St. John's.

Renouf, M.A.P., and T. Bell. 2001. Gould Site, Port au Choix, 2000 Report of Field Activities. On file, Provincial Archaeology Office, Government of Newfoundland and Labrador, St. John's.

Renouf, M.A.P., and T. Bell. 2002. Report of the 2001 Field Season at Port au Choix. On file, Provincial Archaeology Office, Government of Newfoundland and Labrador, St. John's.

Renouf, M.A.P., and T. Bell. 2006. Maritime Archaic site location patterns on the island of Newfoundland. In *The Archaic of the far northeast*, ed. D. Sanger and M.A.P. Renouf, 1–46. Orono: Maine University Press.

Renouf, M.A.P., T. Bell, and J. Macpherson. 2009. Hunter-gatherer impact on Subarctic vegetation: Amerindian and Palaeoeskimo occupations of Port au Choix, northwestern Newfoundland. *Arctic Anthropology* 46(1–2): 176–190.

Robinson, B.S. 2006. Burial ritual, technology, and cultural landscape in the far northeast: 8600-3700 B.P. In *The Archaic of the far northeast*, ed. D. Sanger and M.A.P. Renouf, 341–381. Orono: Maine University Press.

Rothchild, N.A. 1983. The recognition of leadership in egalitarian societies of the northeast. In *The development of political organization in Native North America*, ed. E. Tooker, 165–182. Philadelphia: The American Ethnological Society.

Schlanger, S.H. 1992. Recognizing persistent places in Anasazi settlement systems. In *Space, time and archaeological landscapes*, ed. J. Rossignol and L. Wandsnider, 91–112. New York: Plenum.

Schwarz, F., and R. Skanes. 2005. Stage 2 Assessment of DjBl-04, Woody Point, Bonne Bay, Newfoundland: Interim Report on Site Significance. On file, Provincial Archaeology Office, Government of Newfoundland and Labrador, St. John's.

Stuiver, M., and P.J. Reimer. 1993. Extended 14C Database and Revised CALIB radiocarbon Calibration Program. *Radiocarbon* 35: 215–230.

Temple, B. 2007. The 1966–69 Archaeological Excavations at Back Harbour, North Twillingate Island, Newfoundland. On file, Provincial Archaeology Office, Government of Newfoundland and Labrador, St. John's.

Thibaudeau, P.A. 1993. The Curtis Site: Its Place within the Maritime Archaic. Honour's dissertation, Department of Sociology and Anthropology, Carleton University, Ottawa.

Tuck, J.A. 1970. An Archaic Indian cemetery in Newfoundland. *Scientific American* 226: 112–121.

Tuck, J.A. 1971. An Archaic cemetery at Port au Choix, Newfoundland. *American Antiquity* 36: 343–358.

Tuck, J.A. 1976. *Ancient people of Port au Choix*. Newfoundland Social and Economic Studies 17, Institute of Social and Economic Research, Memorial University of Newfoundland, St. John's.

Tuck, J.A. 1978. Excavations at Cow Head, Newfoundland: An interim report. *Études/Inuit/Studies* 2(1): 138–141.

Walthall, J.A. 1999. Mortuary behaviour and early Holocene land use in the North American midcontinent. *North American Archaeologist* 20(1): 1–30.

Wells, P.J. and M.A.P. Renouf. 2008. Archaeological Survey Report, Back Harbour, North Twillingate Island 2007. On file, Provincial Archaeology Office, Government of Newfoundland and Labrador, St. John's.

Wilmeth, R. 1978. *Canadian archaeological radiocarbon dates (revised version)*. Mercury Series Archaeology Paper 77, National Museums of Canada, Ottawa.

Zedeño, M.N., D. Austin, and R. Stoffle. 1997. Landmark and landscape: A contextual approach to the management of American Indian resources. *Culture and Agriculture* 19(3): 123–129.

# Chapter 4
# Ritual Activity and the Formation of Faunal Assemblages at Two Groswater Palaeoeskimo Sites at Port au Choix

**Patricia J. Wells**

## Introduction

Faunal assemblages from archaeological contexts are usually viewed as the remains of consumption that offer insights into past cultural behaviour such as hunting, processing, transporting and scavenging. However, faunal remains can also provide information about the ritual treatment of animal remains (Jones O'Day et al. 2004; Muir and Driver 2004; Murray 2000; Renouf 2000). This paper presents a comparative examination of faunal remains from middens at two nearby Groswater Palaeoeskimo sites on the Point Riche peninsula in northwestern Newfoundland, Phillip's Garden West (EeBi-11) and Phillip's Garden East (EeBi-1) (Renouf 2005; Wells 2002). Animal exploitation at these temporally overlapping settlements was focused almost exclusively on seal hunting, particularly of the huge harp seal populations that still frequent these waters each late fall and spring (see Murray, Chap. 11). The stone tool assemblage at Phillip's Garden West deviates substantially from typical forms. Qualitative and quantitative descriptions of this particular morphological variant have been presented by Renouf (2005) and Ryan (Chap. 5; see also Melnik 2007 for further discussion). Renouf hypothesizes that the variant has its origins in rituals enacted around seal hunting. A thorough examination of the faunal remains from this site contributes to an understanding of variation at an intra-site level, and a comparison with the faunal assemblage from Phillip's Garden East allows for discussion of similarities and differences at an inter-site level. The faunal analysis explores explanations of the assemblage configurations by quantitatively addressing differential deposition of seal body parts at the sites and differential survival of bones based on their structural density. The results suggest that the frequency of different seal body parts can, with equal reliability, be explained as a consequence of the transport of some meat-rich portions of the skeleton away from the site, as well as from the differential survival of bone based on its structural density.

P.J. Wells (✉)
Department of Archaeology, Memorial University, St. John's, NL, Canada, A1C 5S7
e-mail: pwells@mun.ca

M.A.P. Renouf (ed.), *The Cultural Landscapes of Port au Choix*, Interdisciplinary Contributions to Archaeology, DOI 10.1007/978-1-4419-8324-4_4,

It is argued here that both explanations are operating to account for the frequency of seal body parts. Furthermore, this analysis shows some significant differences in the frequency of seal body parts between the two sites. Specifically, during the time when the two sites were contemporary, there were almost no cranial elements present at Phillip's Garden West, while they were the most frequently represented elements at Phillip's Garden East. This difference is unexpected since the sites are very similar in other ways. Both yielded a full range of domestic and hunting tools as well as the remains of dwelling structures. Furthermore, both are the same distance from the sea, the occupants hunted the same species at the same time of year, and there are excellent and similar preservation conditions. The absence of cranial elements from Phillip's Garden West appears to be intentional. In addition, this absence is interpreted as supporting Renouf's hypothesis that the site functioned, at least partially, as an important location for ritual activities relating to the harp seal hunt.

## Groswater Palaeoeskimos at Port au Choix

The location of Phillip's Garden West and Phillip's Garden East on the outer coast of the Point Riche peninsula offers an excellent view of the Strait of Belle Isle and the migrating sea mammal populations (Fig. 4.1). The sites are approximately 1 km apart, on either side of the large Dorset site of Phillip's Garden, and can be seen from each other. Both sites were excavated by Renouf in the 1980s and 1990s and numerous radiocarbon dates demonstrate that, while Phillip's Garden East was occupied earlier, and Phillip's Garden West later, for most of their occupation period these sites were contemporaneous (Table 4.1; see Appendix) (Renouf 1985, 1986, 1990, 1991, 1992, 2005).

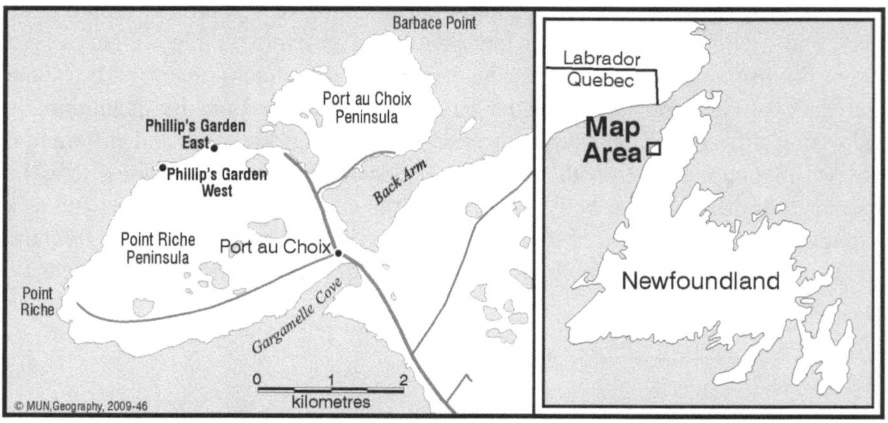


**Fig. 4.1** Location of Phillip's Garden West and Phillip's Garden East

**Table 4.1** Radiocarbon dates from Phillip's Garden West and Phillip's Garden East with calibrations using Calib 6.0html (Stuiver and Reimer 1993)

| Context | C14 years BP | Lab no. | Calibrated median age BP | Cal BP range 1σ | Cal BP range 2σ |
|---|---|---|---|---|---|
| Phillip's Garden West, midden F5 | 2460±120 | Beta 49761 | 2540 | 2700–2360 | 2780–2180 |
| Phillip's Garden West, midden F5 | 2340±100 | Beta 49760 | 2400 | 2680–2160 | 2720–2150 |
| Phillip's Garden West, midden F5 | 2240±70 | Beta 66437 | 2230 | 2340–2160 | 2360–2050 |
| Phillip's Garden West, midden F5 | 1960±80 | Beta 66438 | 1910 | 2000–1820 | 2110–1720 |
| Phillip's Garden East, storage pit F55 | 2500±60 | Beta 50021 | 2570 | 2720–2490 | 2740–2360 |
| Phillip's Garden East, storage pit F53 | 2260±70 | Beta 50022 | 2240 | 2340–2160 | 2460–2060 |

Phillip's Garden West is situated on a terrace 13 m above sea level (asl) and covers an area of approximately 500 m$^2$. The site has an upper terrace living area and a hillside over which debris was deposited. Upper terrace excavations revealed a circular living structure outlined by five postholes. There was one internal hearth and a number of external hearths scattered throughout the upper terrace (Renouf 1994). The terrace edge drops steeply toward the beach. This hillside and lower terrace area contained a large midden deposit. Despite the mixing that would have taken place as faunal and other refuse was thrown over the hillside, excavators were able to distinguish separate dumping episodes. Stone tools at the site reflect a range of domestic and hunting activities. These artefacts include endblades, bifaces, burin-like tools, microblades, cores, scrapers, preforms, hammerstones, abraders, and a few axes.

Phillip's Garden East is located close to shore on a terrace 12.5 m asl and covers an area of approximately 1,500 m$^2$. Renouf identified one Groswater dwelling structure at this site in addition to a second dwelling or pit structure which could be attributable to Dorset or Groswater (Renouf 2003:384). The identified Groswater dwelling was a shallow depression outlined by a perimeter of small stones and debris (Renouf 2003:384). A number of small, shallow midden deposits were located in association with this structure (Wells 2002:69). Similar to Phillip's Garden West, a range of Groswater material culture was recovered at this site, reflecting a variety of domestic and hunting activities (Kennett 1991; LeBlanc 1996, 2000).

## Faunal Preservation

Organic preservation at both Phillip's Garden West and Phillip's Garden East is very good. The limestone bedrock of the Port au Choix and Point Riche peninsulas neutralizes Newfoundland's otherwise acidic soils. Furthermore, the sandy soil at these sites allows good drainage. One possible means of determining the comparative

**Table 4.2** Ratio of MNE to NISP for Phocid elements from Phillip's Garden West and Phillip's Garden East

| Sample | MNE | NISP | MNE:NISP |
|---|---|---|---|
| Phillip's Garden West F18 | 2,055 | 4,026 | 0.51 |
| Phillip's Garden West F5A–D | 371 | 639 | 0.58 |
| Phillip's Garden West F5E | 301 | 656 | 0.46 |
| Phillip's Garden East | 454 | 1,020 | 0.46 |

preservation rates between the two sites is to compare the ratio of minimum number of elements (MNE) to the number of identifiable specimens (NISP). Table 4.2 shows the ratios among the samples to be very similar.

## Faunal Samples

Three midden features were selected from Phillip's Garden West with dates that span the occupation of the site, each representing a separate dumping episode. They are Features 18, 5A–D and 5E, ordered from oldest to youngest. Examining chronologically separate deposits allows an exploration of changes in subsistence and processing over time at Phillip's Garden West (Wells 2005). Because midden features associated with the Groswater dwelling at Phillip's Garden East were fairly small, they have been combined as one sample in this research.

Feature 18 at Phillip's Garden West returned two dates, 2460 ± 120 BP[1] and 2340 ± 100 BP (Table 4.1) (Renouf 1994, 2005). A total of 20,081 bone fragments (specimens) were examined, representing approximately half the faunal material in the feature. Feature 5A–D returned one date of 2240 ± 70 BP (Renouf 1994). The entire faunal assemblage of 3,647 specimens was examined. Feature 5E returned one of the most recent Groswater dates, 1960 ± 80 BP (Renouf 1993). The entire faunal collection of 3,025 specimens was examined.

The faunal sample from Phillip's Garden East consisted of 4,155 specimens representing all the bone recovered from deposits associated with the Groswater dwelling feature. Two uncalibrated radiocarbon dates were returned for the Phillip's Garden East sample. They are 2260 ± 70 BP and 2500 ± 60 BP (Table 4.1) (Renouf 1992).

Seals, in particular harp seals, made up the vast majority of species in the Phillip's Garden West and Phillip's Garden East assemblages (Wells 2002, 2005). Cod fish, a variety of birds and some terrestrial mammals including caribou, red fox, wolf, black bear and beaver contributed to the Groswater economy at these two sites; however, as Table 4.3 shows, seal was of considerable importance.

Faunal analysis of seals is hampered by the difficulty in identifying them to species. There is a great deal of variability in the morphology of seal bones between

[1] All radiocarbon dates in this chapter are uncalibrated years before present (BP); the dates are calibrated in Table 4.1. See also Appendix.

**Table 4.3** Relative proportions of faunal categories at Phillip's Garden West (PGW) and Phillip's Garden East (PGE) (Wells 2002)

| | PGW Feature 18 | | PGW Feature 5A–D | | PGW Feature 5E | | PGE features | |
|---|---|---|---|---|---|---|---|---|
| Faunal category | NISP | %NISP | NISP | %NISP | NISP | %NISP | NISP | %NISP |
| Seal | 4,026 | 97.2 | 639 | 90.4 | 654 | 99.5 | 1,020 | 98.6 |
| Bird | 69 | 1.7 | 39 | 5.5 | 1 | 0.2 | 11 | 1.1 |
| Fish | 29 | 0.7 | 2 | 0.3 | – | – | – | – |
| All other | 19 | 0.5 | 25 | 3.5 | 2 | 0.3 | 3 | 0.3 |
| Mammal total | – | – | – | – | – | – | – | – |
| Total identified | 4,143 | – | 707 | – | 657 | – | 1,034 | – |
| Total unidentified | 15,938 | – | 2,940 | – | 2,368 | – | 3,121 | – |
| Total specimens | 20,081 | – | 3,647 | – | 3,025 | – | 4,155 | – |

individuals of the same species and strong similarities between individuals of different species (Hodgetts 1999:112; Storå 2001). A few elements can indicate species such as the cranium (including the auditory bulla and maxilla), as well as the mandible; but for the most part, precise identification is difficult. For this reason, most of the seal remains in this study are identified to the family (Phocidae) level only, with just a few elements identified to species. Almost all the phocid remains that could be identified to species were harp seal, which likely reflects the case for the majority of bones in the general seal category.

## Body Part Frequency

The relative frequency of different body parts is calculated by determining the MNE that can be accounted for by the fragments and whole elements in each sample (Binford 1978; Lyman 1994). In this study, to avoid over-representing fragmentary elements, particular landmarks on elements are given zone numbers and the most frequent zone defines the MNE for that element. Zonal designation is given only when half or more of the zone is present, and MNE is generated regardless of side or state of fusion (Hodgetts 1999).

The frequency of different elements within one individual animal varies. For instance, there are five cervical vertebrae and only two humeri in seals, potentially resulting in the impression that cervical vertebrae are more frequently deposited than humeri. Minimum animal units (MAU) divide the MNE by the number of times it occurs in the individual (Binford 1978, 1984:50; Lyman 1994), allowing a direct comparison among the frequencies of different elements. In order to compare the relative frequencies of body parts in archaeological samples of different sizes, %MAU is used. This value is calculated by expressing the highest value (MAU) as the standard, dividing all other MAU values by this standard, and multiplying by 100 (Binford 1978:72, 1981; Bunn et al. 1988).

The frequency of different body parts for species can indicate cultural and natural processes that affect the configuration of faunal assemblages recovered from archaeological contexts. Distinguishing what factors, or combination of factors, result in the body part frequency of a faunal assemblage can be challenging (Munro and Bar-Oz 2004). Bones can be removed from a site or they can be destroyed by scavenging carnivores or the chemical and physical agents of weathering (e.g. Behrensmeyer 1983; Binford 1978; Lyman 1994; Stiner 2002). In addition, variation in the frequency witnessed is affected by the structural density of the elements (Lyman 1984, 1994). Furthermore, human processing, including ritual handling, transporting, butchering, cooking, consuming, and disposing of elements, can influence the configuration of specific elements present on a site (e.g. Binford 1978; Lyman 1994; Wells 2002). The following section describes how transport of different body parts and the relative structural density of bones can determine the frequency of different elements in a faunal assemblage.

## *Meat Utility of Faunal Assemblages*

Zooarchaeologists have long recognized that human decision-making processes and butchering practices can explain variability in faunal assemblages (White 1952, 1956). Furthermore, ethnoarchaeological work such as that conducted by Binford (1978) and others (Bunn et al. 1988; O'Connell et al. 1988; O'Connell and Marshall 1989) demonstrates that cultural mechanisms help to structure faunal assemblages. Binford (1978) suggests that the formation of faunal assemblages is influenced by the way that different animal parts are treated, particularly with regard to the relative amount of meat, fat and marrow they provide. He relates the character of the faunal assemblage to site function by proposing that animal parts of higher value would be transported to residential sites, while those of lower value would be deposited at kill sites. Binford develops a utility index as a means of quantitatively evaluating the relative meat, marrow and fat value for each element or set of elements in an animal skeleton. He plots the relationship between economic utility and the relative frequency of particular skeletal elements on a site. The curves generated by this relationship are used to demonstrate the kinds of animal processing activities at sites and therefore site function. Binford (1978) goes on to adjust his index to accommodate the expected possibility that elements of low value attached to those of high value are more likely to be transported to residential sites from kill sites as "riders." He refers to this as the modified meat utility index (MMUI). His ethnographic field work generally agrees with this hypothesis. Others have adopted and adjusted this method of understanding faunal assemblage configurations by developing utility indices for a variety of species (e.g. Lyman et al. 1992; Metcalfe and Jones 1988; Savelle and Friesen 1996).

## *Bone Mineral Density of Faunal Assemblages*

The structural density of bone is a crucially important variable in the survival of bone over time and can influence the relative frequency of skeletal portions in a faunal assemblage (Lyman 1984). The conditions of the burial environment, particularly the pH of the soil, can be destructive to bone. Both alkaline and acidic soils tend to hasten destruction of bone the further they are from neutral (Gordon and Buikstra 1981; Lyman 1994). The hardest, densest bones, teeth, and shell are more likely to withstand destruction. Bone is not usually heterogeneous, and structural density measures an average characteristic of the sample. All skeletal elements are composed of spongy and compact bone. Their ratio will differ according to the element and the location on the element. Researchers calculate density in different ways depending on how they derive the volume of porosity, thus making their measurement results differentially controlled (Lyman 1984:263).

A technique called photon absorptiometry was developed to derive the mineral densities for a number of locations on skeletal elements (Kreutzer 1992;

Lyman 1984; Stiner 2004). A photon beam of known strength is passed through a number of points on an element and the strength of the beam is measured. The higher the mineral content of the scan sites on the element, the weaker the beam, or the fewer photons that will pass through that site (Lyman 1994:238). Sites on elements are chosen for scanning that will reflect known structural variation within each bone, that are easy to locate and describe on the basis of anatomical features, and that include portions often found on archaeological sites. The resulting structural density values are intended to be used as a frame of reference for comparison to archaeological assemblages of the same taxa. Lyman (1994:252) warns that, at best, these values constitute an ordinal scale. He points out that structural density values are averages of a number of individuals and that variation can exist in structural density with respect to age, sex, nutritional status and genetics. Lyman (1992:12) summarizes the various studies of bone density as showing that (a) density is greatest in bone portions that have the greatest compressive and tensile strengths, (b) density is greatest in bones subjected to the greatest weight bearing stresses and (c) increasing porosity (decreasing bulk density) of bone reduces bone strength.

## Data Presentation: Phillip's Garden West and Phillip's Garden East

### *Phocid Body Part Frequency*

Table 4.4 presents the MAU and %MAU values for elements in each of the faunal samples. Features 18 and 5A–D from Phillip's Garden West are dominated by limb and flipper elements, with relatively few ribs, vertebrae and cranial and mandibular bones. Feature 5E at Phillip's Garden West is similar to these in that there are few ribs and vertebral elements and that limbs and flippers are well represented; however, it differs from Features 18 and 5A–D in that crania and mandibles are well represented. At Phillip's Garden East, crania and mandibles dominate with relatively few other elements, and ribs and vertebral elements are the least well represented.

To see how larger articulated portions of the carcass are represented, elements are grouped into seven body portions (Wells 2002). To calculate the MAU value for each body portion, the MNE values for each element in a group are summed and this number is divided by the total number of these elements in one skeleton. The head consists of the cranium and mandible, the vertebrae are included as a group, ribs remain as one segment, front limbs include the scapula, humerus, radius, and ulna, while the front flipper includes the carpals, metacarpals, and front first and second phalanges. The third phalanges were largely fragmented in all the samples from this study making it difficult to determine whether they derive from front or hind limbs in most cases. For this reason, they are not included in the frequency study. The hind limb includes the innominate, femur, tibia and fibula, and the hind flipper includes the astragalus, calcaneus, tarsals, metatarsals, and hind first and second phalanges.

**Table 4.4** Phocid body part frequency as MAU and %MAU for samples from Phillip's Garden West (PGW) and Phillip's Garden East (PGE)

| | PGW Feature 18 | | PGW Feature 5A–D | | PGW Feature 5E | | PGE features | |
|---|---|---|---|---|---|---|---|---|
| Element | MAU | %MAU | MAU | %MAU | MAU | %MAU | MAU | %MAU |
| Cranium | 5 | 17.2 | 2 | 33.9 | 7 | 100 | 11 | 100 |
| Mandible | 9 | 31 | 1 | 17 | 1 | 14.3 | 4.5 | 40.9 |
| Atlas | 2 | 6.9 | 1 | 17 | 1 | 14.3 | 1 | 9.1 |
| Axis | 3 | 10.4 | 2 | 33.9 | 1 | 14.3 | 1 | 9.1 |
| Cervical | 4.4 | 15.2 | 0.4 | 6.8 | 0.4 | 5.7 | 0.4 | 3.6 |
| Thoracic | 2.3 | 7.8 | 0.1 | 2.2 | 0.3 | 4.8 | 0.2 | 1.8 |
| Lumbar | 3.8 | 13.1 | 0.2 | 3.4 | 0.2 | 2.9 | 0.4 | 3.6 |
| Sacrum | 6 | 20.7 | 0 | 0 | 0 | 0 | 0 | 0 |
| Rib | 3 | 10.2 | 0.2 | 2.9 | 0.1 | 1.4 | 0.6 | 5.5 |
| Scapula | 12.5 | 43.1 | 2 | 33.9 | 2 | 28.6 | 1 | 9.1 |
| Humerus | 19 | 65.5 | 3 | 50.9 | 3 | 42.9 | 2 | 18.2 |
| Radius | 29 | 100 | 5 | 84.8 | 6 | 85.7 | 3.5 | 31.8 |
| Ulna | 15.5 | 53.5 | 2 | 33.9 | 3 | 42.9 | 2 | 18.2 |
| Carpal | 21.9 | 75.5 | 2.8 | 66.1 | 2.5 | 35.7 | 3.6 | 33.1 |
| Metacarpal | 18.5 | 63.8 | 5.9 | 100 | 2.2 | 31.4 | 3.2 | 29.1 |
| Phalange 1f | 18.8 | 64.8 | 3.5 | 59.3 | 1.8 | 25.7 | 2.7 | 24.6 |
| Phalange 2f | 22.3 | 76.9 | 3.8 | 64.4 | 1.5 | 21.4 | 4 | 36.4 |
| Phalange 3f and h | 12.6 | 43.5 | 2.7 | 45.8 | 1.9 | 26.4 | 2 | 17.7 |
| Innominate | 3.5 | 12.1 | 0.5 | 8.5 | 1 | 14.3 | 1 | 9.1 |
| Femur | 13.5 | 46.6 | 1.5 | 25.4 | 3 | 42.9 | 2.5 | 22.7 |
| Tibia | 11.5 | 39.7 | 4.5 | 76.3 | 5 | 71.4 | 1 | 9.1 |
| Fibula | 20 | 69 | 5.5 | 93.2 | 3 | 42.9 | 0.5 | 4.6 |
| Tarsal | 11.7 | 40.3 | 2.1 | 49.2 | 1.6 | 22.4 | 2 | 18.2 |
| Metatarsal | 13.2 | 45.5 | 2.4 | 40.7 | 4.1 | 58.6 | 6.8 | 61.8 |
| Phalange 1 hind | 11.1 | 38.3 | 1.9 | 32.2 | 3.1 | 44.3 | 7.1 | 64.6 |
| Phalange 2 hind | 11.5 | 39.7 | 2.3 | 39 | 1.1 | 16.1 | 4 | 36.4 |
| Total | 304.6 | – | 58.3 | – | 56.8 | – | 68 | – |

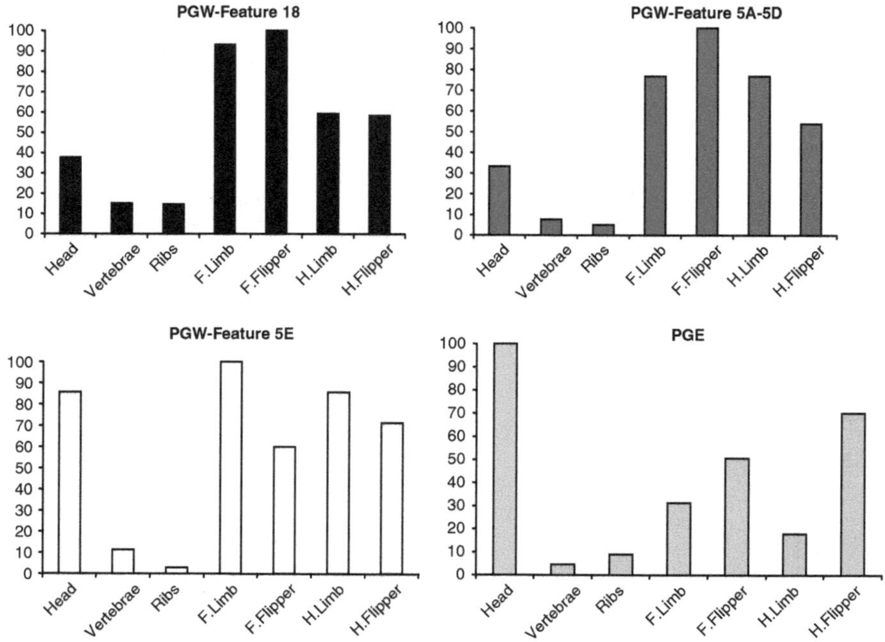


**Fig. 4.2** %MAU frequency of phocid body portions for each sample from Phillip's Garden West (PGW) and Phillip's Garden East (PGE). *F* front; *H* hind

Figure 4.2 presents the %MAU values for the summed body portions for each faunal sample. Ribs and vertebrae are poorly represented in all samples. Features 18 and 5A–D from Phillip's Garden West are dominated by limb and flipper elements and relatively few cranial elements. Feature 5E from Phillip's Garden West has a good representation of most elements, apart from ribs and vertebrae. The Phillip's Garden East sample is dominated by head and flipper elements.

## *Meat Utility Against Phocid Body Part Frequency*

In order to understand how differential frequency of skeletal parts may have been influenced by utility value, I employ the meat utility index (MUI) for phocid seals developed by Lyman et al. (1992) and plot this value against the MAU for each sample.[2] In their research, Lyman et al. (1992) base their utility index on the average weight of meat per

[2]Spearman's rank order correlation coefficient was used in this study to test significance of relationships. The correlation coefficient is a value that ranges from 1 to −1 and is expressed as $r_s$. Significance values are given for each $r_s$ in order to assess confidence that this value is not the result of sampling vagaries and is expressed as *p* (Drennan 1996:231). As *p* values decrease, confidence rises. A *p* value of 0.05 or lower is considered significant in social sciences and is used here.

skeletal portion from three harp seals and one hooded seal. The results indicate that the rib cage is of greatest food utility, followed in order by the pelvis, vertebrae and proximal limb elements; distal limb elements rank lowest in food value. Lyman et al. (1992) went on to develop a MMUI for seals to account for riders.

When the MAU values for the faunal samples from both Phillip's Garden West and Phillip's Garden East are plotted against the %MUI and %MMUI derived by Lyman et al. (1992), there is an L-shaped curve which indicates that elements of high meat value are relatively low in number, while those of low meat value are more frequent (Wells 2002). This configuration is commonly associated with kill sites. There is a negative correlation when MAU values are plotted against %MUI[3]; however, when elements are grouped, the L-shaped appearance of the scatterplot becomes more apparent and the correlation becomes stronger (Wells 2002). According to Lyman et al. (1992:548), the stronger correlation with the MMUI suggests that seal carcasses may have been transported in units that include a number of elements. Figure 4.3 presents the scatterplots of

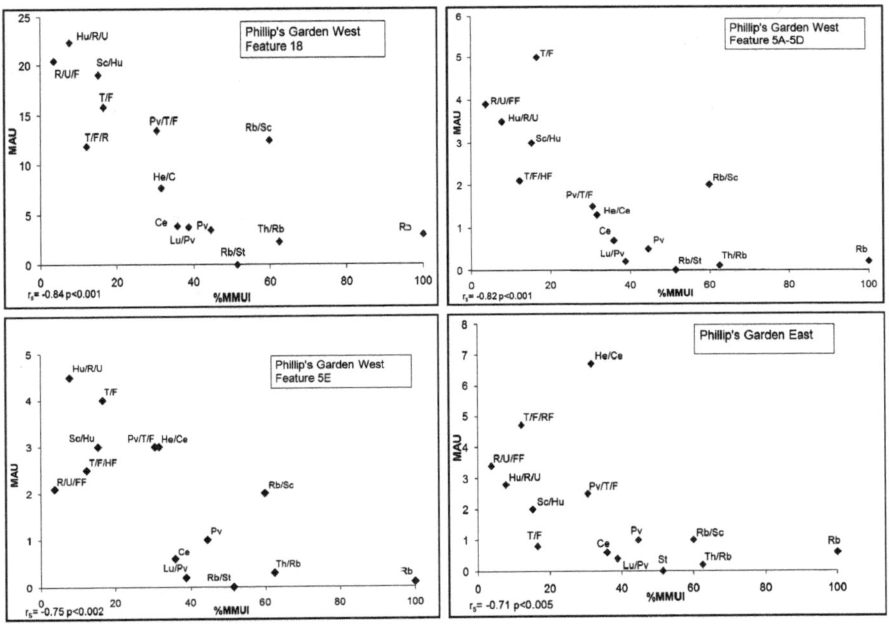


**Fig. 4.3** MAU frequencies of phocids from faunal assemblages from Phillip's Garden West and Phillip's Garden East against %MMUI. *He* head; *Ce* cervical vertebra; *Th* thoracic vertebra; *Lu* lumbar vertebra; *Pv* pelvis; *St* sternum; *Rb* rib; *Sc* scapula; *Hu* humerus; *R* radius; *U* ulna; *T* tibia; *F* fibula; *FF* front flipper; *HF* hind flipper

[3] MAU values against %MUI (ungrouped elements)
Feature 18 $r_s = -0.5$, $p < 0.05$
Feature 5A–D $r_s = -0.4$, $p < 0.01$
Feature 5E $r_s = -0.4$, $p < 0.2$
Phillip's Garden East $r_s = -0.2$, $p < 0.5$

the relationship of MAU against %MMUI for each sample. It is clear that the more meat-rich portions of the skeleton are the least frequent, and that these portions may have been removed from the sites.

## *Bone Mineral Density Values Against Phocid Body Part Frequency*

Structural density values are compared to the frequency of skeletal elements from the Phillip's Garden West and Phillip's Garden East samples to determine if a correlation exists between the relative structural density and the survival of the bones. In the present study, the bone mineral density values for the scan sites correspond to the most frequently occurring zone for each element in each sample (Wells 2002:140). Generally, the results of these comparisons demonstrate that the density of bones is positively correlated with their frequency.

Figure 4.4 illustrates this positive correlation; however, there are a number of anomalies. This figure shows the frequency of phocid elements against their bone mineral density values for the assemblage from Feature 18. This scatterplot suggests

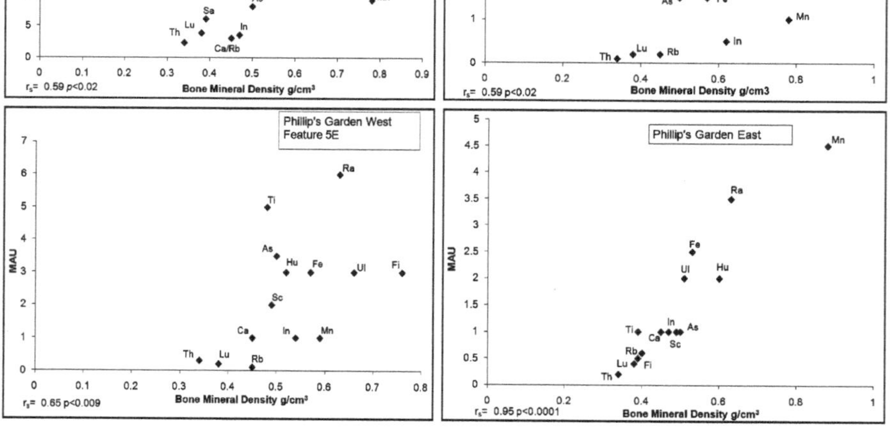


**Fig. 4.4** MAU frequencies of phocids from faunal assemblages from Phillip's Garden West and Phillip's Garden East against bone mineral density values for seals. *He* head; *Ce* cervical vertebra; *Th* thoracic vertebra; *Lu* lumbar vertebra; *Pv* pelvis; *St* sternum; *Rb* rib; *Sc* scapula; *Hu* humerus; *R* radius; *U* ulna; *T* tibia; *F* fibula; *FF* front flipper; *HF* hind flipper

that variability in density of phocid elements may account for the configuration of body parts in the assemblage from this feature. A few of the elements do not fit the overall trend. The ulna is relatively frequent despite the fact that it has a relatively low bone mineral density value. Conversely, although the mandible has the highest bone mineral density value, it is infrequently represented in the assemblage. Similar results are noted for Feature 5A–D, with a general trend toward greater numbers of denser bones. Again, there are a few exceptions. The ulna and tibia are more frequently represented than would be expected from their relatively low bone mineral density values, and as in Feature 18, there is a relatively low frequency of mandibles despite their high density values. There is also a very low representation of innominates despite having bone mineral densities that are about the same as the radius, which is highly represented. For Feature 5E at Phillip's Garden West, there is a positive and significant correlation between variables. Nevertheless, the frequency of innominate, ulna, fibula and mandible is relatively low despite their high density values. The sample from Phillip's Garden East shows a strong and highly significant positive correlation between density and relative frequency. No exceptions are noted here.

## Discussion: Body Part Frequency

When both transport (utility indices) as well as bone mineral density values are used in comparisons to MAU values from a site, it may not be possible to differentiate which, or to what extent of each, best accounts for the frequency of faunal remains if both show significant results (Lyman 1994, 2004; Munro and Bar-Oz 2004). This is the case in the present study. Lyman (1994:258) reports a tendency toward a negative correlation between utility indices and bone mineral density. This trend indicates that bones with low structural density tend to rank high in utility, while bones with high structural density tend to rank low in utility. Since utility is assumed to be directly related to decisions about transport, this situation presents a problem of interpreting whether transport or structural density more strongly influences the body part frequency seen in the faunal assemblages.

Issues of equifinality, or the circumstance where final states can result from different initial conditions, have been the focus of increased attention by taphonomists (Lyman 2004). A symposium held in 2004 and a subsequent issue of the *Journal of Taphonomy* is dedicated to resolving problems of equifinality (Munro and Bar-Oz 2004). Most contributors focus on gaining greater precision with existing techniques for identification and quantification. Marean et al. (2004) argue that more precise methods of identification and quantification of bone shafts in particular will reduce the likelihood of equifinality (see also Outram 2004). In addition, Bar-Oz and Munro (2004) suggest a three-part analytical approach that involves investigating various taphonomic variables such as: primary quantification and sources of damage; analyses that assess the degree of fragmentation in samples; and comparisons among subgroups of assemblages such as taxa, size and age.

In an earlier work, Lyman (1984:258) suggests that other lines of evidence are necessary to sort out what process best accounts for a particular faunal assemblage. He states that the nature of tool assemblages along with other evidence from constructed features on sites should contribute to any interpretation of faunal assemblage configuration (Lyman 1992:19). I would add that it is important to describe the context of this type of faunal analysis, both cultural and natural. The ecology of the prey species, proximity to the kill location, hunting technology, and the possibility of damage caused by carnivores must be assessed for each study.

A review of the evidence for both transport and density-mediated destruction follows for each feature, with an evaluation of each. In addition, evidence regarding site features and artefact configuration, site location, hunting practices and species morphology will be considered in the interpretation of these faunal assemblages. Available evidence suggests that transport away from the sites and the structural density of the bones both influence the character of the faunal assemblages.

A reverse utility configuration was generated when meat utility was compared to element frequency in each of the samples. This configuration suggests that the sites functioned as kill and/or butchery locations from which seal portions of high meat value were removed. The exclusive designation of the sites as functioning solely as kill/butchery locations is not supported by the other archaeological evidence; nevertheless, butchery was an important activity at the sites and it seems likely that some portion of the seal catch was processed for transport and later consumption.

Since harp seals in this region were moving north and south in their seasonal migrations, they were likely killed in open water from boats, or along the ice edge (Sergeant 1985; Stenson et al. 1995). Given that the sites were located on the coast, it is likely that seal carcasses were returned to the sites whole for butchering. The shape of a seal lends itself to being transported whole (Lyman et al. 1992). Pinnipeds have torpedo-shaped, streamlined bodies with only short appendicular protrusions, making them fairly easy to drag whole. This condition suggests that initial butchering to facilitate easy transport at the kill site was unlikely, and removal of portions may not have had the same practical considerations as with terrestrial mammals. The transport of heavy, low value parts of terrestrial mammals, particularly of ungulates, would have to be considered during primary butchery, as these parts tend to be rather large and cumbersome and have low meat value (Binford 1978). This would not be a practical consideration for the butchering of seal as it could be handled easily from any side without the removal of either cranium or limbs. Furthermore, there is no practical reason to abandon portions of the seal carcass at the kill since the consistent quality of the fur covering the animal and a consistent blubber layer would make no part of this animal particularly extraneous.

While it appears likely that the sites functioned as phocid butchering locations, interpreting Phillip's Garden West and Phillip's Garden East as kill/butchery sites rather than residential sites based on the results of comparisons between skeletal element frequency and utility is simplistic and relies on the assumption that sites must be designated either as kill/butchery location or residential sites. The location of both sites near the shore where seals could have been hunted or landed is a natural choice for the primary butchery of seal for some immediate, limited consumption and processing for transport of high meat value portions. Nevertheless, Phillip's Garden West and Phillip's Garden East had many functions, including the butchering and processing of seals, the monitoring of game as well as more domestic tasks.

The features and the range of artefacts present at both sites reflect residential occupations where numerous activities took place. Both sites had at least one dwelling feature. At Phillip's Garden West, the dwelling was defined by a circle of five post holes. Five hearths were identified, four outside the dwelling, and one situated in the centre of the house (Renouf 1992). At Phillip's Garden East, the dwelling was a circular arrangement of stones enclosing an area of approximately 5 m in diameter (Renouf 1992, 1994). The artefacts recovered from both sites include the full range of Groswater material culture, representing a variety of domestic and hunting implements. Tools include endblades, bifaces and harpoon heads and also more domestic implements such as scrapers, needles and burin-like tools. The sites and middens were strewn with stone flakes and charcoal as well as a number of preforms and cores. Strict designation of sites as either kill/butchery or residential sites is inappropriate and does not reflect the multitude of activities that were performed. It is likely that Phillip's Garden West and Phillip's Garden East were residential sites where, because they were located close to sealing locations, hunting and processing activities were carried out around a social life that involved other domestic tasks such as food preparation, tool manufacture, and shelter construction and maintenance.

Some portion of the phocid assemblage was likely processed for transport and later consumption. The ecology of the harp seals that migrate along the coast of the Point Riche and Port au Choix peninsulas suggests the likelihood that some degree of processing for transport away from the site was conducted. The harp seal populations that passed the coast of Port au Choix arrived in huge numbers for restricted periods from early spring to early summer and again for a short period in the early winter. These concentrations provided Groswater hunters at Port au Choix the opportunity to capture numerous seals over relatively short time periods. Unlike catches of individual ringed seal, available to the north, the sudden availability of large quantities of harp seal meat provided the opportunity for processing for transport and later consumption. This makes sense in light of the temporal restriction of this abundant resource.

The consistent paucity of the rib, innominate and vertebral elements in these samples suggests the possibility that these high meat value elements were transported from the sites. Ethnographic and archaeological references state that seal meat was dried for later transport and consumption by Arctic populations (Birket-Smith 1945:96; Mathiassen 1928:206; Nelson 1983:299; Petersen 1984). Birket-Smith (1929:144) describes seal meat being cut into strips and laid in the sun to create a dark crust around the soft interior meat. Otto Fabricius (1962:108–109) states that the Inuit of Greenland processed seal for storage: "For drying purposes it is cut into flat slices as far as this can be done on account of the bones, which are allowed to remain; the slices are then laid upon bare rocks with sun and wind in summer... and afterwards they have this wind-dried meat for winter supplies."

While transport of skeletal parts may have played a significant role in the configuration of bones in these samples, the relative frequency of elements may be equally the result of differential bone mineral densities. A comparison of bone mineral density to element frequency demonstrates that, with few exceptions, denser elements are more frequent. The bone mineral density of the ribs and vertebrae in all the samples goes some way toward explaining their relatively low numbers on the site. The vertebrae ranked lowest in density followed by the ribs. While it is possible that the ribs and vertebrae were transported from the site, it is also likely that the low density of the bones contributed to their under-representation.

Lyman et al. (1992) suggest that variability in frequencies of phocid bones is likely to be a result of taphonomic processes that take place at the residential site, including feeding dogs and natural post-depositional processes. There is no evidence that the Groswater Palaeoeskimo had dogs and there were no signs of large carnivore gnawing on the bones from these sites. Since it is likely that the harp seals were returned whole to Phillip's Garden West and dogs were not likely to have contributed to the under-representation of faunal remains, density-mediated post-depositional destruction is a likely contributor to the variability in the faunal assemblage noted for Phillip's Garden West.

Taken alone, the correlation observed between the meat utility indices devised by Lyman et al. (1992) and the MAUs from the samples at Phillip's Garden West suggests that this site functioned as a butchering station from which packages of high meat value were removed for consumption elsewhere. However, the features on the site, the range of artefacts present, the location of the site, and the evidence of structural density of seal bones all challenge this interpretation. Taken together, the evidence suggests that Phillip's Garden West and Phillip's Garden East were residential sites from which hunting, game monitoring and domestic activities were initiated. Seals were hunted from boats in open water or along the ice edge and returned whole to the site for processing and consumption. It is possible that some portion of this meat was transported elsewhere, as suggested by the low frequency of the high meat value parts of the skeleton. While exportation may explain the low frequency of the relatively dense innominate in all samples, it is impossible to discount the destruction of elements due to natural post-depositional forces, especially for the less dense but meaty elements including the vertebrae and ribs. Both interpretations have validity and are probably operating to a greater or lesser extent.

## Intra-Site and Inter-Site Comparisons

The faunal assemblages from Phillip's Garden West and Phillip's Garden East are alike in the species exploited and, for the most part, in phocid body part frequency (Table 4.5; Wells 2002). These commonalities indicate similarities of site function

**Table 4.5** Spearman's rho calculations of MAU values for phocids in features from Phillip's Garden West and Phillip's Garden East

| | PGW Feature 18 | PGW Feature 5A–D | PGW Feature 5E |
|---|---|---|---|
| PGW Feature 18 | – | $r_s=0.80, p<0.001$ | $r_s=0.61, p<0.001$ |
| PGW Feature 5A–D | – | – | $r_s=0.68, p<0.001$ |
| Phillip's Garden East | $r_s=0.45, p=0.02$ | $r_s=0.45, p=0.02$ | $r_s=0.61, p=0.001$ |

including the range of activities that took place. Nevertheless, an examination of the variation among the faunal assemblages indicates some important differences in the activities performed at the two sites; specifically, the distribution and handling of seal skulls. The following section compares samples and offers an interpretation of results. It is proposed that seal skulls may have had particular ideological importance that manifested itself in ritual practices surrounding the harp seal hunt.

Archaeological evidence and ethnographic literature document the special treatment given to seal cranial elements. They are often overly abundant on archaeological sites where these animals are hunted. Murray (1992) finds that cranial elements were the most frequent phocid bones in her sample from Phillip's Garden, the Dorset site sitting between the two Groswater sites in this study (Murray, Chap. 11; see also Renouf and Murray 1999). A similar relative frequency of skulls was found elsewhere in Phillip's Garden by Linehan (1990). Stewart (1979) notes the high representation of seal skulls in the faunal assemblage from the Groswater site of Factory Cove, south of Port au Choix (Auger 1985). Elsewhere, Savelle (1984) notes a relatively large number of seal skulls on a historic Inuit site on Somerset Island in the Canadian Arctic. Hodgetts (1999) describes an over-abundance of seal skulls on Younger Stone Age sites in northern Norway, and Lyman (1991) notes a similar situation for sea lion skulls from the Pacific Northwest. Likewise, Giddings (1967:145) excavated a cache of animal heads, including a large number of seals from Ipiutak Beaches on Cape Krusenstern, Alaska. The auditory bulla, an extremely dense and recognizable element, contributes to this high frequency of crania. Murray (2000) suggests that the over-abundance of seal skulls may also be explained, at least partially, as a result of their retention on sites for ritual practices.

Ethnographic sources from northern regions note that seal skulls have special ritual functions in a number of contexts. Rasmussen (1931) notes that when the Netsilik residents moved, seal heads were laid on clean snow or sea ice pointed in the direction of a new camp so that the souls of the seals could follow the people and ensure good hunting (see also Søby 1970). Murdoch (1892) states that the Inupiat avoided fracturing or throwing seal skulls into the sea, keeping them in piles in front of their houses. This was done to keep the souls of the seals content. Lantis' (1947) ethnographic work among the Alaskan hunters of Nunivak Island describes hunters keeping seal skulls on shelves facing the door of their dwellings. In the spring, the skulls and bones were buried in special disposal sites. Fienup-Riordan (1994:105) describes the care taken with seal heads by the people of Nelson Island and the Yukon Delta. To ensure the return of seal in the future, the women of Nelson Island place the head of seals inside the house facing the door, while the people of the Yukon Delta place seal heads facing toward the interior of dwellings to encourage other seals to follow them into the human world.

Figure 4.5a shows the summed %MAU values for the assemblages from Phillip's Garden West. There is generally little difference in the treatment of harp seal carcasses over the occupation of the site; however, it is significant to note that during the early

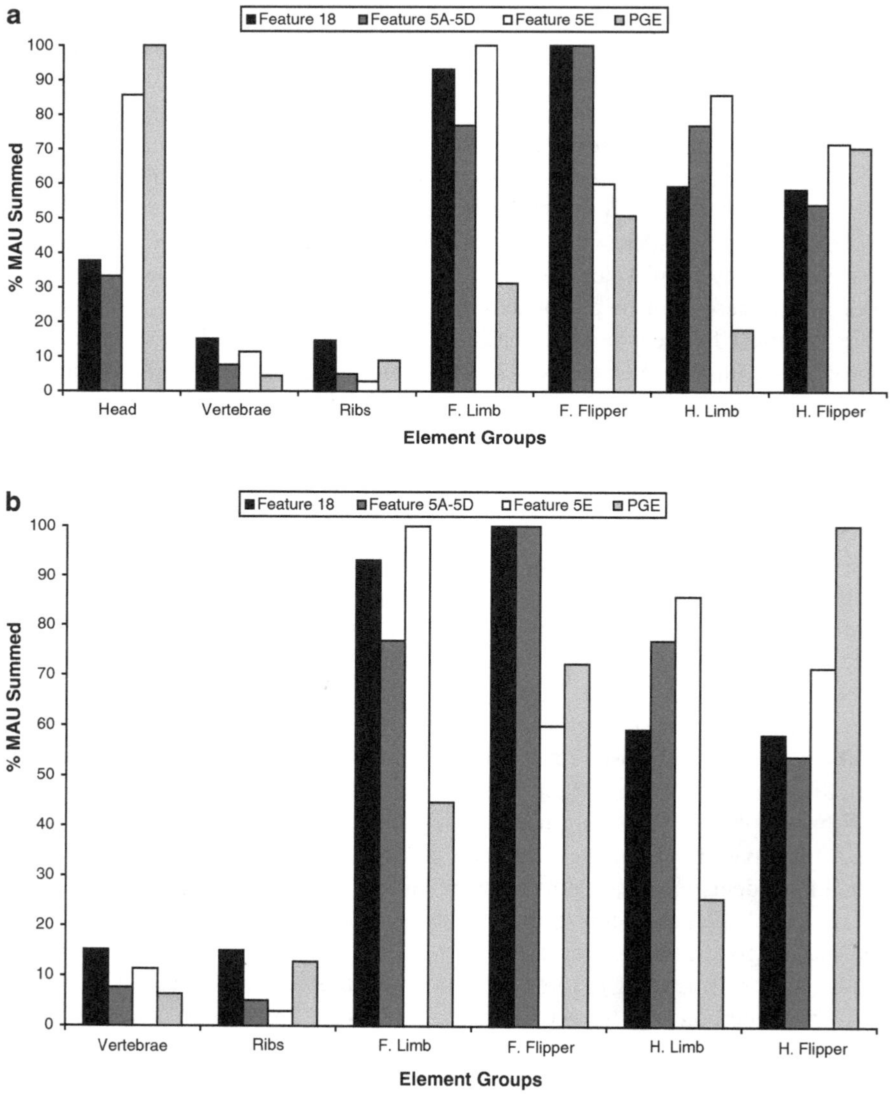


**Fig. 4.5** (**a**) %MAU frequency of phocid body portions (summed) in faunal assemblages from Phillip's Garden West and Phillip's Garden East (PGE). (**b**) %MAU frequency of phocid body portions (summed), excluding cranial elements in faunal assemblages from Phillip's Garden West and Phillip's Garden East (PGE)

occupation of the site (Features 18 and 5A–D) seal skulls are poorly represented, while later (Feature 5E) they become a much more frequent element.[4]

At the inter-site level, the summed %MAU values from Phillip's Garden East are significantly correlated with those from Phillip's Garden West, particularly with Feature 5E (Table 4.5). There are a number of similarities as well as some noteworthy differences between the sites' assemblages. The samples from both sites show a low relative frequency of vertebrae and ribs. Hind flippers are also almost equally represented at the sites, with slightly greater representation during the later occupation at Phillip's Garden West (Feature 5E) and at Phillip's Garden East. The front and hind limbs, as well as the front flippers, are poorly represented in the Phillip's Garden East sample compared to the Phillip's Garden West features. For the proximal limb bones, there are striking differences between the Phillip's Garden East sample and Feature 5E, where the latter has numerous front and hind limbs compared to the former. The crania also show conspicuous differences among the samples. They are extremely well represented in the samples from Phillip's Garden East and Feature 5E and poorly represented in the other samples from Phillip's Garden West. The stronger positive correlation between Phillip's Garden East and Feature 5E may be related to the fact that both are largely dominated by cranial elements.

An interesting pattern emerges when comparing the samples from Phillip's Garden East and Features 18 and 5A–D at Phillip's Garden West. The dates from these sites overlap and the species exploited and season of occupation are similar, suggesting that these two sites were occupied at the same time and season (Table 4.1) (Wells 2002). As mentioned above, Phillip's Garden West and Phillip's Garden East are within a kilometre of each other on the same shoreline and have very similar soil conditions. In addition, all the faunal remains were excavated from middens. It is expected that natural post-depositional conditions would be essentially the same for faunal material on both sites. As the sites were occupied by people of the same culture, one would expect similar treatment of seal carcasses on two contemporary, residential sites. This is clearly not the case. Indeed, in some instances the faunal assemblages resemble a mirror image of each other. For instance, there are many crania at Phillip's Garden East and few in the temporally overlapping samples from Phillip's Garden West. While there are few front limbs, front flippers and hind limbs in the Phillip's Garden East sample, there are greater numbers in the temporally overlapping samples from Phillip's Garden West. Thus, holding soil conditions, culture and chronology constant, variability in the frequency of body parts at the two residential sites suggests some degree of functional difference in the processing of seal between the sites. A few scenarios can be explored to explain this pattern.

It is possible that these two sites were directly connected to one another. For instance, they could have been involved in the processing of seal carcasses in some

[4] I confirmed the low cranial numbers for the unidentified portion of Feature 18 by quickly examining the whole assemblage. Furthermore, I viewed the remaining faunal material from the rest of the site, which had been excavated in its entirety, and found very few cranial elements.

cooperative fashion which resulted in the differential disposal of body parts. It is only the disposal that we can see at this stage, making it difficult to demonstrate how processing was coordinated between the sites. Phillip's Garden East may have functioned as a hunting camp where seals were initially butchered and parts of low meat value were deposited (skulls and flippers), while other parts were transported to Phillip's Garden West. This suggests that consumption may have taken place at Phillip's Garden West and only butchery at Phillip's Garden East. The presence of hearths, a dwelling feature and a wide range of artefact functional categories at Phillip's Garden East, all of which indicate a residential site, is difficult to explain in light of this interpretation. Nevertheless, there were far more harpoon heads found at Phillip's Garden East ($n=13$) compared to Phillip's Garden West ($n=1$). This implies a slight difference in the activities performed at the two sites. Perhaps, hunting and preliminary butchery were initiated out of Phillip's Garden East, with some members of the group using this site as a residence while processing the seal carcasses. It is possible that the residents of the two sites confined their consumption and disposal of most edible parts of the carcass to Phillip's Garden West.

Continuing with the assumption that the occupants of both sites cooperated in seal hunting, it is possible that, while seal carcasses may have been returned to both sites, some aspect of the activities at Phillip's Garden West may have necessitated the removal of skulls. Renouf (2005) argues that the Phillip's Garden West lithic assemblage represents a distinct Groswater variant which she connects to possible ritual activities carried out at the site. In particular, a number of the endblades recovered from Phillip's Garden West were so finely serrated and extremely elongated that they may not have been functional (Renouf 2005, Fig. 13). These particularly elegant pieces have been found singly in a few other Groswater sites, and in small numbers at the Dorset site of Phillip's Garden (Renouf 2005, Fig. 12), but are common only at Phillip's Garden West.

The ritual treatment associated with the hunting and processing of game by northern hunter gatherers has been, and continues to be, a widespread and cross-cultural phenomenon (e.g. Balikci 1970; Fienup-Riordan 1994; Nuttall 1992, 2000; Søby 1970; Tanner 1979). Preparations for hunts and the treatment of carcasses after capture involve carefully performed rituals to show respect for animals and continued success in harvesting them in the future. It is certain that the close relationship between humans and animals was very important in the past and would have had a series of associated ritual behaviours. The relative frequency of some skeletal elements, along with the lithic evidence from Phillip's Garden West, may be a tangible indication of ritual activities.

Alternatively, it is possible that the sites were independent of one another and that both were simply contemporary settlements with slight differences in the focus of activities. Because of the greater emphasis on hunting at Phillip's Garden East evident in the number of harpoon heads, this site may have functioned primarily as a hunting, butchering, and processing camp from which some meat was transported (LeBlanc 1996). The presence of 13 harpoon heads at Phillip's Garden East in contrast to the single specimen found at Phillip's Garden West may indicate a site with only marginal domestic activity and a greater focus on hunting. The under-representation of most appendicular elements at Phillip's Garden East cannot be entirely explained as resulting from differential survival, as nearby Phillip's Garden West has good preservation of some bones largely missing from Phillip's Garden East. Their absence could be interpreted as the result of transport. However, the apparent under-representation of appendicular elements at Phillip's Garden East is partly a product of the way %MAU is calculated. The great number of cranial elements can make the number of other elements appear relatively small. When %MAU (summed) values were re-calculated for all samples excluding the cranial elements, the characterization of Phillip's Garden East as a site focused primarily on initial butchering remained unchanged. Although the appendicular skeleton was now better represented at Phillip's Garden East, it remains low as do the limb elements, while parts such as the hind flipper are still highly represented compared to Phillip's Garden West (Fig. 4.5b). Phillip's Garden West, on the other hand, appears to represent a more focused residential location where hunting, butchering, processing, consumption and disposal of seals were undertaken with no connection to the hunting camp at Phillip's Garden East.

While it may be true that Phillip's Garden East represents a hunting camp with fewer activities compared to Phillip's Garden West, it is unlikely that two contemporary sites of the same culture located within a short distance of one another would not have had some degree of contact. It is interesting to note that after the abandonment of Phillip's Garden East, there is a shift in the frequency of various phocid elements at Phillip's Garden West. This may be due to a change in the activities at Phillip's Garden West once there was no longer a settlement at Phillip's Garden East. Feature 5E post-dates the occupation at Phillip's Garden East. With the exception of vertebrae and ribs (highly subject to taphonomic processes), the elements identified in this sample are relatively equally represented. All element groups (%MAU) are between 60 and 100%. It appears that there may have been less transport of meat packages from Phillip's Garden West during this later period.

These suggestions are not necessarily exclusive of one another. It is possible that some combination of these factors is at work. It is conceivable that the sites of Phillip's Garden East and Phillip's Garden West were related to one another and that some activities performed at one site were excluded from the other. The variability in the frequency of phocid body parts at the two sites suggests some degree of functional difference between them.

## Conclusions

Faunal assemblages from two Groswater Palaeoeskimo sites, Phillip's Garden West and Phillip's Garden East, were presented; these sites temporally overlap for a large part of their occupations. Approximately 90% of the assemblages comprised seal, demonstrating the importance of this animal to the site's inhabitants. In addition, an examination of phocid body part frequency indicated slight differences between the sites. Explanations of the relative frequency were explored, focusing on how meat utility and structural density of the elements may have influenced this configuration.

Problems with distinguishing between human transport and the differential survival of bone in understanding body part representation were outlined and other factors such as site locations, features and artefacts, as wells as hunting practices, seal ecology, and the ritual treatment of body parts were introduced to develop a comprehensive picture of seal hunting, butchering and processing at the sites.

The results indicate slight differences in the activities at the sites, in particular during the period of temporal overlap, the most remarkable of which was connected to the treatment of seal cranial elements. Since it was likely that killed seals were returned to these sites whole, it was unusual that the very dense and recognizable cranial elements were not returned to Phillip's Garden West during the early part of site occupation (Features 18 and 5A–D). In contrast, they were well represented at Phillip's Garden East. The absence of cranial elements from Phillip's Garden West cannot be explained as a result of the transport of meat-rich portions or low structural density since these low meat value elements are among the densest in the phocid body. The exclusion of cranial elements is clearly intentional and likely related to ideological beliefs about the seal skull and its possible role in ritual practice.

**Acknowledgements** I wish to acknowledge the generous financial support of the Institute of Social and Economic Research (MUN) and the Smallwood Centre for Newfoundland and Labrador Studies (MUN). I am very grateful to my supervisor, Dr. M.A.P. Renouf, and to Dr. Lisa Hodgetts, both of whom assisted in all aspects of my research. Darlene Balkwill and Anne Rick, both formerly of the Canadian Museum of Nature, contributed to the identification of some of the Phillip's Garden East seal material. Thanks to Charlie Conway, Memorial University Geography, who drafted the map. I am also grateful for the helpful comments of two anonymous reviewers.

## References

Auger, R. 1985. Factory Cove: Recognition and Definition of the Early Palaeo-Eskimo Period in Newfoundland. Master's thesis, Department of Anthropology, Memorial University of Newfoundland, St. John's.

Balikci, A. 1970. *The Netsilik Eskimo*. New York: The Natural History Press.

Bar-Oz, G., and N.D. Munro. 2004. Beyond cautionary tales: A multivariate taphonomic approach for resolving equifinality in zooarchaeological studies. *Journal of Taphonomy* 2(4): 201–221.

Behrensmeyer, A.K. 1983. Patterns of natural bone distribution on recent land surfaces: Implications for archaeological site formation. In *Animals and archaeology: 1. Hunters and their prey*, ed. J. Clutton-Brock and C. Grigson, 93–106. British Archaeological Reports International Series 163. British Archaeological Reports, Oxford.

Binford, L.R. 1978. *Nunamiut ethnoarchaeology*. New York: Academic Press.

Binford, L.R. 1981. *Bones: Ancient men and modern myths*. New York: Academic Press.

Binford, L.R. 1984. *Faunal remains from Klasies River Mouth*. New York: Academic Press.

Birket-Smith, K. 1929. *The Caribou Eskimos: Material and social life and their cultural position*. Report of the Fifth Thule Expedition 1921–24, Vol. 5(1). Gyldeddalske Boghandel, Nordisk Forlag, Copenhagen.

Birket-Smith, K. 1945. *Ethnographical collections from the Northwest Passage*. Report of the Fifth Thule Expedition 1921–24, Vol. 6(2). Gyldeddalske Boghandel, Nordisk Forlag, Copenhagen.

Bunn, H., L.E. Bartram, and E.M. Kroll. 1988. Variability in bone assemblage formation from Hadza hunting, scavenging and carcass processing. *Journal of Anthropological Archaeology* 7: 412–457.

Drennan, R.D. 1996. *Statistics for archaeologists: A common sense approach*. New York: Plenum Press.

Fabricius, O. 1962. Full description of the Greenland seals. First section: The Harp Seal (*Phoca groenlandica*). In *Otto Fabricius' ethnographical works*, ed. E. Holtved, 103–117. Meddelelser om Grønland 140(2). Copenhagen: C.A. Reitzel.

Fienup-Riordan, A. 1994. *Boundaries and passages: Rule and ritual in Yup'ik Eskimo oral tradition*. Norman: University of Oklahoma Press.

Giddings, J.L. 1967. *Ancient men of the Arctic*. New York: Knopf.

Gordon, C.C., and J.E. Buikstra. 1981. Soil pH, bone preservation, and sampling bias at mortuary sites. *American Antiquity* 46: 566–571.

Hodgetts, L.M. 1999. Animal Bones and Human Society in the Late Younger Stone Age of Arctic Norway. PhD dissertation, Department of Archaeology, University of Durham, Durham.

Jones O'Day, S., Van Neer, W., and A. Ervynck (eds.), 2004. *Behaviour behind bones: The zooarchaeology of ritual, religion, status, and identity*, Proceedings of the 9th Conference of the International Council of Archaeozoology, Durham, August 2002. Oxford: Oxbow Books.

Kennett, B.L. 1991. Phillip's Garden East: An Examination of the Groswater Palaeo-Eskimo Phase. Master's thesis, Department of Anthropology, Memorial University of Newfoundland, St. John's.

Kreutzer, L.A. 1992. Bison and deer bone mineral densities: Comparisons and implications for the interpretation of archaeological faunas. *Journal of Archaeological Science* 19: 271–294.

Lantis, M. 1947. Social culture of the Nunivak Eskimo. *Transactions of the American Philosophical Society* 35(3): 153–323.

LeBlanc, S. 1996. A Place with a View: Groswater Subsistence-settlement Patterns in the Gulf of St. Lawrence. Master's thesis, Department of Anthropology, Memorial University of Newfoundland, St. John's.

LeBlanc, S. 2000. Groswater technological organization: A decision-making approach. *Arctic Anthropology* 37(2): 23–37.

Linehan, D. 1990. The Analysis and Description of Four Midden Squares at Phillip's Garden, Port au Choix, Newfoundland. Honour's dissertation, Department of Anthropology, Memorial University of Newfoundland, St. John's.

Lyman, R.L. 1984. Bone density and differential survivorship of fossil classes. *Journal of Archaeological Science* 3: 259–299.

Lyman, R.L. 1991. Taphonomic problems with archaeological analyses of animal carcass utilization and transport. In *Beamers, bobwhites, and blue-points: Tributes to the career of Paul W. Parmalee*, ed. J.R. Purdue, W.E. Klippel, and B.W. Styles, 125–138. Illinois State Museum Scientific Papers, Vol. 23, Springfield.

Lyman, R.L. 1992. Anatomical considerations of utility curves in zooarchaeology. *Journal of Archaeological Science* 19: 7–22.

Lyman, R.L. 1994. *Vertebrate taphonomy*. Cambridge Manuals in Archaeology. Cambridge: Cambridge University Press.

Lyman, R.L. 2004. The concept of equifinality in taphonomy. *Journal of Taphonomy* 2(4): 15–26.

Lyman, R.L., J.M. Savelle, and P. Whitridge. 1992. Derivation and application of a meat utility index for phocid seals. *Journal of Archaeological Science* 19: 531–555.

Marean, C.W., M. Domínguez-Rodrigo, and T.R. Pickerking. 2004. Skeletal element equifinity in zooarchaeology begins with method: The evolution and status of the "Shaft Critique". *Journal of Taphonomy* 2(1): 69–98.

Mathiassen, T. 1928. *Material culture of the Iglulik Eskimos*. Report of the Fifth Thule Expedition 1921–24, Vol. 6(1). Gyldeddalske Boghandel, Nordisk Forlag, Copenhagen.

Melnik, M.E. 2007. Salmon Net (EfAx-25): An Investigation of Groswater Material Culture Stylistic Variability in Newfoundland. Master's thesis, Department of Anthropology, Memorial University of Newfoundland, St. John's.

Metcalfe, D., and K.T. Jones. 1988. A reconsideration of animal body-part utility indices. *American Antiquity* 53(3): 486–504.

Muir, R.J., and J.C. Driver. 2004. Identifying ritual use of animals in the northern American Southwest. In *Behaviour behind bones: The zooarchaeology of ritual, religion, status, and identity,* Proceedings of the 9th Conference of the International Council of Archaeozoology, Durham, August 2002, ed. S. Jones O'Day, W. Van Neer, and A. Ervynck, 128–143. Oxford: Oxbow Books.

Munro, N.D., and G. Bar-Oz. 2004. Debating issues of equifinality in ungulate skeletal part studies. *Journal of Taphonomy* 2(1): 1–13.

Murdoch, J. 1892. *Ethnological results of the Point Barrow expedition.* Reprinted from the Ninth Annual Report of the Bureau of Ethnology 1887–1888. Washington, D.C.: Smithsonian Institution Press.

Murray, M.S. 1992. Beyond the Laundry List: The Analysis of Faunal Remains from a Dorset Dwelling at Phillip's Garden (EeBi 1), Port au Choix, Newfoundland. Master's thesis, Department of Anthropology, Memorial University of Newfoundland, St. John's.

Murray, M.S. 2000. A zooarchaeological approach to Arctic prehistory. In *Animal bones, human societies*, ed. P. Rowley-Conwy, 58–64. Oxford: Oxbow Books.

Nelson, E.W. 1983. *The Eskimo about Bering Strait*, (1899). Washington, D.C.: Smithsonian Institution Press.

Nuttall, M. 1992. *Arctic homeland: Kinship, community and development in northwest Greenland.* Toronto: University of Toronto Press.

Nuttall, M. 2000. Becoming a hunter in Greenland. *Études/Inuit/Studies* 24(2): 33–45.

O'Connell, J.F., and B. Marshall. 1989. Analysis of kangaroo body part transport among the Alyawara of Central Australia. *Journal of Archaeological Science* 16: 393–405.

O'Connell, J.F., K. Hawkes, and N.B. Jones. 1988. Hadza hunting, butchering, and bone transport and their archaeological implications. *Journal of Anthropological Research* 44(2): 113–161.

Outram, A.K. 2004. Applied models and indices vs. high-resolution, observed data: Detailed fracture and fragmentation analyses for the investigation of skeletal part abundance patterns. *Journal of Taphonomy* 2(1): 167–184.

Petersen, R. 1984. East Greenland before 1950. In *Handbook of North American Indians: Arctic*, ed. D. Damas, 622–639. Washington, D.C.: Smithsonian Institution Press.

Rasmussen, K. 1931. *The Netsilik Eskimos: Social life and spiritual culture.* Report of the Fifth Thule Expedition 1921–1924, Vol. 8(1–2). Glydendalske Boghandel, Nordisk Forlag, Copenhagen.

Renouf, M.A.P. 1985. Archaeology of the Port au Choix National Historic Park: Report of 1984 Field Activities. On file, Parks Canada, Archaeology, Atlantic Region, Halifax.

Renouf, M.A.P. 1986. Report of 1985 Excavations at Point Riche and Phillip's Garden, Port au Choix National Historic Park. On file, Parks Canada, Archaeology, Atlantic Region, Halifax.

Renouf, M.A.P. 1990. The Transitional Period in Eastern Arctic Prehistory: Another Mesolithic Interlude? Paper given at the Sixth International Conference on Hunting and Gathering Societies, Fairbanks. On file, Department of Archaeology, Memorial University of Newfoundland, St. John's.

Renouf, M.A.P. 1991. Archaeological Investigations at the Port au Choix National Historic Park: Report of the 1990 Field Season. On file, Parks Canada, Archaeology, Atlantic Region, Halifax.

Renouf, M.A.P. 1992. The 1991 Field Season, Port au Choix National Historic Park: Report of Archaeological Excavations. On file, Parks Canada, Archaeology, Atlantic Region, Halifax.

Renouf, M.A.P. 1993. The 1992 Field Season, Port au Choix National Historic Park: Report of Archaeological Excavations. On file, Parks Canada, Archaeology Atlantic, Region, Halifax.

Renouf, M.A.P. 1994. Two transitional sites at Port au Choix, northwestern Newfoundland. In *Threads of Arctic prehistory: Papers in honour of William E. Taylor Jr.*, ed. D. Morrison and J.-L. Pilon, 166–195. Mercury Series Archaeology Paper 149, Canadian Museum of Civilization, Ottawa.

Renouf, M.A.P. 2000. Symbolism and subsistence: Seals and caribou at Port au Choix, northwestern Newfoundland. In *Animal bones, human societies*, ed. P. Rowley-Conwy, 65–73. Oxford: Oxbow Books.

Renouf, M.A.P. 2003. A review of Palaeoeskimo dwelling structures in Newfoundland and Labrador. *Études/Inuit/Studies* 27(1–2): 375–416.

Renouf, M.A.P. 2005. Phillip's Garden West: A Newfoundland Groswater variant. In *Contributions to the study of the Dorset Palaeo-Eskimos*, ed. P.D. Sutherland, 57–80. Mercury Series Archaeology Paper 167, Canadian Museum of Civilization, Gatineau.

Renouf, M.A.P., and M.S. Murray. 1999. Two winter dwellings at Phillip's Garden, a Dorset site in northwestern Newfoundland. *Arctic Anthropology* 36(1–2): 118–132.

Savelle, J.M. 1984. Cultural and natural formation processes of a historic Inuit snow dwelling site, Somerset Island, Arctic Canada. *American Antiquity* 49(3): 508–524.

Savelle, J.M., and M.T. Friesen. 1996. An odontocete (cetacea) meat utility index. *Journal of Archaeological Science* 23(5): 713–721.

Sergeant, D.E. 1985. *Underwater world: The hooded seal.* Department of Fisheries and Oceans, Marine Fish Division, Dartmouth.

Søby, R.M. 1970. The Eskimo animal cult. *Folk* 11–12: 43–78.

Stenson, G.B., Hammill, M.O., Kinsley, M.C.S., Sjare, B., Warren, W.G., and R.A. Myers. 1995. *Pup Production of Harp Seals, Phoca groenlandica, in the Northwest Atlantic during 1994.* Department of Fisheries and Oceans, Atlantic Fisheries Research Document 95/20, St. John's.

Stewart, F. L. 1979. Faunal Remains from the Factory Cove Site (DlBk-3), Newfoundland. In Factory Cove: Recognition and Definition of the Early Palaeo-Eskimo Period in Newfoundland, R. Auger, 217–222. Master's thesis, Department of Anthropology, Memorial University of Newfoundland, St. John's.

Stiner, M.C. 2002. On in situ attrition and vertebrate body part profiles. *Journal of Archaeological Science* 29: 979–991.

Stiner, M.C. 2004. A comparison of photon densitometry and computed tomography parameters of bone density in ungulate body part profiles. *Journal of Taphonomy* 2(1): 117–145.

Storå, J. 2001. *Reading bones: Stone age hunters and seals in the Baltic.* Stockholm Studies in Archaeology 21, Stockholm.

Stuiver, M., and P.J. Reimer. 1993. Extended 14C database and revised CALIB radiocarbon calibration program. *Radiocarbon* 35: 215–230.

Tanner, A. 1979. *Bringing home animals: Religious ideology and mode of production of the Mistassini Cree hunters.* Social and Economic Studies No. 23, Institute of Social and Economic Research, Memorial University of Newfoundland, St. John's.

Wells, P.J. 2002. An Analysis of Faunal Remains from Two Groswater Palaeoeskimo Sites at Port au Choix, Northwestern Newfoundland: Phillip's Garden West (EeBi-11) and Phillip's Garden East (EeBi-1). Master's thesis, Department of Anthropology, Memorial University of Newfoundland, St. John's.

Wells, P.J. 2005. Animal exploitation and season of occupation at the Groswater Palaeoeskimo site of Phillip's Garden West. *Newfoundland and Labrador Studies* 20(1): 75–90.

White, T.E. 1952. Observations on the butchering technique of some aboriginal peoples: No. 1. *American Antiquity* 4: 337–338.

White, T.E. 1956. The study of osteological materials in the plains. *American Antiquity* 21: 401–404.

# Chapter 5
# Mobility, Curation, and Exchange as Factors in the Distribution of the Phillip's Garden West Groswater Toolkit

**Karen Ryan**

## Introduction

The Groswater Palaeoeskimo site of Phillip's Garden West (EeBi-11) is situated on the Point Riche Peninsula, on the west coast of Newfoundland's Northern Peninsula (Fig. 5.1). Located on an approximately 500 $m^2$ terrace, 13 m above sea level (asl), the site was found in 1982 by Fitzhugh (1983) and excavated between 1990 and 1992 by Renouf (1991, 1992, 1993a, 1994). Although the lithic assemblage from the site is Groswater in affiliation, several of the artefact types are demonstrably different from Groswater toolkits reported elsewhere in Newfoundland, Labrador, and Quebec (Renouf 1994, 2005; Ryan 1997). These distinctive and atypical artefacts pose a number of culture-historical questions relating to the end of the Groswater period in Newfoundland (identified in this chapter as the Phillip's Garden West phase), while the geographic distribution of this toolkit hints at possible inter- and intra-cultural relations between late Groswater Phillip's Garden West phase populations and people of other cultural groups who migrated to the island of Newfoundland.

In this chapter, I first provide a brief overview of the Groswater period in Newfoundland (2990–1820 cal BP[1]) and of cultural developments in Newfoundland between 2110 and 1820 cal BP. Following this, a discussion of the metric and non-metric attributes used to distinguish Phillip's Garden West-type artefacts from more typical Groswater materials is presented. Unlike Renouf (2005), who demonstrated that differences exist in several classes of artefact between Phillip's Garden West and Phillip's Garden East (EeBi-1), the present paper focuses only on metric analysis of endblades, tying changes in this tool form to chronological developments.

[1]Unless otherwise indicated, all dates are calibrated to 1σ using Calib5.0 (Stuiver and Reimer 1993). See Appendix for further details of dates.

K. Ryan (✉)
Department of Anthropology, University of Toronto, 19 Russell Street, Toronto, ON, Canada, M5S 2S2
e-mail: karen.ryan@utoronto.ca

M.A.P. Renouf (ed.), *The Cultural Landscapes of Port au Choix*, Interdisciplinary Contributions to Archaeology, DOI 10.1007/978-1-4419-8324-4_5,

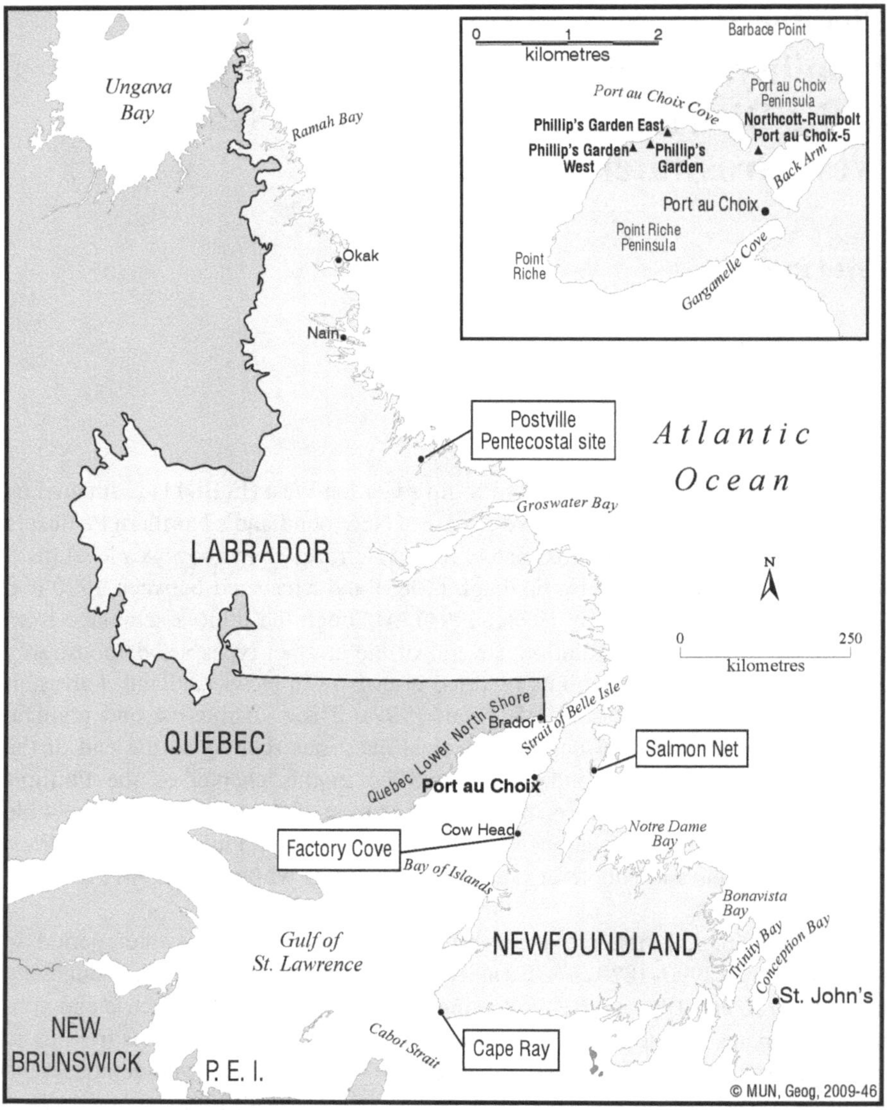


**Fig. 5.1** Location of Groswater, Dorset, and Recent Indian sites referred to in the text

Recognising the temporally sensitive nature of changing endblade styles is important because it allows isolated Phillip's Garden West style artefacts found elsewhere in Newfoundland to be chronologically situated, and also permits the identification of several mechanisms that may account for the geographic distribution of this distinctive artefact style. Viable explanations for the occurrence of Phillip's Garden West-type artefacts centre on aspects of group mobility, collection and possible curation of particular artefact forms, as well as the potential for interaction between terminal Phillip's Garden West Groswater groups and other populations.

## The Groswater Period (2990–1820 Cal BP)

Groswater was first defined as a regional variant of Dorset on the basis of archaeological work in the Groswater Bay area of central Labrador (Fitzhugh 1972:148–151; see also Tuck and Fitzhugh 1986) (Fig. 5.1). Since that initial research Groswater has been recognised as a distinct "transitional" population situated between earlier Pre-Dorset and later Dorset occupations and is now known from Diana and Ungava Bays in Nunavik (Gendron 1990, 1999; Plumet 1994), along virtually the entire coast of Labrador (Fitzhugh 1980; Tuck and Fitzhugh 1986), the Quebec Lower North Shore (LeBlanc 1996; Martijn 1974; Pintal 1994) and Newfoundland (Auger 1984; Kennett 1990; Renouf 1994; Tuck and Fitzhugh 1986). Uncalibrated radiocarbon dates indicate Groswater emerged as a recognisable cultural entity at approximately 2800 BP and became archaeologically invisible in Labrador by 2200 BP and by 1900 BP in Newfoundland (Loring and Cox 1986:66; Tuck 1988:112; Tuck and Fitzhugh 1986:164).

As with other "transitional" period cultures identified in the eastern Arctic, it is unclear how Groswater relates to the succeeding Dorset tradition, especially considering Early Dorset (2500–2400 BP, uncalibrated) is contemporaneous with Groswater in Labrador. Initial field research had indicated that Early Dorset was restricted to northern Labrador and Groswater to areas further south, with the intervening area, populated by Intermediate and Recent Indians, avoided by both Palaeoeskimo groups (Fitzhugh 1980:27–28; Nagle 1978). However, the identification of Newfoundland-sourced soapstone in Early Dorset sites from the Okak area by Cox (1977:37) called both this population distribution, and the assumed lack of opportunity for contact, into question. Subsequent work outside the community of Nain led Fitzhugh (1981) to suggest that contact between Groswater and Early Dorset populations in Labrador had not only occurred, but was more extensive than originally thought. However, Anton (2004) subsequently concluded that, although Groswater and Early Dorset did live in the same general areas at the same time, any contact was minimal and did not result in the level of influence envisioned by Fitzhugh (1981).

Although Groswater populations are now known to have inhabited much of coastal Newfoundland, Groswater materials were not initially distinguished from larger, typically Middle Dorset site assemblages (Renouf 1994:169; Tuck and Fitzhugh 1986:165). This is because Groswater sites, like other eastern Arctic Palaeoeskimo sites dating to the "transitional" period, are small and do not display a deep stratigraphy typically indicative of longer-term occupations (Fitzhugh 1976:114; Loring and Cox 1985; Maxwell 1985:121; McGhee 1981:39; Schledermann 1978:48). This suggests a high degree of residential mobility (LeBlanc 1996:121) that seems to be linked to a generalised economic strategy geared to exploiting both marine and terrestrial resources (Fitzhugh 1972:149, 1980:24; Marshall 1990:216; Tuck 1988:110; Tuck and Fitzhugh 1986:176).

This potentially relates to two circumstances. First, recognition of an Early Palaeoeskimo presence in Newfoundland did not occur until the mid-to-late 1970s (Bishop 1974; Tuck 1978), with the result that Groswater contexts were not expected and therefore Groswater artefacts were generally identified as finely made Dorset oddities (e.g. Harp 1964:46). Second, Groswater occupations can be difficult to discern, a consequence of the apparently high mobility strategy practised in many areas (LeBlanc 1996). At multi-component sites in particular, the ephemeral nature of a Groswater occupation typically meant that it was subsumed by much more intensive, generally Middle Dorset, occupations (Renouf 1994:169). Given these circumstances, information pertaining to Groswater comes primarily from four large unmixed sites (Fig. 5.1): the Postville Pentecostal site (GfBw-4) in Labrador (Loring and Cox 1986), and the western Newfoundland sites of Factory Cove (DlBk-3) (Auger 1982, 1984, 1986), Phillip's Garden East (Kennett 1990; Renouf 1994), and Phillip's Garden West (Renouf 1994, 2000, 2005; Ryan 1997; Wells 2002).[2]

The material excavated from these sites is consistent with the list of material culture traits for the Groswater culture first identified by Fitzhugh (1972:148–151, 1976:109; also Cox 1978:104) at sites in Groswater Bay. These include a variety of distinctive plano-convex side-notched endblades dominated by the "box-based type," large oval or circular sideblades, burins and burin-like tools, sidescrapers, stemmed or side-notched bifaces, "eared" endscrapers, microblades (some of which are stemmed or have side-notches), minimal production of ground slate tools, and round or oval soapstone vessels (Fig. 5.2a). It has been suggested that the use of Ramah chert increased through time at Labrador Groswater sites (Loring and Cox 1986:78), and that it is present in small quantities in most Groswater assemblages in Newfoundland (Auger 1986:113; Kennett 1990; Renouf 1994:174). However, the preferred raw material for Groswater groups in Newfoundland and the areas adjacent to the Strait of Belle Isle was the fine-grained and colourful chert that outcrops on the Cow Head Peninsula (Fig. 5.1) on Newfoundland's west coast (LeBlanc 1996:3; Nagle 1986:100–101; Pintal 1994:151; Tuck 1978). Faunal remains suggest that Groswater primarily exploited sea mammals, although birds, caribou, and other terrestrial mammals were also hunted (e.g. Auger 1984; Kennett 1990).

## Cultural Developments in Newfoundland 2110–1820 Cal BP

The disappearance of the Maritime Archaic Indians at approximately 3340 cal BP was originally thought to have created a population vacuum on Newfoundland which existed until the first Groswater groups crossed the Strait of Belle Isle by 2995 cal BP. However, recent research (see below) suggests that a very small Intermediate Indian population persisted on the island during this supposed hiatus. It was not until circa 2000 BP (Renouf et al. 2000, Table 6), when Middle Dorset and Recent Indians (the latter representing either an immigrant population or an in situ development) appeared, that Groswater may have become aware of relatively large numbers of "strangers."

[2] since this paper was written another unmixed Groswater site, Salmon Net (EfAx-25), was excavated on the east coast of the Northern Peninsula by Melnik (2007) (Fig. 5.1).

**Fig. 5.2** (**a**) A typical Groswater tool assemblage, represented by this assemblage from Phillip's Garden East. (**b**) Groswater tool assemblage from Phillip's Garden West

Middle Dorset (1990–1020 cal BP) were, like the Groswater, part of the Arctic Small Tool tradition (see Irving 1957) and were focused on marine resources to an even greater extent than the Groswater (Renouf 1999:408; but see Hodgetts et al. 2003), leaving behind larger sites that may have been occupied on a year-round basis (Murray 1992; Renouf 1999). The majority of substantial Middle Dorset sites, including Phillip's Garden (EeBi-1) (Murray 1992; Renouf 1993b, Chap. 7) and Cape Ray (CjBt-1) (Fogt 1998; Linnamae 1975) (Fig. 5.1), are located on headlands. Site location and analysed faunal remains from Phillip's Garden indicate that harp seal herds were the economic focus. Material culture remains include tip-fluted triangular endblades, burin-like tools, several types of ground slate tool, an absence of drilled holes, and the widespread use of rectangular soapstone vessels (Harp 1964). Although initially characterized as a singularly homogenous "Newfoundland Dorset" occupation (e.g. Fitzhugh 1980:22–23, 26; Harp 1969/1970:123; Linnamae 1975:93), it is now recognised that a strong degree of regionalism existed that most probably resulted from decreased group mobility and an attendant intensification in the use of local resources (LeBlanc 2000:102, 2008; Robbins 1986).

Unlike Groswater and Middle Dorset Palaeoeskimos, the people of the Cow Head complex (2110–930 cal BP) are Amerindians. Few sites of this period are known, with the result that settlement and subsistence strategies are poorly understood (Renouf et al., Chap. 13; Teal 2001:17; Tuck 1978); limited data suggest that, like later Recent Indian populations, Cow Head complex groups were marine-oriented generalists (Cridland 1998; Holly 1997; Pastore 1986; Rowley-Conwy 1990; Schwarz 1994). In terms of material culture, the complex is characterized by stemmed and notched projectile points in a number of forms, small endscrapers, large sidescrapers, ovate, lanceolate, and bi-pointed bifaces, blade-like flakes, and high numbers of bifacial preforms, as well as small amounts of pottery (Teal 2001:13; Tuck 1988:158).

The origins of the Cow Head complex are unclear. Similarities in material culture between contemporary Amerindian populations on either side of the Strait of Belle Isle, including the occurrence of Ramah chert in Newfoundland and Newfoundland cherts in Labrador and the Quebec Lower North Shore (Loring 1992; Pintal 1998, 2000), suggested to Hartery (2001, 2007) that the Cow Head complex represents a non-resident population that originated in the Quebec-Labrador area and migrated to Newfoundland. However, identification of materials identified as Intermediate Indian (e.g. Beaton 2004) in Newfoundland, albeit limited, lends credence to Tuck's (1988:159, 160) hypothesis of in situ development from Maritime Archaic into Recent Indian.

In summary, radiocarbon dates indicate that three distinct populations inhabited Newfoundland in the period between 2110 and 1820 cal BP. Given Newfoundland's low ecological productivity (Bergerud 1983) and the fact that the two chief mammalian species, harp seals and caribou, are only seasonally available, it was no doubt necessary for each of these groups to carefully schedule their movements so that these resources could be procured where and when available (see Tuck and Pastore 1985:74–77). Perhaps because of this, Schwarz (1994:65, Fig. 3) suggested that the only major difference in subsistence and settlement patterns occurred during the winter period, when Recent Indians moved to the near interior or near coastal zone from which they could access both coastal and interior resources

(Rowley-Conwy 1990), while the Palaeoeskimos remained on the coast. Considering that for the remainder of the year settlement patterns were broadly similar (Schwarz 1994:59), inter-cultural contact, as has been identified between Middle Dorset and Recent Indians at Port au Choix (Renouf 2003; Renouf et al. 2000; Teal 2001: 82–83), must have been virtually unavoidable at some points.

Terminal Groswater populations may also have been interacting and trading with the Recent Indians, as implied by the identification of Cow Head chert in Recent Indian Flèche Littorale (2500–1500 BP, uncalibrated) sites on the Quebec Lower North Shore (Pintal 1998:117). Fitzhugh (1980:25, 28) alludes to additional indicators of contact near Brador, on the Lower North Shore, while Hartery (2001:129) suggests Groswater-Recent Indian interaction on Newfoundland's west coast. It is apparent from these findings that the period between 2110 and 1820 cal BP was a dynamic and culturally complex time on Newfoundland and that the cultural developments that occurred there also played out on the other side of the Strait of Belle Isle. How these events may help account for the distribution of Phillip's Garden West-type artefacts is explored in the following sections of this paper.

## The Phillip's Garden West Site and Toolkit

The Phillip's Garden West site was initially interpreted as a short-term warm weather occupation based on an absence of soapstone, a small amount of fire-cracked rock, and a lightly constructed circular tent-like structure, in combination with a relatively low number of artefacts (Renouf 1994:188). Analysis of the excavated faunal remains has, however, indicated that the site was actually occupied during the late winter/spring period, with occupation possibly extending into the early summer (Wells 2002, Chap. 4). Based on a series of ten uncalibrated radiocarbon dates indicating the site was revisited over a period of 600 years (Fig. 5.3), Renouf (1994:184, Table 4) originally partitioned the site into three chronologically discrete loci: a late occupation on the terrace top; early deposits in the lower part of the hillside midden; and intermediate-aged deposits on the upper hillside midden (Renouf 1994:184, Table 4). A small activity area was also identified near the base of the sloping midden on the 6.5 m asl lower terrace (Renouf 1993a:11–12). Renouf (2005) subsequently combined the three site divisions into two for the purposes of comparing the Phillip's Garden West and Phillip's Garden East chronologies. Phillip's Garden West-1 (PGW1) included the total midden area and the activity area at the midden base and Phillip's Garden West-2 (PGW2) included the terrace top deposits. Calibrated and uncalibrated radiocarbon dates showed considerable but not complete overlap between PGW1 and PGW2; the oldest dates were within PGW1 and, with one exception, the youngest dates were within PGW2 (Renouf 2005:61–65). Since the intention here is to examine chronological trends within Phillip's Garden West, the original and more spatially detailed (lower hillside, middle hillside and upper terrace) (Fig. 5.4) site divisions are used.

Renouf (1994:183, 2005) previously identified several general characteristics that typify the Phillip's Garden West lithic toolkit (Fig. 5.2b). These include a

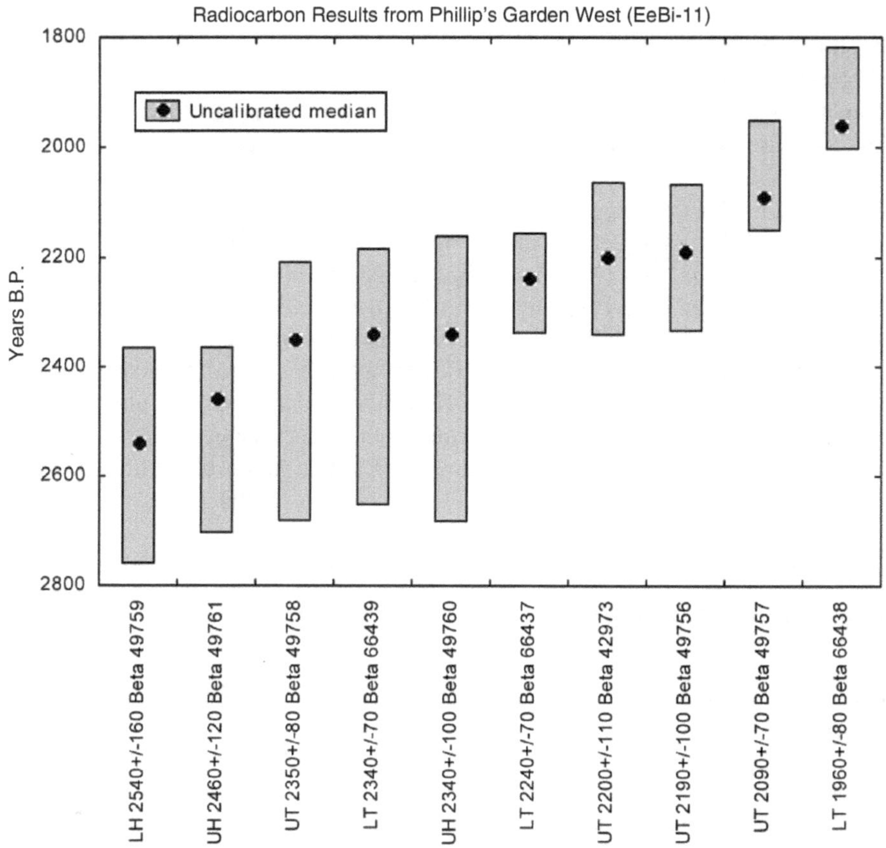


**Fig. 5.3** Radiocarbon results from Phillip's Garden West (EeBi-11) by locus (*LH* lower hillside midden; *UH* upper hillside; *UT* upper terrace; *LT* lower terrace)

preference for high-quality colourful Cow Head cherts, extremely fine pressure (often parallel) flaking, a high frequency of surface grinding and polishing, and finely executed edge serration. In reference to the endblades from the site, Renouf (1994:184, 2005:65) noted a preference for side-notches that were narrower than those more commonly seen in Groswater contexts, the occurrence of multiple side-notches on some examples, more elongated blades, and bases that are often concave and may also have lateral tangs. Renouf (1994:184) also grossly sorted the endblades into two impressionistic categories, an "exquisite" class and a less well-crafted group. Chronological influences were suggested to account for this variability, where less finely made specimens were produced early in the sequence and the best made examples were created during later periods (Renouf 1994:184).

Ryan (1997) conducted a comparative analysis of the Phillip's Garden West endblades and confirmed Renouf's impression that the endblades associated with the Phillip's Garden West phase (Fig. 5.2b) were morphologically distinguishable from classic Groswater examples (Fig. 5.2a). As shown in Fig. 5.5a, when the

**Fig. 5.4** Phillip's Garden West site annotated to show (**a**) the upper terrace, (**b**) the lower hillside and (**c**) the lower terrace

length and width measurements of Phillip's Garden West endblades are compared to those reported from Phillip's Garden East, the Postville Pentecostal site and Factory Cove, the Phillip's Garden West endblades display the highest length-to-width ratio. Additionally, a comparison of the side-notch measurements of endblades from Phillip's Garden West with those from the Postville Pentecostal site (the only other Groswater site for which these measurements have been published) demonstrate in a quantifiable manner that the side-notches of the endblades discovered at the Postville Pentecostal site are much shallower and broader than those from Phillip's Garden West (Fig. 5.5b).

Several additional criteria also proved to be useful for distinguishing Phillip's Garden West endblades from more common Groswater examples (Fig. 5.5c). Of these, the high frequency of endblades made from Cow Head chert, as well as endblades exhibiting plano-convex cross-sections, are less useful as these are attributes shared by all Groswater endblades. However, concave bases (sometimes with lateral tangs), the prevalence of grinding or more finely accomplished polishing (both of which can be quite extensive), and lateral edge serration are frequently observed on Phillip's Garden West-type endblades while remaining comparatively rare on more typical Groswater specimens. As such, a visual assessment of these attributes, in combination with the metric analyses described above, is quite helpful for the identification of Phillip's Garden West-type endblades, especially in instances where the artefact under examination is fragmentary.

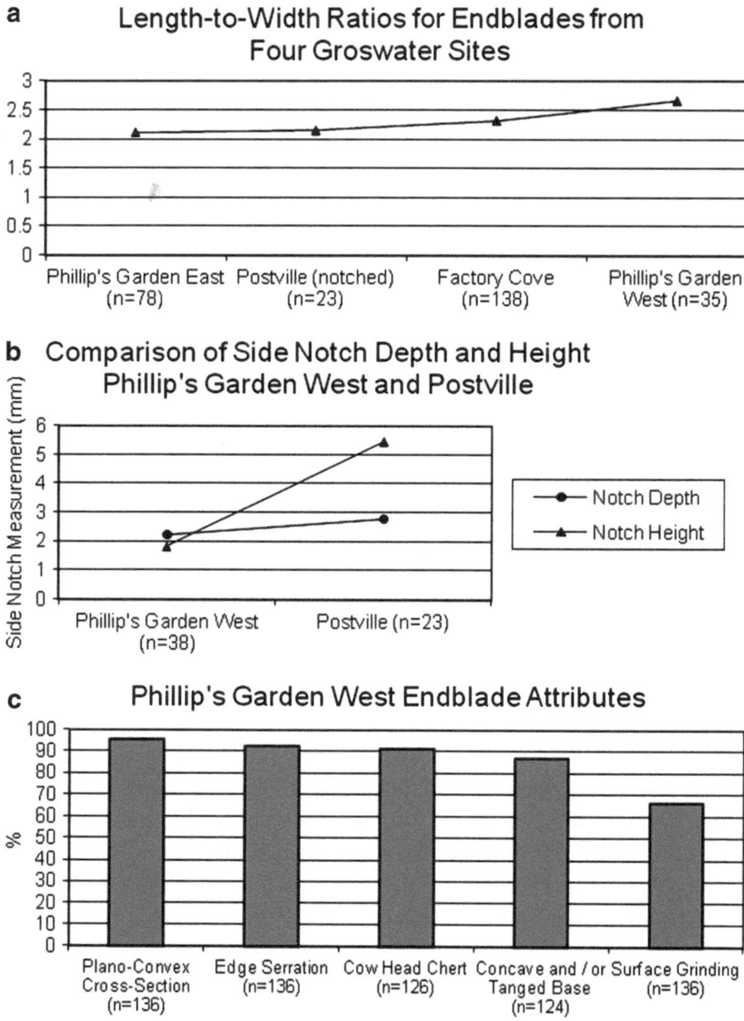


**Fig. 5.5** Chronological development of the Phillip's Garden West-type endblade shown through: (**a**) comparison of endblade measurements among four Groswater sites, (**b**) comparison of endblade side-notch measurements between the Phillip's Garden West and Postville sites, and (**c**) additional attributes used to define Phillip's Garden West-type endblades

The remaining artefacts making up the Phillip's Garden West toolkit are: triangular or ovate sideblades, unifacial and bifacial concave sidescrapers, and side-notched bifacial knives (Fig. 5.2b). Typically, Phillip's Garden West sideblades can be distinguished by their small and distinctive form, which is plano-convex in comparison to typical ovate to triangular Groswater sideblades, and also by the frequent appearance of edge serration, surface grinding, and the precise pressure flaking typical of the Phillip's Garden West toolkit (see also Renouf 2005:67). Five bifacial knives are also included as part of the Phillip's Garden West inventory despite the fact that all were recovered from sites outside the Port au Choix area (Ryan 1997). These knives are made on Cow Head chert using exacting parallel pressure flaking, display edge serration, are remarkably similar in outline (unlike typical Groswater bifaces, which Tuck (1988:109) notes are variable in form) and exhibit narrow elongated blades. The single complete example also has an asymmetrically pointed distal end and a bifacially bevelled base. None are ground, and although the only complete example has relatively broad side-notches when compared to the notches on Phillip's Garden West-type endblades, the notches are much narrower than on more typical Groswater bifaces. The final artefact type, unifacial and bifacial concave sidescrapers, have no analogous tool type in classic Groswater assemblages.

In summary, four artefact types – endblades, sideblades, bifacial knives, and unifacial and bifacial concave sidescrapers – have thus far been clearly linked to the terminal Groswater Phillip's Garden West phase in Newfoundland (Renouf 2005 also distinguishes additional tool categories). As discussed previously, changes in the appearance of endblades from the Phillip's Garden West site (including an increasing frequency of edge serration, increased elongation, and more finely accomplished pressure flaking) have been linked to temporal considerations (Renouf 1994:184; Ryan 1997), a suggestion which will be more fully investigated in the following section.

## Chronological Development of the Phillip's Garden West Endblade

The ten radiocarbon dates from Phillip's Garden West indicate that the site was subject to repeated, if not intensive, occupation by Groswater for approximately 600 years (Fig. 5.3). As interpreted by Renouf (1994:181, 188), site occupants lived primarily on the upper terrace, throwing their garbage down the midden slope. Periodic activities also took place on the lower terrace and the resultant refuse of these episodes was deposited near the base of the midden (Renouf 1993a:11–12). Based on radiocarbon dates, the most recent site occupation is the upper terrace and its contemporary activity area is on the lower terrace (Fig. 5.3). Not surprisingly, some dates from the midden overlap the range of occupation on the upper terrace; however, it is evident that the midden deposits become progressively older as they decrease in elevation down the slope (Figs. 5.3 and 5.4).

When analysing the cultural material from the site, Renouf (1994:184) observed that the highest frequency of finely made endblades occurred on the terrace top, a ratio that became progressively smaller when analysis moved to the upper and lower portions of the hillside midden. Linking these stylistic changes to the well-dated terrace and midden areas of the site, she proposed that temporal considerations might account for the changing frequencies of finely-crafted endblades (Renouf 1994:184). Although she is not explicit concerning the criteria used to differentiate between the two groups of endblades, it appears that less adept flaking skill, greater spacing between individual lateral serrations, an overall tendency towards

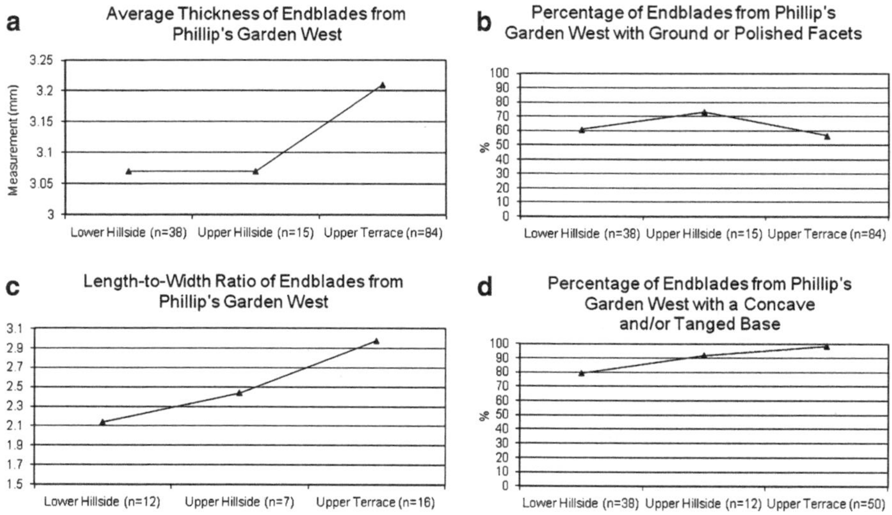


**Fig. 5.6** Chronological development of the Phillip's Garden West-type endblade shown through: (**a**) changes through time in endblade thickness at Phillip's Garden West, (**b**) changes through time in the occurrence of grinding or polishing facets on endblades from Phillip's Garden West, (**c**) changes through time in the length-to-width ratio for endblades from Phillip's Garden West, and (**d**) changes through time in the frequency of endblades with concave and/or tanged proximal bases at the Phillip's Garden West site

wider and thicker dimensions, and less elongated blades all serve to distinguish the less aesthetically pleasing examples from those produced at the end of the sequence (Renouf 1994:184).

An examination of the endblades from Phillip's Garden West using a variety of metric attributes confirms the proposal that there is a progression through time at the site towards the production of increasingly better-crafted endblades (Fig. 5.6a–d). For example, Fig. 5.6a shows the average thickness for the endblades associated with each of the three areas of the site. As can be seen, mean endblade thickness is consistent from the lower hillside through upper hillside periods at 3.07 mm, before increasing to 3.21 mm during the upper terrace period. While this might be taken as an indication that endblades made during the latest occupation of the site were more robust than those produced during earlier periods, it should be kept in mind that the increase in thickness is less than one-fifth of one millimetre. Given such a minor rise in value, it seems more likely that the increased thickness resulted from a coincident decrease in the frequency with which endblades were ground. As shown in Fig. 5.6b, grinding decreases by approximately 16% from the upper hillside period to the upper terrace period, or from roughly 73 to 57%.

It is argued that the decrease in grinding seen during the upper terrace phase of site occupation serves as a useful proxy indicator for increased skill levels among tool-makers at the site. As discussed by Whittaker (1994:167, 170), it takes a high degree of control to remove flakes from a biface in a parallel-patterned fashion, such as that seen on many of the upper terrace endblades. If the techniques used by the tool-makers in this period were sufficiently developed (as they certainly appear to have been), grinding might not have always been a necessary production step for controlling flaking platforms and eliminating flaws such as step and hinge fractures. Indeed, the overall decrease in grinding noted on upper terrace endblades may be taken as an indication that bifacial thinning techniques had advanced to a point where toolmakers were less dependent upon grinding to produce the desired product. One might also infer, given the obvious preference for better-crafted endblades during the latest period of occupation at the site, that the practise of grinding endblades may have been discouraged as it might have detracted from the final appearance of the artefact.

More evidence that the techniques employed to create endblades were changing with time can be found when the length-to-width ratios of endblades from each of the three site areas are compared. The length-to-width ratio was previously used to illustrate the difference between Phillip's Garden West-type endblades and those from typical Groswater assemblages, where the higher ratio of the Phillip's Garden West endblades was taken to indicate a more advanced level of production, or at least a preference for endblades to take this form (Fig. 5.4a). As shown in Fig. 5.6c, when the length and width ratios for the endblades from each of the three Phillip's Garden West site areas are compared, the lower hillside midden endblades have the lowest value (i.e. they are less elongated), while the upper terrace endblades display the highest ratio (i.e. these are the most elongated). Upper hillside endblades are intermediate between the two. These results verify that a definite trend towards increasingly elongated endblades from the earliest to latest components at the Phillip's Garden West site exists. Using this as an indicator for fineness of production, it is clear that endblades at the site become increasingly finer through time.

A final attribute, base form, can also be used to corroborate Renouf's (1994:184) original hypothesis that morphological differences identified in the endblades from the Phillip's Garden West site have chronological significance. As noted earlier (Fig. 5.5c), when differentiating between Phillip's Garden West-type and classic Groswater endblades, 87% of Phillip's Garden West-type endblades have a base that can be characterized as concave (with or without lateral tangs) (Fig. 5.2b) in contrast to typical Groswater endblade bases which are more or less straight from lateral edge to lateral edge (Fig. 5.2a). Interestingly, when the percentage of concave and/or tanged bases is calculated separately for each of the three chronologically discrete site areas, the frequency of this attribute varies (Fig. 5.6d). Endblades in the earliest deposits (the lower hillside midden) have the lowest percentage of concave and/or tanged bases at approximately 79%, while almost 92% of the endblades from the upper hillside midden deposits have this distinctive base. Finally, 98% of endblades excavated from the upper terrace, the most recently occupied area of the site, are concave and/or tanged.

Thus, it is apparent, based on the analysis of endblade attributes, that Phillip's Garden West endblades are not only morphologically distinguishable from typical Groswater artefacts, but are also a chronologically sensitive time marker of the general Phillip's Garden West phase (Figs. 5.5a–c, 5.6a–d, and 5.7). This latter point is a key development in terms of identifying and attempting to temporally situate Phillip's Garden West-type endblades identified at locations outside the type site, especially when one realises that this material is often the only Groswater artefact at such sites (refer to Ryan 1997).

**Fig 5.7** Endblades from the lower hillside (*bottom row*), upper hillside (*middle row*), and upper terrace (*top row*) components of the Phillip's Garden West site

## The Distribution of Phillip's Garden West-Type Artefacts

Prior to Ryan's (1997) examination of site collections housed at the Newfoundland Museum (now The Rooms Provincial Museum) which were supplemented in 2006 by inspection of materials stored at the Canadian Museum of Civilization in Gatineau, Phillip's Garden West material had only been identified from five

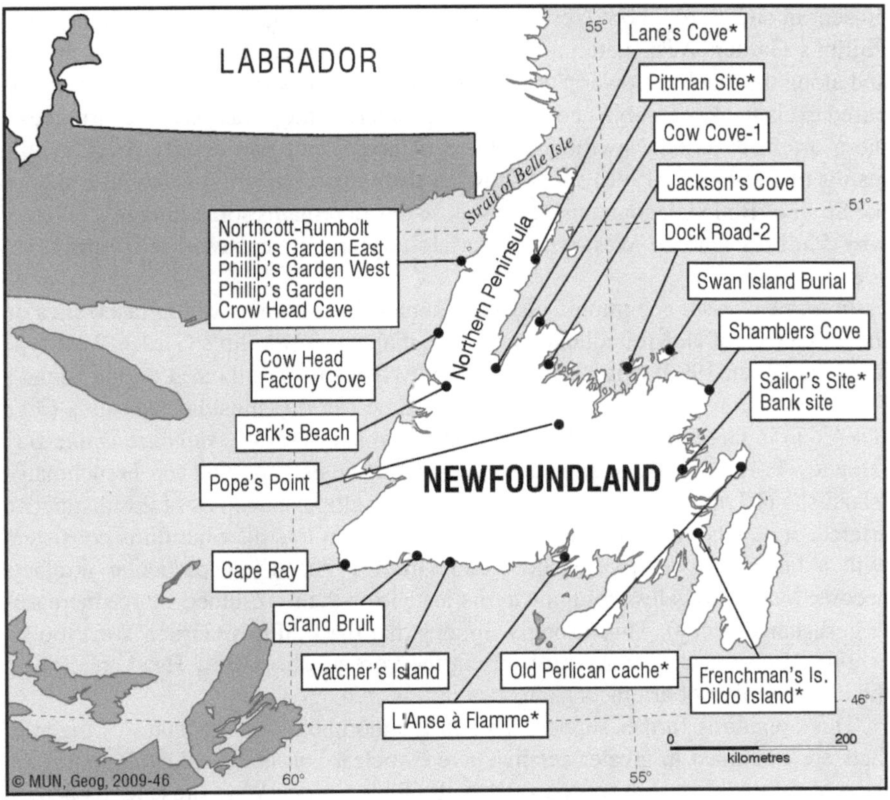


**Fig. 5.8** Known distribution of Phillip's Garden West-type artefacts. Site names with an *asterisk* indicate tentative identification

locations: Phillip's Garden West, Phillip's Garden (EeBi-1) (Harp 1964), the Port au Choix-5 or Northcott/Rumbolt site (EeBi-5, -7) (Harp 1964; Renouf 1985), the Cow Head site (DlBk-1) (Tuck 1978) and Frenchman's Island (ClAl-1) (Evans 1982) (Fig. 5.8).[3] The identification of Phillip's Garden West material at the Port au Choix and Cow Head sites was not surprising given their proximity to both the type site and the main raw material source. However, the fifth locale, Frenchman's Island, is situated at the bottom of Trinity Bay on the east coast of Newfoundland (Figs. 5.1 and 5.8) and as such was an unexpectedly isolated find.

But when site collections throughout Newfoundland and Labrador were examined in order to determine whether unrecognised Phillip's Garden West-type materials occurred in other sites, it quickly became obvious that elements of the toolkit were

[3]material from the Salmon Net site (Fig. 5.1) is similar to, but not exactly the same as, Phillip's Garden West-type material (Melnik 2007). Similar material comprised the Groswater component of the Recent Indian St. Pauls Bay-2 site (DlBk-6) at St. Pauls Inlet (Fig. 5.1) (Lavers and Renouf 2009).

present in sites throughout Newfoundland (Fig. 5.8, Table 5.1). At the same time, Phillip's Garden West material was conspicuously absent from sites in Labrador and along the Quebec Lower North Shore (although my survey of Quebec sites was based exclusively on published reports). Considering the widespread occurrence of these artefacts along Newfoundland's west coast, and particularly recalling the results of LeBlanc's (1996) study showing that Groswater groups on either side of the Strait of Belle Isle maintained very close ties, it remains something of a mystery why Phillip's Garden West-type materials appear not to have been transported across the Strait.

In terms of both geographic distribution and sheer number of artefacts, sites on the west coast of Newfoundland contain the majority of Phillip's Garden West-type artefacts (Ryan 1997) (Fig. 5.8). In fact, only five sites not located on the island's west coast: Cow Cove-1 (EaBa-14) on the Baie Verte Peninsula; Jackson's Cove (DkAx-1) in Green Bay; the Swan Island burial site (DiAs-9) in Notre Dame Bay (Howley 1915); the Bank site (DdAk-5) in Bonavista Bay; and the Frenchman's Island site in Trinity Bay (Evans 1982); contain multiple examples of this distinctive artefact style (Ryan 1997). The pattern evident from this distribution is consistent with a fall-off model (see Hodder and Orton 1976) where particular artefacts become increasingly less common as distance from their presumed source increases (e.g. distance-decay). This strongly implies that the Phillip's Garden West toolkit originated on the west coast, probably in the Port au Choix-Cow Head area, given the large number of artefacts from sites in this area.

This opinion is further supported when the morphological attributes of the artefacts are examined in greater detail. There is a clear chronological evolution in the form of the endblades associated with the Phillip's Garden West site (Fig. 5.6a–d), a developmental sequence that should logically be applicable to specimens located outside the type site. Employing the identified chronological markers as a guide, it is possible to at least broadly place individual endblades within the temporal sequence identified at Phillip's Garden West. From this comparative analysis it becomes clear that while endblades from each of the three time periods identified at the type site occur on the west coast, only endblades falling within upper hillside and upper terrace ranges have been identified at sites elsewhere in Newfoundland (Table 5.1).

These findings pose a number of new questions regarding the distribution of this distinctive toolkit. What mechanisms can be proposed to account for the geographic distribution of Phillip's Garden West-type artefacts in Newfoundland? Why is it that the earliest form of the Phillip's Garden West endblade is only found on the west coast of Newfoundland, while those that were apparently produced later in the sequence are so widely distributed? Why are so many of the sites at which these artefacts are found either dominated by non-Groswater assemblages or contain Groswater assemblages that are almost completely of the typical Groswater style? What, if any, significance should be attached to the Jackson's Cove site, which contains both a large number of Phillip's Garden West-type artefacts (11 endblades and two bifacial knives) and more typically-styled Groswater tools? Finally, why is it that no specimens occur in sites across the Strait of Belle Isle? Some of these questions will be explored in the following section.

**Table 5.1** Distribution of known Phillip's Garden West-type artefacts by proposed chronological period

| | Lower hillside (1820–2650 cal BP) | Upper hillside (2160–2760 cal BP) | Upper terrace (1950–2680 cal BP) |
|---|---|---|---|
| Phillip's Garden West (EeBi-11) (Groswater) | Present | Present | Present |
| Phillip's Garden (EeBi-1) (Middle Dorset) | Present | Present | Present |
| Crow Head Cave (EeBi-4) (Middle Dorset) | | Present | |
| Northcott/Rumboldt (EeBi-5, -7) (Groswater, Middle Dorset) | Present | | |
| Cow Head, Band 5–6 (DlBk-1) (Groswater) | Present | | |
| Parke's Beach (DgBm-1) (Groswater, Middle Dorset, Recent Indian) | Present | Unclear | Unclear |
| Grand Bruit (CkBn-1) (Middle Dorset) | | | Present |
| Vatcher's Island (CjBj-8) (Groswater, Middle Dorset) | | | Present |
| Frenchman's Island (ClAl-1) (Middle Dorset) | | | Present |
| Bank (DdAk-5) (Groswater, Middle Dorset, Recent Indian) | | | Present |
| Shambler's Cove (DgAj-1) (Groswater, Middle Dorset, Recent Indian) | | | Present |
| Swan Island Burial (DiAs-9) (Recent Indian) | | | Present |
| Dock Road-2 (DjAq-8) (Groswater, Middle Dorset) | | | |
| Jackson's Cove (DkAx-1) (Groswater) | | Present | Present |
| Cow Cove-1 (EaBa-14) (Groswater) | | Unclear | Present |
| Pope's Point (DfBa-1) (Recent Indian) | | Unclear | |

Only endblades and bifaces exhibit sufficient morphological change through time for temporal placement to be possible. Those specimens where the identification is tentative are called "unclear." In many cases this tentative designation was a result of the incompleteness of the artefact. In other cases the artefact lacked a particular attribute, for example a concave base or edge serration. See Ryan (1997) for more a detailed discussion

## The Mechanisms Behind the Distribution of the Phillip's Garden West Toolkit

Three propositions, not necessarily mutually exclusive, are offered to explain the distribution of Phillip's Garden West-type artefacts. The first of these, mobility, centres on the idea that the manufacturers of the Phillip's Garden West artefacts were the same people who transported them throughout Newfoundland as a part of their seasonal round. The second explanation, curation, postulates that artefacts of this toolkit were collected archaeologically and transported throughout Newfoundland as curiosities by some combination of Groswater, Middle Dorset, and Recent Indian populations. The third suggestion, exchange, supposes that Phillip's Garden West materials were transported to sites throughout Newfoundland by means of their purposeful exchange with other cultural populations, either non-Phillip's Garden West Groswater, Middle Dorset or Recent Indian. Each of these proposals is evaluated in turn.

### *Mobility*

As mentioned above, Groswater populations were highly mobile, moving from one location to another on a regular basis (LeBlanc 1996). Given this tendency, it is possible that a small group of toolmakers, who occasionally returned to the Phillip's Garden West site over several generations, travelled along the north-eastern and southern coasts of Newfoundland for the remainder of the year(s), depositing the occasional Phillip's Garden West-type artefact at sites along the way. However, even considering the large amount of territory Groswater populations appear to have covered, it is difficult to accept that the apparently small population that produced the Phillip's Garden West toolkit would traverse the entire coast of Newfoundland, leaving only the slimmest indications of its passing.

Additionally, with the exception of the Cow Cove-1, Jackson's Cove, Swan Island, Bank, and Frenchman's Island sites, all Phillip's Garden West-type artefacts identified outside the Port au Choix-Cow Head Peninsula area occur as single finds (Ryan 1997). It is possible that the Groswater occupation of some sites may have been brief and associated debris so ephemeral that components were not identified, as suggested by the fact that seven of the 16 sites containing Phillip's Garden West-type material have no other recognisable Groswater elements (Ryan 1997) (Table 5.1). However, the fact that in most cases only a single Phillip's Garden West artefact is present argues against the idea of direct transportation by Phillip's Garden West toolmakers, despite the highly mobile seasonal round that Groswater groups are thought to have practised.

## *Curation*

Radiocarbon dates suggest that terminal Groswater may have shared Newfoundland with Middle Dorset and Recent Indian Cow Head complex populations (see Appendix) although it is unclear whether Groswater populations overlapped in both time and space with either group (as Dorset and Recent Indian populations did, Renouf et al. 2000, Table 6). Whether or not Groswater groups were still present in Newfoundland when the other two cultural groups arrived, it is conceivable that Phillip's Garden West artefacts were collected from abandoned Groswater sites by Dorset or Recent Indian individuals who practised what Bielawski (1979:105) termed "antiquarianism." In this manner, a site such as Phillip's Garden West, visible from the Middle Dorset site of Phillip's Garden (Fig. 5.9) and certainly within the range of Recent Indians at the Gould (EeBi-42) site in modern Port au Choix (Fig. 5.1), could have been visited and some artefacts collected as curiosities. This may help to explain why so many Phillip's Garden West-type artefacts were excavated from Phillip's Garden (see Harp 1964:45). Ryan (1997) located 86 in her initial, preliminary examination. This is the largest number of Phillip's Garden West-type artefacts found outside the site.

Unfortunately, there is no easy way to distinguish between an artefact brought to a site as a curated object (or "salvaged," see Schiffer 1987:104) vs. one that is

**Fig. 5.9** View of Phillip's Garden (the meadow in the background) looking east from) the hill that also overlooks Phillip's Garden West (*midground* with light patch)

present because it was an in situ element of an earlier occupation. One possible method is reported by Park (1993:222–223) who argues that curated artefacts are usually eye-catching (e.g. harpoon heads and carvings) while earlier non-curated artefacts include more mundane or commonplace forms. Using this observation as a basis for examining the Phillip's Garden West assemblage from Phillip's Garden site, presents an interesting finding. Although the Phillip's Garden West assemblage from the site also includes sideblades and unifacial and bifacial concave sidescrapers, almost 84% of all artefacts are endblades (Ryan 1997).[4] Considering that Phillip's Garden West-type endblades are quite distinctive and are the most remarkable and "collectible" element of the toolkit, there is at least a superficial agreement with Park's (1993) observation.

Curation is therefore a logical possible explanation for the occurrence of Phillip's Garden West artefacts elsewhere in Newfoundland. Coupling the discontinuous distribution of Phillip's Garden West artefacts with the fact that seven of 16 sites containing the material have no other evidence of a Groswater component (Table 5.1) suggests that Phillip's Garden West Groswater individuals did not visit these sites and could not have been responsible for depositing the identified artefacts.

## *Inter- and Intra-Cultural Trade*

Radiocarbon dates from Port au Choix sites indicate that Groswater, Middle Dorset, and Recent Indians may have overlapped chronologically (Renouf et al. 2000, Table 6). There is evidence that at least some of these occupations may have taken place in the same season. The longest and most intensively occupied locale is Phillip's Garden, which Middle Dorset inhabited on a multi-seasonal basis within the context of a primary focus on late winter/early spring harp seal hunting (Harp 1976:132; Hodgetts et al. 2003:114–116; Renouf 1994:190–191, Chap. 7). Although Teal (2001:97; see also Renouf et al., Chap. 13) notes that insufficient faunal material was found in the Cow Head complex component of the Gould site to determine seasonality, its apparently strategic location, offering easy access to the interior and ocean, can be used to suggest a possible fall–winter occupation (Rowley-Conwy 1990; Schwarz 1994). Wells' (2002) analysis of faunal material from the Phillip's Garden West site indicates that the site was occupied during the late winter and early spring harp seal migration, and perhaps into the early summer, making it feasible that Middle Dorset and Recent Indians may have been present in the Port au Choix area when the Phillip's Garden West site was occupied.

---

[4] since this paper was written Lavers (2006) studied the Groswater artefacts found in the complete Phillip's Garden artefact assemblage of >35,000 tools. Almost all of the 280 Groswater artefacts from the assemblage are Phillip's Garden West-type. Endblades comprised 121 (42%) of that total (Lavers 2006:14).

Unlike the Gould site where Middle Dorset and Cow Head complex groups appear to have interacted (Renouf et al. 2000), there is no direct evidence of extra-cultural contact involving Groswater populations in the Port au Choix area (although see Fitzhugh 1980; Hartery 2001; Pintal 1998). However, if contact did occur, it most probably would have happened on the Point Riche Peninsula, where the late winter/early spring harp seal hunting sites of Phillip's Garden West, Phillip's Garden East, and Phillip's Garden are located (Fig. 5.1). If Groswater and Middle Dorset populations were occupying these sites contemporaneously (a possibility suggested by radiocarbon dates) then some Phillip's Garden West-type material may have been incorporated into the Phillip's Garden Dorset deposits as a result of contact and possible trading activities. The recovery of a Middle Dorset-like whalebone sled runner from the Phillip's Garden West midden (Renouf 1992:40) may suggest one reciprocal trade item.

## Conclusions

I began this chapter by posing a number of questions relating to how and why the Phillip's Garden West artefact assemblage came to be so widely distributed throughout Newfoundland. A single definitive answer has not been found and the suggestions offered here are tentative and exploratory. However, based on more recent research it is becoming increasingly apparent that the Strait of Belle Isle was a very dynamic place in the years bracketing 2000 BP, with three possibly contemporary and culturally distinct populations living there. That this chronological and spatial overlap appears not to have resulted in any significant or archaeologically visible enculturation is not surprising (although Marshall (1978:152) suggests borrowing may have occurred), particularly if that contact was limited. In such situations, as discussed by McGhee (1997:212) with reference to the Late Dorset and Thule Inuit, any evidence of interaction can be expected to be "slight, ambiguous [and] difficult to interpret."

On morphological grounds I suggested that the Phillip's Garden West toolkit was first conceived and created in the Port au Choix-Cow Head area and spread from there to the rest of Newfoundland, although the underlying reasons for its inception may never be known. If the Phillip's Garden West phase did originate on Newfoundland's west coast, and considering the long history of communication and movement over the Strait of Belle Isle (Howley 1915:257, 270; Marshall 1996:54–60, 277; Martijn 1990:229–230; Pastore 1987:57–58, 1989:61; although see Robbins 1989), it is very surprising that no Phillip's Garden West-type artefacts have been identified in Labrador or Quebec. It is likely that the Groswater occupation of Labrador and Quebec ended prior to 2000 BP and that the Groswater population in Newfoundland after this point was extremely isolated. Following this scenario, this remnant Groswater population may have been motivated to interact with the Middle Dorset and Recent Indian groups (who first occupied Newfoundland in the centuries around 2000 BP) in an attempt to survive cultural extinction.

One final and fundamental question must be addressed before more in-depth analysis of the Phillip's Garden West assemblage is possible. There is a continuity of elaboration associated with Phillip's Garden West endblades, where increasingly greater amounts of time and energy were invested in the creation of individual artefacts. This investment surpassed the functional requirements of the endblade, suggesting that the production sequence had transcended the practical considerations of artefact form and function to encompass more esoteric concerns (e.g. Geneste and Plisson 1990:304–316). In order to properly evaluate this observation, it is necessary to conduct use-wear and fracture pattern studies to determine whether Phillip's Garden West artefacts were broken through use, and were therefore intended to be functional tools, or were created for some other, perhaps symbolic, purpose (e.g. Renouf 2005; Wells, Chap. 4). Such studies will add another layer of meaning to the already intriguing Phillip's Garden West toolkit.

**Acknowledgements** This paper is based on the results of research carried out as part of my Honour's paper and was presented in the session "Archaeology at Port au Choix, Northwestern Newfoundland" at the 36th Annual Meeting of the Canadian Archaeological Association in Hamilton. I am indebted to a number of people for making the original research, and this paper, possible. Kevin McAleese, Curator of Archaeology and Ethnology at the Newfoundland Museum (now The Rooms Provincial Museum), gave me free rein to explore the provincial collections, helped me dig the Frenchman's Island collection out of the artefact purgatory that is storage, and allowed me to borrow some of the Phillip's Garden West material housed at the Museum for study and photos. In the Collections Room, Delphina Mercer and Katherine Scott were patient and helpful souls who let me go through the newly-returned Phillip's Garden material in 1996 before and during its re-cataloguing with nary a word of complaint. Elaine Anton and Lori Balson graciously permitted me to borrow artefacts in 2003 when the Museum was ostensibly researcher-free. Priscilla Renouf suggested the whole project in the first place, answered my numerous questions about the Phillip's Garden West site, allowed me uncounted hours with the site assemblage, gave me permission to photograph artefacts from Phillip's Garden West, and assisted with the production of some figures. My appreciation goes to Charles Conway, Memorial University Geography Department, who drafted the maps for Figs. 5.1 and 5.8 and annotated Fig. 5.4. David Keenlyside, formerly of the Canadian Museum of Civilization, kindly allowed me access to the Newfoundland and Labrador materials housed there. Thanks to Priscilla and Patty Wells for organising the 2003 CAA session. I am grateful for the comments of Matthew Betts, two anonymous reviewers, as well as the editor, on an earlier version of this text. Photographs are from The Port au Choix Archaeology Project.

## References

Anton, E.P. 2004. St. John's Harbour 5 (HeCi-30) and an Examination of Groswater and Early Dorset Relationships in Labrador. Master's thesis, Department of Anthropology, Memorial University of Newfoundland, St. John's.

Auger, R. 1982. A preliminary report on Early Dorset occupations on the west coast of Newfoundland. In *Archaeology in Newfoundland and Labrador 1981*, ed. J. Sproull Thomson and C. Thomson, 130–151. Historic Resources Division, Government of Newfoundland and Labrador, St. John's.

Auger, R. 1984. Factory Cove: Recognition and Definition of the Early Palaeo-Eskimo Period in Newfoundland. Master's thesis, Department of Anthropology, Memorial University of Newfoundland, St. John's.

Auger, R. 1986. Factory Cove: An early Palaeo-Eskimo component from the west coast of Newfoundland. In *Palaeo-Eskimo cultures in Newfoundland, Labrador and Ungava*, 111–118. Reports in Archaeology No. 1, Memorial University of Newfoundland, St. John's.

Beaton, G. 2004. A Chip off the Old Block: Investigations of a Maritime Archaic Lithic Workshop/Quarry Site in Big Brook (EjBa-2), Northwestern Newfoundland. Master's thesis, Department of Anthropology, Memorial University of Newfoundland, St. John's.

Bergerud, A.T. 1983. Prey switching in a simple ecosystem. *Scientific American* 24(9): 130–141.

Bielawski, E. 1979. Contactual transformation: The Dorset-Thule succession. In *Thule Eskimo culture: An anthropological retrospective*, ed. A.P. McCartney, 100–109. Mercury Series Archaeology Paper 88, National Museums of Canada, Ottawa.

Bishop, P. 1974. Final report: 1973 excavations at Norris Point, Gros Morne National Park. On file, Parks Canada, Archaeology, Atlantic Region, Halifax.

Cox, S.L. 1977. Prehistoric Settlement and Culture Change at Okak, Labrador. PhD dissertation, Department of Anthropology, Harvard University, Boston.

Cox, S.L. 1978. Palaeo-Eskimo occupations of the north Labrador coast. *Arctic Anthropology* 15(2): 96–118.

Cridland, J. 1998. Late Prehistoric Indian Subsistence in Northeastern Newfoundland: Faunal Analysis of Little Passage Complex Assemblages from the Beaches and Inspector Island Sites. Master's thesis, Department of Anthropology, Memorial University of Newfoundland, St. John's.

Evans, C.O. 1982. Frenchman's Island site (ClAl-1) preliminary field report. In *Archaeology in Newfoundland and Labrador 1981*, ed. J. Sproull Thomson and C. Thomson, 210–225. Historic Resources Division, Government of Newfoundland and Labrador, St. John's.

Fitzhugh, W.W. 1972. *Environmental archeology and cultural systems in Hamilton Inlet, Labrador: Survey of the central Labrador coast from 3000 B.C. to the present.* Washington, D.C.: Smithsonian Institution Press.

Fitzhugh, W.W. 1976. Preliminary culture history of Nain, Labrador: Smithsonian fieldwork, 1975. *Journal of Field Archaeology* 3: 123–142.

Fitzhugh, W.W. 1980. A review of Paleo-Eskimo culture history in southern Quebec-Labrador and Newfoundland. *Études/Inuit/Studies* 4: 21–31.

Fitzhugh, W.W. 1981. Smithsonian archaeological surveys, central and northern Labrador, 1980. In *Archaeology in Newfoundland and Labrador 1980*, ed. J. Sproull Thomson and B. Ransom, 26–47. Historic Resources Division, Government of Newfoundland and Labrador, St. John's.

Fitzhugh, W.W. 1983. Archaeological surveys in the Strait of Belle Isle. In *Archaeology in Newfoundland and Labrador 1982*, ed. J. Sproull Thomson and C. Thomson, 118–132. Historic Resources Division, Government of Newfoundland and Labrador, St. John's.

Fogt, L.M. 1998. The Excavation and Analysis of a Dorset Palaeoeskimo Dwelling at Cape Ray, Newfoundland. Master's thesis, Department of Anthropology, Memorial University of Newfoundland, St. John's.

Gendron, D. 1990. The JgEj-3 site: A Groswater occupation in northwestern Ungava Bay. Paper presented at the 23rd Annual Meeting of the Canadian Archaeological Association, Whitehorse.

Gendron, D. 1999. The Early Palaeo-Eskimo Boulder Field Sites: Methodological and Interpretative Issues. Paper presented at the 32nd Annual Meeting of the Canadian Archaeological Association, Whitehorse.

Geneste, J.-M., and H. Plisson. 1990. Technologie fonctionelle des pointes à cran Solutrénnes: l'apport des nouvelles donnėes de la Grotte de la Combe Saunière. In *Feuilles de Pierre 42*, ed. J. Kozlowski, 293–320. ERAUL: Liège.

Harp, E. 1964. *The cultural affinities of the Newfoundland Dorset Eskimo*. National Museums of Canada Bulletin 200, Ottawa.

Harp, E. 1969/1970. Late Dorset art from Newfoundland. *Folk* 11/12: 109–124.

Harp, E. 1976. Dorset settlement patterns in Newfoundland and southeastern Hudson Bay. In *Eastern Arctic prehistory: Paleoeskimo problems*, ed. M.S. Maxwell, 119–138. Washington, D.C.: Society for American Archaeology.

Hartery, L.J. 2001. The Cow Head Complex. Master's thesis, Department of Archaeology, University of Calgary, Calgary.

Hartery, L.J. 2007. *The Cow Head complex and the Recent Indian period in Newfoundland, Labrador and the Quebec Lower North Shore*. Occasional Papers in Northeastern Archaeology 17. St. John's: Copetown Press.

Hodder, I., and C. Orton. 1976. *Spatial analysis in archaeology*. Cambridge: Cambridge University Press.

Hodgetts, L.M., M.A.P. Renouf, M.S. Murray, D. McCuaig-Balkwill, and L. Howse. 2003. Changing subsistence practices at the Dorset Paleoeskimo site of Phillip's Garden, Newfoundland. *Arctic Anthropology* 40(1): 106–120.

Holly, D.H. 1997. Revisiting Marginality: Settlement Patterns on the Island of Newfoundland. Master's thesis, Department of Anthropology, Brown University, Providence.

Howley, J.P. 1915. *The Beothucks or Red Indians: The aboriginal inhabitants of Newfoundland*. Cambridge: Cambridge University Press.

Irving, W. 1957. An archaeological survey of the Susitna Valley. *Anthropological Papers of the University of Alaska* 6(1): 37–52.

Kennett, B.L. 1990. Phillip's Garden East: An Examination of the Groswater Palaeo-Eskimo Phase. Master's thesis, Department of Anthropology, Memorial University of Newfoundland, St. John's.

Lavers, D. 2006. The Groswater Palaeoeskimo Component at Phillip's Garden (EeBi-1), Port au Choix, Newfoundland. Honour's dissertation, Department of Anthropology, Memorial University of Newfoundand, St. John's.

Lavers, D., and M.A.P. Renouf. 2009. St. Pauls archaeology project: St. Pauls Bay-2 (DlBk-6). On file, Provincial Archaeology Office, Government of Newfoundland and Labrador, St. John's.

LeBlanc, S. 1996. A Place with a View: Groswater Subsistence-Settlement Patterns in the Gulf of St. Lawrence. Master's thesis, Department of Anthropology, Memorial University of Newfoundland, St. John's.

LeBlanc, S. 2000. Groswater technological organization: A decision-making approach. *Arctic Anthropology* 37(2): 23–37.

LeBlanc, S. 2008. Middle Dorset Variability and Regional Cultural Traditions: A Case Study from Newfoundland and Saint- Pierre and Miquelon. PhD dissertation, Department of Anthropology, University of Alberta, Edmonton.

Linnamae, U. 1975. *The Dorset culture: A comparative study in Newfoundland and the Arctic*. Technical Papers of the Newfoundland Museum 1, St. John's.

Loring, S.G. 1992. Princes and Princesses of Ragged Fame: Innu Archaeology and Ethnohistory in Labrador. PhD dissertation, Department of Anthropology, University of Massachusetts, Amherst.

Loring, S.G., and S. Cox. 1986. The Postville Pentecostal Groswater site, Kaipokok Bay, Labrador. In *Palaeo-Eskimo cultures in Newfoundland, Labrador and Ungava*, 65–95. Reports in Archaeology No.1, Memorial University of Newfoundland, St. John's.

Marshall, I. 1978. The significance of Beothuk carved bone pendants. *Canadian Journal of Archaeology* 2: 139–154.

Marshall, I. 1990. Evidence for two Beothuk subsistence economies. In *Papers of the twenty-first Algonquian conference*, ed. W. Cowan, 216–226. Ottawa: Carleton University Press.

Marshall, I. 1996. *A history and ethnography of the Beothuk*. Montreal: McGill-Queen's University Press.

Martijn, C.A. 1974. Archaeological research on the Lower North Shore, Quebec. In *Archaeological salvage projects 1972*, ed. D. Byrne, 112–130. Mercury Series Archaeology Paper 15, National Museums of Canada, Ottawa.

Martijn, C.A. 1990. Innu (Montagnais) in Newfoundland. In *Papers of the twenty-first Algonquian conference*, ed. W. Cowan, 227–246. Ottawa: Carleton University Press.

Maxwell, M.S. 1985. *Prehistory of the eastern Arctic*. New York: Academic Press.

McGhee, R. 1981. *The Dorset occupations in the vicinity of Port Refuge, High Arctic Canada*. Mercury Series Archaeology Paper 105, National Museums of Canada, Ottawa.

McGhee, R. 1997. Meetings between Dorset culture Palaeo-Eskimos and Thule culture Inuit: evidence from Brooman Point. In *Fifty years of Arctic research: Anthropological studies from Greenland to Siberia*, ed. R. Gilberg and H.C. Gulløv, 209–213. Copenhagen: Department of Ethnography, National Museum of Denmark.

Melnik, M.E. 2007. Salmon Net (EfAx-25): An Investigation of Groswater Material Culture Stylistic Variability in Newfoundland. Master's thesis, Department of Anthropology, Memorial University of Newfoundland, St. John's.

Murray, M.S. 1992. Beyond the Laundry List: The Analysis of Faunal Remains from a Dorset Dwelling at Phillip's Garden East (EeBi 1), Port au Choix, Newfoundland. Master's thesis, Department of Anthropology, Memorial University of Newfoundland, St. John's.

Nagle, C. 1978. Indian occupations of the Intermediate period on the central Labrador coast: A preliminary synthesis. *Arctic Anthropology* 15(2): 119–145.

Nagle, C. 1986. Flaked stone procurement and distribution in Dorset culture sites along the Labrador coast. In *Palaeo-Eskimo cultures in Newfoundland, Labrador and Ungava*, 95–110. Reports in Archaeology No. 1, Memorial University of Newfoundand, St. John's.

Park, R.W. 1993. The Dorset-Thule succession in Arctic North America: Assessing claims for culture contact. *American Antiquity* 63(3): 417–435.

Pastore, R.T. 1986. The spatial distribution of late Palaeo-Eskimo sites on the island of Newfoundland. In *Palaeo-Eskimo cultures in Newfoundland, Labrador and Ungava*, 125–134. Reports in Archaeology No. 1, Memorial University of Newfoundland, St. John's.

Pastore, R.T. 1987. Fishermen, furriers, and Beothuks: The economy of extinction. *Man in the Northeast* 33: 47–62.

Pastore, R.T. 1989. The collapse of the Beothuk world. *Acadiensis* 19(1): 52–71.

Pintal, J.-Y. 1994. A Groswater site at Blanc-Sablon, Quebec. In *Threads of Arctic prehistory: Papers in honour of William E. Taylor, Jr.*, ed. D. Morrison and J.-L. Pilon, 145–164. Mercury Series Archaeology Paper 149, Canadian Museum of Civilization, Ottawa.

Pintal, J.-Y. 1998. *Aux frontières de la mer: La préhistoire de Blanc-Sablon*. Les Publications du Québec, Dossiers 102. Collections Patrimoines et Municipalité de Blanc-Sablon, Québec.

Pintal, J.-Y. 2000. On the (Early) Origins of the Beothuk. Paper presented at the 33rd Annual Meeting of the Canadian Archaeological Association, Ottawa.

Plumet, P. 1994. Le Paléoesquimau de la baie du Diana (Arctique Québécois). In *Threads of Arctic prehistory: Papers in honour of William E. Taylor, Jr.*, ed. D. Morrison and J.-L. Pilon, 103–144. Mercury Series Archaeology Paper 149, Canadian Museum of Civilization, Ottawa.

Renouf, M.A.P. 1985. Archaeology of the Port au Choix National Historic Park: Report of 1984 Field Activities. On file, Parks Canada, Archaeology, Atlantic Region, Halifax.

Renouf, M.A.P. 1991. Archaeological Investigations at the Port au Choix National Historic Park: Report of the 1990 Field Season. On file, Parks Canada, Archaeology, Atlantic Region, Halifax.

Renouf, M.A.P. 1992. The 1991 Field Season, Port au Choix National Historic Park: Report of Archaeological Excavations. On file, Parks Canada, Archaeology, Atlantic Region, Halifax.

Renouf, M.A.P. 1993a. The 1992 Field Season, Port au Choix National Historic Park: Report of Archaeological Excavations. On file, Parks Canada, Archaeology, Atlantic Region, Halifax.

Renouf, M.A.P. 1993b. Palaeoeskimo seal hunters at Port au Choix, northwestern Newfoundland. *Newfoundland Studies* 9(2): 188–210.

Renouf, M.A.P. 1994. Two transitional sites at Port au Choix, northwestern Newfoundland. In *Threads of Arctic prehistory: Papers in honour of William E. Taylor, Jr.*, ed. D. Morrison and J.-L. Pilon, 165–196. Mercury Series Archaeology Paper 149, Canadian Museum of Civilization, Ottawa.

Renouf, M.A.P. 1999. Prehistory of Newfoundland hunter-gatherers: Extinctions or adaptations? *World Archaeology* 30(3): 403–420.

Renouf, M.A.P. 2000. Symbolism and subsistence: Seals and caribou at Port au Choix, northwestern Newfoundland. In *Animal bones, human societies*, ed. P. Rowley-Conwy, 65–72. Oxford: Oxbow Books.

Renouf, M.A.P. 2003. Hunter-gatherer interactions: Mutualism and resource partitioning on the island of Newfoundland. *Before Farming* 1(4): 84–95.

Renouf, M.A.P. 2005. Phillip's Garden West: A Newfoundland Groswater variant. In *Contributions to the study of the Dorset Palaeo-Eskimos*, ed. P.D. Sutherland, 57–80. Mercury Series Archaeology Paper 167, Canadian Museum of Civilization, Gatineau.

Renouf, M.A.P., T. Bell, and M. Teal. 2000. Making contact: Recent Indians and Palaeoeskimos on the island of Newfoundland. In *Identities and cultural contacts in the Arctic*, ed. M. Appelt, J. Berglund, and H.C. Gulløv, 106–119. Copenhagen: The Danish National Museum and Danish Polar Center.

Robbins, D.T. 1986. "Newfoundland Dorset" culture? In *Palaeo-Eskimo cultures in Newfoundland, Labrador and Ungava*, 119–124. Reports in Archaeology No. 1, Memorial University of Newfoundland, St. John's.

Robbins, D.T. 1989. Regards archéologiques sur les Béothuks de Terre-Neuve. *Recherches Amérindiennes au Québec* 19(2–3): 21–32.

Rowley-Conwy, P. 1990. Settlement patterns of the Beothuk Indians of Newfoundland: A view from away. *Canadian Journal of Archaeology* 14: 13–32.

Ryan, K. 1997. Groswater Palaeoeskimo Toolmakers: Phillip's Garden West and Beyond. Honour's dissertation, Department of Anthropology, Memorial University of Newfoundland, St. John's.

Schiffer, M. 1987. *Formation processes of the archaeological record*. Salt Lake City: University of Utah Press.

Schledermann, P. 1978. Prehistoric demographic trends in the Canadian High Arctic. *Canadian Journal of Archaeology* 2: 43–58.

Schwarz, F.A. 1994. Paleo-Eskimo and Recent Indian subsistence and settlement patterns on the island of Newfoundland. *Northeast Anthropology* 47: 55–70.

Stuiver, M., and P.J. Reimer. 1993. Extended 14C database and revised CALIB radiocarbon calibration program. *Radiocarbon* 35: 215–230.

Teal, M.A. 2001. An Archaeological Investigation of the Gould Site (EeBi-42) in Port au Choix, Northwestern Newfoundland: New Insight into the Recent Indian Cow Head Complex. Master's thesis, Department of Anthropology, Memorial University of Newfoundland, St. John's.

Tuck, J.A. 1978. Excavations at Cow Head, Newfoundland: An interim report. *Études/Inuit/Studies* 2(1): 138–141.

Tuck, J.A. 1988. Prehistory of Atlantic Canada. Manuscript on file, Department of Archaeology, Memorial University of Newfoundland, St. John's.

Tuck, J.A., and R.T. Pastore. 1985. A nice place to visit, but...prehistoric extinctions on the island of Newfoundland. *Canadian Journal of Archaeology* 9(1): 69–80.

Tuck, J.A., and W.W. Fitzhugh. 1986. Palaeo-Eskimo traditions of Newfoundland and Labrador: A re-appraisal. In *Palaeo-Eskimo cultures in Newfoundland, Labrador and Ungava*, 161–167. Reports in Archaeology No. 1, Memorial University of Newfoundland, St. John's.

Wells, P.J. 2002. An Analysis of Faunal Remains from Two Groswater Palaeoekimo Sites at Port au Choix, Northwestern Newfoundland: Phillip's Garden West (EeBi-11) and Phillip's Garden East (EeBi-1). Master's thesis, Department of Anthropology, Memorial University of Newfoundland, St. John's.

Whittaker, J.C. 1994. *Flintknapping: Making and understanding stone tools*. Austin: University of Texas Press.

# Chapter 6
# A Sheltered Life: Inner Cove Groswater Palaeoeskimo Occupation at Port au Choix

**Kendra D. Stiwich**

## Introduction

This chapter summarizes data from the Groswater Palaeoeskimo occupation at the Party site (EeBi-30) situated in the sheltered inner reaches of Back Arm, Port au Choix (Fig. 6.1). This location contrasts with better known Groswater Palaeoeskimo sites on the Point Riche headland, in particular Phillip's Garden East (EeBi-1) and Phillip's Garden West (EeBi-11) (Harp 1951, 1964; Kennett 1990; LeBlanc 1996; Renouf 1985, 1986, 1987, 1991, 1992, 1993, 1994, 2005; Wells 2002; Wintemberg 1939, 1940; see also Chaps. 4 and 5) (Fig. 6.1). This chapter presents the artefact assemblage, intra-site spatial patterning, and radiocarbon data from two areas of the Party site and concludes that these represent distinct occupations with different economic foci. These data, along with site location and available resources, suggest that the Party site was occupied during the warmer months, in contrast with the late winter/early spring focus of the Phillip's Garden East and Phillip's Garden West sites. The Party site expands our knowledge about the range of Groswater site locations and activities at Port au Choix.

The Party site was excavated in the summer of 2003 (Wheatley 2004a, 2004b). Based on the results of previous survey work (Renouf 2002; Renouf and Bell 2001, 2002) two excavation areas were opened during that field season. Area 1 is on the lower terrace, 4–6 m above sea level (asl). Area 2 is located on the upper terrace, 6–8 m asl and 5 m to the southwest of Area 1 (Figs. 6.1 and 6.2). The site was hidden by dense, stunted spruce and fir forest and a thick layer of peat. Not only did this make the site difficult to find (Renouf 1991) but for excavation it required a considerable effort to clear trees and remove the upper layer of peat (Wheatley 2004a). Description of the two excavation areas and their associated material is presented below.

K.D. Stiwich (✉)
#20, 3208 Gibbins Road, Duncan, BC, Canada, V9L 1G8
e-mail: kendrastiwich@hotmail.com

M.A.P. Renouf (ed.), *The Cultural Landscapes of Port au Choix*, Interdisciplinary Contributions to Archaeology, DOI 10.1007/978-1-4419-8324-4_6,

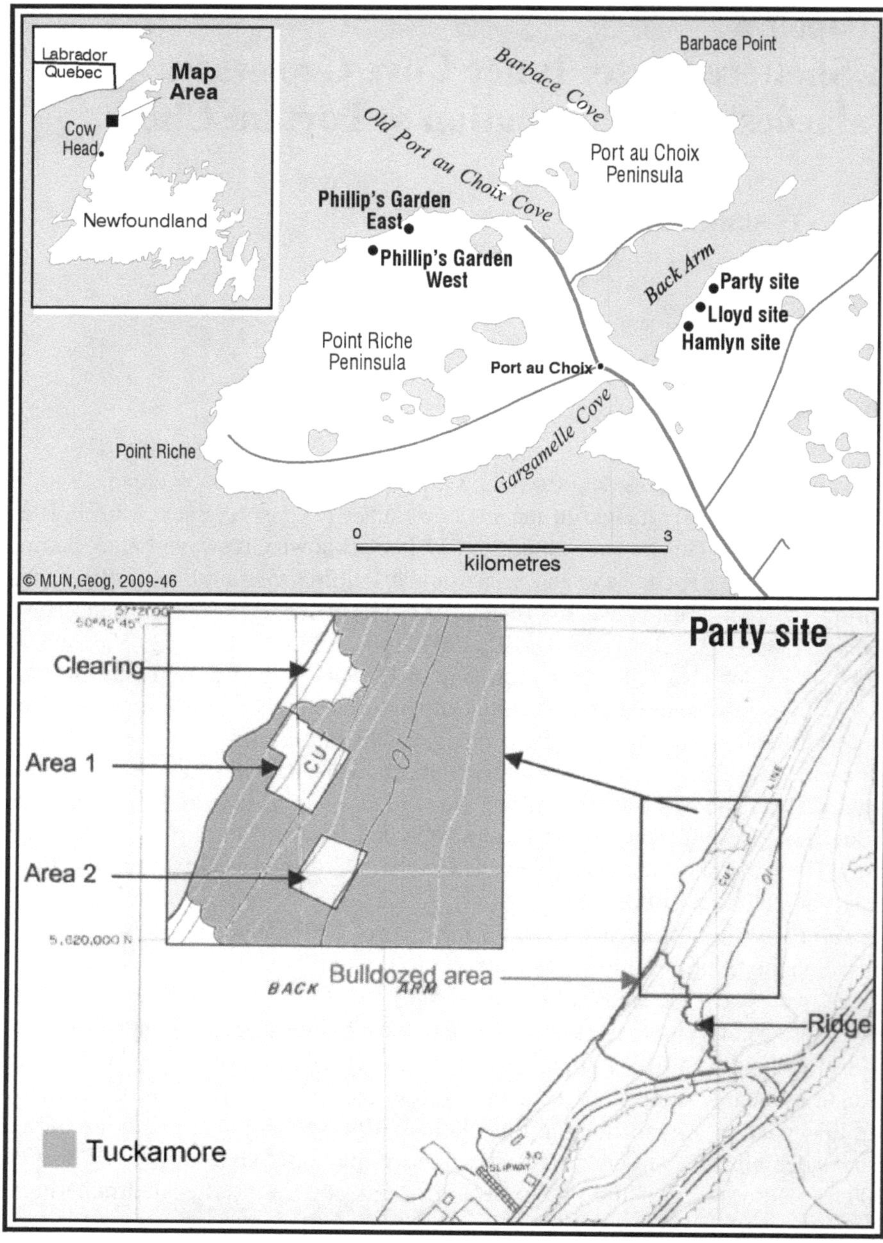


**Fig. 6.1** Location of the Party site on a 1:2,500 community map including the 2003 excavation. *Inset* map of Port au Choix shows location of sites mentioned in the text

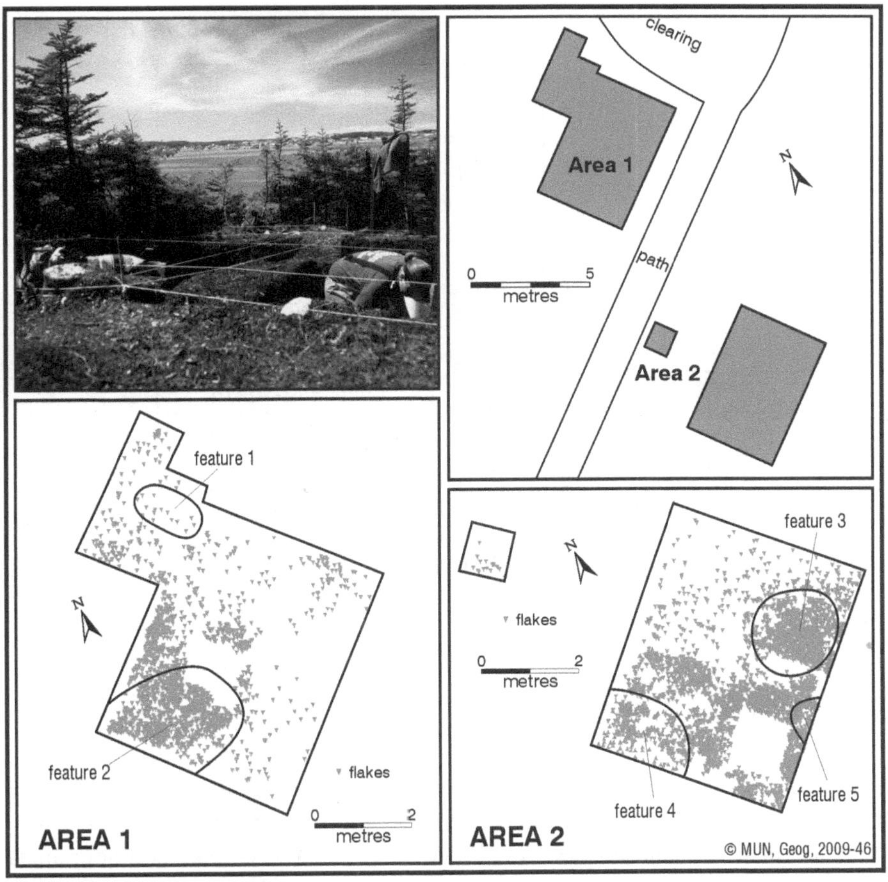


**Fig. 6.2** Areas 1 and 2 showing lithic concentrations and location of features

# Area 1

## *Stratigraphy*

The Party site stratigraphy is defined on the basis of natural and cultural deposits. Area 1 has three natural strata, Levels 1, 2, and 3 (Fig. 6.3). Level 1 begins at ground surface and is a continuous 66–80 cm thick stratum of organic orange-brown peat which becomes more compact and darker with depth. Level 2 is a 1–10 cm dark clay-like sediment which contains over 95% of the cultural material. Level 2 is likely the result of acidic peat reacting with the basic limestone beach, creating a separate sediment layer. If so, this indicates that the cultural material was originally deposited on the beach itself. The remaining cultural material is found at the interface between Level 1 and Level 2. Level 3 is the limestone beach.

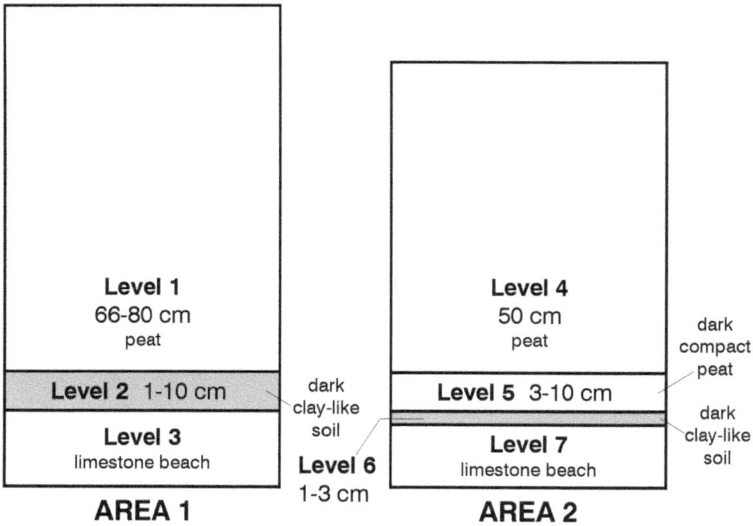


**Fig. 6.3** Stratigraphy of Areas 1 and 2 at the Party site

## *Raw Material*

The Area 1 artefact assemblage shows a predominance of chert use over any other lithic raw material. At least 42 different colour varieties were identified at the Party site, most of which appear to be Cow Head chert, based on colour and fine-grained texture. Cow Head cherts originate along the west coast of Newfoundland (Fig. 6.1) and were used extensively by Groswater groups (LeBlanc 1996). A subset of chert artefacts from Area 1 does not appear to be made of typical Cow Head cherts. These cherts are darker and a number have distinctive swirling and banding patterns. These exotic cherts were not found in Area 2.

## *Radiocarbon Dating*

In Area 1 a charcoal sample from a hearth (Feature 1) returned a date of 2710±40 BP[1] (2850–2770 cal BP, see Table 6.1).

[1] All dates used in this paper are calibrated to calendar years before present (cal BP) using the Calib 6.0html program (Stuiver and Reimer 1993).

**Table 6.1** Radiocarbon dates from Areas 1 and 2 of the Party site

| Context | C14 years BP | Lab No. | Median age cal BP | Cal BP range 1σ | Cal BP range 2σ |
|---|---|---|---|---|---|
| Party site, Area 1 | 2710±40 | Beta 183603 | 2810 | 2850–2770 | 2920–2750 |
| Party site, Area 2 | 2570±60 | Beta 146666 | 2860 | 2760–2510 | 2790–2370 |
| Party site, Area 2 | 2460±70 | Beta 183604 | 2540 | 2700–2370 | 2720–2360 |

### *Location and Spatial Organization*

It appears that the original Area 1 occupation occurred on or very near what was the beach at the time of occupation, based on three observations. First, the artefacts and debitage density within Area 1 increased towards the modern shoreline (Fig. 6.2), and the two features, a hearth (Feature 1) and a concentration of retouch flakes (Feature 2) (Fig. 6.2), were also closer to the seaward side of the excavation area. Second, most cultural material occurred in the dark clay layer directly above the limestone substrate or beach. Third, most Feature 1 stones were sitting directly on the beach.

### *Artefacts*

Forty-one artefacts were found in Area 1 (Table 6.2). In her study of Groswater sites on the Northern Peninsula and the Quebec Lower North Shore, LeBlanc (1996:51) separated Groswater artefact classes into one of four functional categories: procurement, processing, maintenance and manufacturing. Her categories are followed here and artefacts that are likely multi-functional are included within more than one functional category (Table 6.3). For example, because the awl found at Area 1 may have been used to process a new hide (manufacture) and/or maintain one brought to the site (maintenance), it is counted in both functional categories.

## Area 2

### *Stratigraphy*

Area 2 is defined on the basis of natural and cultural strata. There are three natural strata, Levels 4, 6, and 7, and one cultural stratum, Level 5 (Fig. 6.3). Level 4 begins at ground surface and is a continuous stratum approximately 50 cm thick composed of organic orange-brown peat which becomes more compact and darker with depth. Level 5 is 3–10 cm thick and is composed of dark, compact peat; all cultural material is found in this level. Level 6 is a 1–5 cm thick dark clay-like sediment. Level 7 is

**Table 6.2** List of artefacts from Area 1 and Area 2 of the Party site, including material from the 2003 excavations (Wheatley 2004a, b) and test pits from 2001 (Renouf and Bell 2002)

| | Area 1 | | Area 2 | |
|---|---|---|---|---|
| Artefact class | Number | % of Total | Number | % of Total |
| Awl | 1 | 2.4 | 0 | 0 |
| Axe | 0 | 0 | 1 | 1.3 |
| Biface | 4 | 9.8 | 9 | 12 |
| Blade | 4 | 9.8 | 5 | 6.7 |
| Burin-like tool | 1 | 2.4 | 5 | 6.7 |
| Core fragment | 6 | 14.6 | 17 | 22.7 |
| Endblade | 1 | 2.4 | 5 | 6.7 |
| Hammerstone | 1 | 2.4 | 0 | 0 |
| Microblade | 14 | 34.1 | 17 | 22.7 |
| Microblade core | 0 | 0 | 1 | 1.3 |
| Preform | 6 | 14.6 | 9 | 12 |
| Scraper | 3 | 7.3 | 3 | 4 |
| Sideblade | 0 | 0 | 2 | 2.7 |
| Unidentified tool fragment | 0 | 0 | 1 | 1.3 |
| Total | 41 | 99.8 | 75 | 100.1 |

**Table 6.3** Artefact functional categories, Areas 1 and 2, the Party site

| Activities | Indicators | Area 1 | Area 2 |
|---|---|---|---|
| Procurement[1] | Endblade; sideblade | 1 | 7 |
| Processing[2] | Scraper; awl; biface; blade/microblade | 26 | 34 |
| Maintenance[3] | Scraper; axe; awl; biface; blade/microblade; burin-like tool | 27 | 43 |
| Manufacturing[4] | Axe; burin-like tool; microblade core; core fragment; preform; hammerstone | 14 | 33 |

[1] Hunting tools
[2] Skin/meat processing tools
[3] Tools used to maintain the working state of other objects
[4] Items that are used to make tools, or tools that are in the process of manufacture

the limestone beach. The main difference between Area 1 and Area 2 stratigraphy is that whereas in Area 1 the cultural level was the dark clay-like sediment above the beach, in Area 2 the cultural level is dark, compact peat above that clay-like sediment.

## *Raw Material*

Like Area 1, Area 2 artefacts are predominantly chert, most likely from Cow Head. However, some yellow and pink cherts are part of the Area 2 assemblage, colour variations that are absent from Area 1.

## Radiocarbon Dating

A charcoal sample from a test pit on the upper terrace near Area 2 (Renouf and Bell 2002) returned a date of 2570±60 BP (2760–2510 cal BP, see Table 6.1). This was associated with lithic debris and preserved animal hide in a stratigraphic context comparable to Level 5. A second sample collected from the upper terrace was obtained from a midden (Feature 4) in Area 2 (Fig. 6.2) and is associated with lithic debris and fire-cracked rocks. This sample is dated to 2460±70 BP (2700–2370 cal BP, see Table 6.1).

## Location and Spatial Organization

Area 2 is on the upper terrace which slopes towards the sea. Three features were excavated in this area, a flake concentration (Feature 3), a midden (Feature 4) and a hearth (Feature 5) (Fig. 6.2). In comparison to the rest of Area 2 a large amount of lithic debitage was recovered from Feature 3. Microblades, a microblade core, endblades, a burin-like tool, red ochre and a small amount of charcoal were associated. Feature 4 is interpreted as a midden based on the high concentration of material (Fig. 6.2), including many flakes and various preforms at different stages of the lithic reduction sequence, the presence of broken tools, the abundance of fire-cracked rocks with no apparent spatial pattern, a large amount of charcoal, and sediment that differs in colour (much darker) and texture (more clay-like) from all other sediments found in Area 2. All materials were intermixed and no identifiable strata within the midden could be identified. Feature 5 is a hearth, composed of an ill-defined circle of fire-cracked rocks and charcoal. No artefacts were recovered in the hearth and no sediment change was evident except for the inclusion of charcoal.

## Artefacts

Area 2 has 75 artefacts (Table 6.2). As with Area 1, these have been placed into one of four functional categories (Table 6.3). As before, certain artefacts are classified in more than one functional category.

# Discussion and Conclusions

A closer look at the raw material distribution, radiocarbon dates, spatial organization and artefacts indicate that Area 1 and Area 2 are not contemporaneous and have different economic foci, both within the warmer months of the year.

## Site Chronology

The cultural material at each of the two areas was found in different soil strata. In Area 1 most cultural material was found in Level 2, a dark clay-like sediment. In Area 2 most cultural material was found in Level 5, dark compact peat lying above a dark clay-like sediment comparable to Level 2 of Area 1. This does not necessarily relate to two chronologically separate occupations and instead likely relates to differences in ground cover at the time of occupation. Therefore site stratigraphy is not used to determine whether the two areas at the site are contemporaneous.

Differences in Area 1 and Area 2 raw material may indicate that the two areas were occupied at different times since yellow and pink cherts were only found in Area 2. However, this may also indicate that some tool production episodes took place at specific locations on the site rather than throughout the site.

The three radiocarbon dates from the Party site (Table 6.1) are within the known range of Groswater in Newfoundland, 2990–1820 cal BP (Renouf 2005). The single date from Area 1 is slightly older than both dates from Area 2, indicating that the two areas were not contemporaneous.

## Economic Function

The Party site is situated on the shore of a sheltered cove 10–15 m from a small stream. Since site function is directly related to site location, this alone would suggest that both Areas 1 and 2 had the same or similar economic function. However, the features and artefacts from both areas indicate some differences in economic focus.

Port au Choix residents tell us that winter pack ice remains in Back Arm until April and sometimes as late as May. Studies based on fossil pollen (see Bell and Renouf, Chap. 2; Macpherson 1981) indicate that during the time of Groswater occupation, Newfoundland climate conditions were colder than present. Therefore it is unlikely that the Area 1 occupation was during the colder months since the beach, on which hearth Feature 1 was built, would have been covered in snow and ice. Possibly the occupation was on top of the snow. However, if this were the case, after the snow melted the hearth would shift downward to lie on top of the beach and there would be a displacement of the rocks. The hearth's spatial integrity (Fig. 6.4) indicates it was originally built on the beach, suggesting that Area 1 was most likely occupied during the warmer months of late spring to early fall. During this period many food resources would have been available such as harbour seals, straggler harp seals, mollusks, fish, birds, and crustaceans (Burrows 1989; Chapman 1966; Collins 1993; Gordon and Weeks 1982; Northcott and Phillips 1976; Squires 1996; Wheatley 2004b:31).

In Area 2, Feature 3 (a flake and artefact concentration) and Feature 5 (a hearth) are located at the upper edge of the excavation area (Fig. 6.2) away from the low terrace edge. Based on the presence of these features, this location appears to be where the site residents chose to spend their time while engaging in domestic

**Fig. 6.4** Area 1, hearth Feature 2

activities such as tool making, hide production, and food consumption. In contrast, the midden is downslope and closer to the terrace edge. Although the presence of a midden indicates a certain degree of refuse organization and therefore site longevity (Chatters 1987), it is not an area where people most likely spent their time.

The spatial patterning indicates that in Area 2 the living areas were farther from the beach than non-living/refuse areas. This spatial pattern may indicate the season and/or function of Area 2 related to the need for shelter. If Area 2 was occupied during the winter, then the group may have wanted to be farther from the beach and the cold winter winds. However, during the winter very few resources would have been available directly at the site aside from a stray caribou or harp seal or the occasional beaver or red fox (Wheatley 2004b:34).

In contrast, as mentioned above, there are many more resources available at this location during the summer months, in particular harbour seals, mollusks, fish, birds, and crustaceans; we observed some of these resources near the site during our summer field season. The best time to hunt harbour seal is during the late spring and early summer, when pups are born on shore (Northcott and Phillips 1976). If Area 2 residents were there to hunt harbour seal they might have preferred to be set back from the beach so as not to frighten their prey.

Both areas contain tools from all four functional categories: procurement, processing, maintenance and manufacturing (Table 6.3). This range of activities indicates that both areas were residential sites as opposed to task-specific sites with a narrower range of functional tool categories (Binford 1980). If both areas are a residential type of occupation, the differences observed in the artefact collections from both areas would be the result of a different economic focus.

Procurement tools are those tools used for hunting. One endblade was found in Area 1 compared to five endblades from Area 2 (Table 6.2). Two sideblades were also recovered from Area 2 and none from Area 1. Although these numbers are small, they nevertheless indicate that the people occupying Area 2 were hunting. There are morphological differences between the Area 2 endblades and the single example from Area 1 (Fig. 6.5). The endblades from Area 2 are virtually identical to the endblades from Phillip's Garden East which, based on faunal and artefactual data, has been interpreted as a focused harp seal hunting site (Kennett 1990; LeBlanc 1996; Renouf 2005). Therefore it is likely that the residents of Area 2 hunted seal. Since Back Arm is not a good harp seal hunting locale, and is instead more appropriate for harbour seals, it is likely that the Area 2 endblades were used for harbour seal hunting. This in turn indicates a late spring or summer hunt. There is only one endblade tip from Area 1 (Fig. 6.5) which is more elongated than the Area 2 endblades and is unlike the typical Groswater endblades from Area 2 and Phillip's Garden East. This suggests that seal hunting was not an activity carried out from Area 1.

In addition, more burin-like tools were found at Area 2 ($n=5$) compared to Area 1 ($n=1$). In Area 2 both finished burin-like tools and burin-like tool preforms ($n=3$) are in the assemblage, which indicates they were manufactured and used by the area occupants. Their use is substantiated by the fact that all of the burin-like tools found in

**Fig. 6.5** Endblades from the 2003 excavation (*top row* – Area 1, *bottom row* – Area 2)

Area 2 are broken. Since burin-like tools were likely used to make organic components of a sealing harpoon and since there were endblades found in the area, it is plausible that harpoon manufacture or maintenance for the purpose of seal hunting was an Area 2 activity.

The Party site findings show that Groswater Palaeoeskimos occupied the sheltered Back Arm area of Port au Choix, likely for harbour seal hunting in addition to other economic activities. Radiocarbon dates indicate that the Party site was occupied at least twice between 2850 and 2370 cal BP.

Regarding the earlier occupation (Area 1), stratigraphic and other evidence suggest that this occurred during the warmer months of the year. There is no evidence of seal hunting associated with this area. Instead, Area 1 occupants might have been exploiting the mussel beds that are today on the beach near the site, or small fish that swim in the nearby stream, or any of the seabirds and ducks that frequent Back Arm in the summer months (Wheatley 2004b:32–40).

Regarding the later occupation (Area 2), the endblades from this area indicate that the occupants were engaged in seal hunting. The fact that Back Arm is ideal for harbour rather than harp seals suggests that this area, like Area 1, was occupied during the warmer months.

Prior to our excavations at the Party site, Groswater research at Port au Choix had focused on sites on the outer coast, most notably Phillip's Garden East and Phillip's Garden West. Extensive faunal and artefactual data from these larger and richer sites demonstrated that late winter/early spring harp seal hunting was an important economic focus for Groswater Palaeoeskimos at Port au Choix (see Wells, Chap. 4). The Party site data broadens this view of Groswater activities within the Port au Choix area, shifting the focus away from the headlands to the sheltered shore of Back Arm. Two other Groswater sites have been identified in that same area, the Hamlyn site (EeBi-39) and the Lloyd site (EeBi-41) (Fig. 6.1) (Renouf and Bell 1998, 1999). Both sites were found during landscaping activities and were consequently very disturbed; at one time they, like the Party site, would have been obscured by woods and a thick layer of peat. Their location along the same Back Arm shoreline as the Party site indicates that this part of Port au Choix was regularly occupied by Groswater peoples. In light of our data from the Party site the Hamlyn and Lloyd sites were probably warm weather occupations.

In conclusion, the Party site provides information about other, lesser known, aspects of Groswater settlement and subsistence in Port au Choix. Along with the other Groswater sites in Back Arm, it increases our appreciation of the importance of the Port au Choix area, both headland and inner cove, to the Groswater people.

**Acknowledgements** My research was made possible by generous financial support from the Institute for Social and Economic Research, the Newfoundland Archaeological Heritage Outreach Program, the Provincial Archaeology Office and the Port au Choix Archaeology Project. The Port au Choix Party site crew (Ainslie Cogswell, Dominique Lavers, Mindy Pitre and Sarah Player) deserve applause. Thanks to Charlie Conway, Memorial University Geography Department, for drafting the figures. Additionally, two anonymous reviewers helped greatly with reading clarity; the readers, I am sure, thank you. And of course, thanks to Dr. M.A.P. Renouf, for helping in so many ways.

# References


Binford, L.R. 1980. Willow smoke and dogs' tails: Hunter-gatherer settlement systems and archaeological site formation. *American Antiquity* 45(1): 4–20.

Burrows, R. 1989. *Birding in Atlantic Canada: Newfoundland.* St. John's: Jesperson Press.

Chapman, F.M. 1966. *Handbook of birds of eastern North America.* New York: Dover.

Chatters, J.C. 1987. Hunter-gatherer adaptations and assemblage structure. *Journal of Anthropological Archaeology* 6: 336–375.

Collins, M. 1993. *Life on the Newfoundland seashore: Seaweeds, invertebrates and fish.* St. John's: Jesperson Press.

Gordon, J., and T.E. Weeks. 1982. *Seashells of the northeast coast from Cape Hatteras to Newfoundland.* Washington: Hancock House.

Harp, E. 1951. An archaeological survey in the Strait of Belle Isle area. *American Antiquity* 16: 203–221.

Harp E. 1964. *The cultural affinities of the Newfoundland Dorset Eskimo.* National Museums of Canada Bulletin 200, Ottawa.

Kennett, B.L. 1990. Philip's Garden East: An Examination of the Groswater Palaeo-Eskimo Phase. Master's thesis, Department of Anthropology, Memorial University of Newfoundland, St. John's.

LeBlanc, S. 1996. A Place with a View: Groswater Subsistence and Settlement Patterns in the Gulf of St. Lawrence. Master's thesis, Department of Anthropology, Memorial University of Newfoundland, St. John's.

Macpherson, J.B. 1981. The development of the vegetation of Newfoundland and climate change during the Holocene. In *The natural environment of Newfoundland, past and present,* ed. A.G. Macpherson and J.B. Macpherson, 189–217. Department of Geography, Memorial University of Newfoundland, St. John's.

Northcott, T.H., and F.R. Phillips. 1976. *The land and sea mammals of Port au Choix National Historic Park, Newfoundland.* On file, Canada and the Department of Indian and Northern Affairs, Ottawa.

Renouf, M.A.P. 1985. Report of the archaeological investigations at the Port au Choix and Point Riche Peninsulas. In *Archaeology in Newfoundland and Labrador 1984,* ed J. Sproull Thomson and C. Thomson, 298–321. Historic Resources Division, Government of Newfoundland and Labrador, St. John's.

Renouf, M.A.P. 1986. Archaeological Investigations at Phillip's Garden and Point Riche, Port au Choix National Historic Park: Report of 1985 Field Activities. On file, Parks Canada, Archaeology, Atlantic Region, Halifax.

Renouf, M.A.P. 1987. Archaeological Excavations at the Port au Choix National Historic Park: Report of 1986 Field Activities. On file, Parks Canada, Archaeology, Atlantic Region, Halifax.

Renouf, M.A.P. 1991. Archaeological Investigations at the Port au Choix National Historic Park: Report of the 1990 Field Season. On file, Parks Canada, Archaeology, Atlantic Region, Halifax.

Renouf, M.A.P. 1992. The 1991 Field Season, Port au Choix National Historic Park: Report of Archaeological Excavations. On file, Parks Canada, Archaeology, Atlantic Region, Halifax.

Renouf, M.A.P. 1993. The 1992 Field Season, Port au Choix National Historic Park: Report of Archaeological Excavations. On file, Parks Canada, Archaeology, Atlantic Division, Halifax.

Renouf, M.A.P. 1994. Two transitional sites at Port au Choix, northwestern Newfoundland. In *Threads of Arctic prehistory: Papers in honour of William E. Taylor, Jr.,* ed D. Morrison and J.L. Pilon, 165–196. Mercury Series Archaeology Paper 149, Canadian Museum of Civilization, Ottawa.

Renouf, M.A.P. 2002. *Archaeology at Port au Choix, northwestern Newfoundland 1990-1992.* Occasional Papers in Northeastern Archaeology 12. St. John's: Copetown Press.

Renouf, M.A.P. 2005. Phillip's Garden West: A Newfoundland Groswater variant. In *Contributions to the study of the Dorset Palaeo-Eskimos,* ed P.D. Sutherland, 57–80. Mercury Series Archaeology Paper 167, Canadian Museum of Civilization, Gatineau.

Renouf, M.A.P., and T. Bell. 1998. Searching for the Maritime Archaic Indian Habitation Site at Port au Choix, Newfoundland: An Integrated Approach using Archaeology, Geomorphology and Sea Level History. On file, Provincial Archaeology Office, Government of Newfoundland and Labrador, St. John's.

Renouf, M.A.P., and T. Bell. 1999. The Port au Choix Archaeology and Sea Level History Project: Preliminary Report of the 1998 Field Season. On file, Provincial Archaeology Office, Government of Newfoundland and Labrador, St. John's.

Renouf, M.A.P., and T. Bell. 2001. Gould Site Excavations, Port au Choix: 2000 Field Season. On file, Provincial Archaeology Office, Government of Newfoundland and Labrador, St. John's.

Renouf, M.A.P., and T. Bell. 2002. Report of the 2001 Site Testing and Survey on the Northern Peninsula, Newfoundland. On file, Provincial Archaeology Office, Government of Newfoundland and Labrador, St. John's.

Squires, H.J. 1996. Decapod crustaceans of Newfoundland, Labrador and the Canadian eastern Arctic. *Canadian Technical Report of Fisheries and Aquatic Sciences* 2359: 1–234.

Stuiver, M., and P.J. Reimer. 1993. Extended 14C database and revised CALIB radiocarbon calibration Program. *Radiocarbon* 35: 215–230.

Wells, P.J. 2002. An Analysis of Faunal Remains from Two Groswater Palaeoeskimo Sites at Port au Choix, Northwestern Newfoundland: Phillip's Garden West (EeBi–11) and Phillip's Garden East (EeBi–1). Master's thesis, Department of Anthropology, Memorial University of Newfoundland, St. John's.

Wheatley, K.D. 2004a. A Report of the 2003 Field Season at the Party Site (EeBi-30) in Port au Choix, Newfoundland. On file, Provincial Archaeology Office, Government of Newfoundand and Labrador, St. John's.

Wheatley, K.D. 2004b. The Party Site (EeBi-30) and Beyond: An Interpretation of Groswater Mobility and Landscape from Port au Choix, NL. Master's thesis, Department of Anthropology, Memorial University of Newfoundland, St. John's.

Wintemberg, W.J. 1939. Eskimo sites of the Dorset culture in Newfoundland, part I. *American Antiquity* 5(2): 83–102.

Wintemberg, W.J. 1940. Eskimo sites of the Dorset culture in Newfoundland, part II. *American Antiquity* 5(4): 309–333.

# Chapter 7
# On the Headland: Dorset Seal Harvesting at Phillip's Garden, Port au Choix

M.A.P. Renouf

## Introduction

This chapter reviews data from the Dorset Palaeoeskimo site of Phillip's Garden (EeBi-1) on the Point Riche headland (Fig. 7.1) that show how harp seal hunting and processing were labour-intensive activities, involving specialized technology and large multi-family dwellings. The geography of Phillip's Garden was ideal for exploiting the large numbers of harp seals that regularly appeared a short distance offshore, offering the Phillip's Garden Dorset an unprecedented economic opportunity. This chapter argues that because the seals were available only for a short time, a large and coordinated labour force was necessary to capitalize on a narrow window of economic opportunity. Phillip's Garden is unique among Dorset sites in Newfoundland and is more comparable to a culturally diverse range of large and intensively occupied northern and Arctic sites which were based on similarly predictable, abundant and accessible resources.

## Phillip's Garden

Phillip's Garden is a 2.17 ha meadow on a series of raised terraces 6–11 m above sea level (Fig. 7.2). Site occupation dates from 1990 to 1180 cal BP.[1] The intensity of past activities is directly reflected in 20–60 cm of dark organically enriched soil laden with artefacts, animal bones and lithic debris. Many shallow depressions, 3–4 m in diameter, are spread throughout the meadow. Others occur in the dense, stunted spruce forest that rings the site's landward perimeter. The Port au Choix

[1] Except where indicated, all calendar dates in this chapter were calibrated using Calib 6.0html (Stuiver and Reimer 1993) and are represented by the one sigma probability range.

M.A.P. Renouf (✉)
Department of Archaeology, Memorial University, St. John's, NL, Canada, A1C 5S7
e-mail: mapr@mun.ca

M.A.P. Renouf (ed.), *The Cultural Landscapes of Port au Choix*, Interdisciplinary Contributions to Archaeology, DOI 10.1007/978-1-4419-8324-4_7,

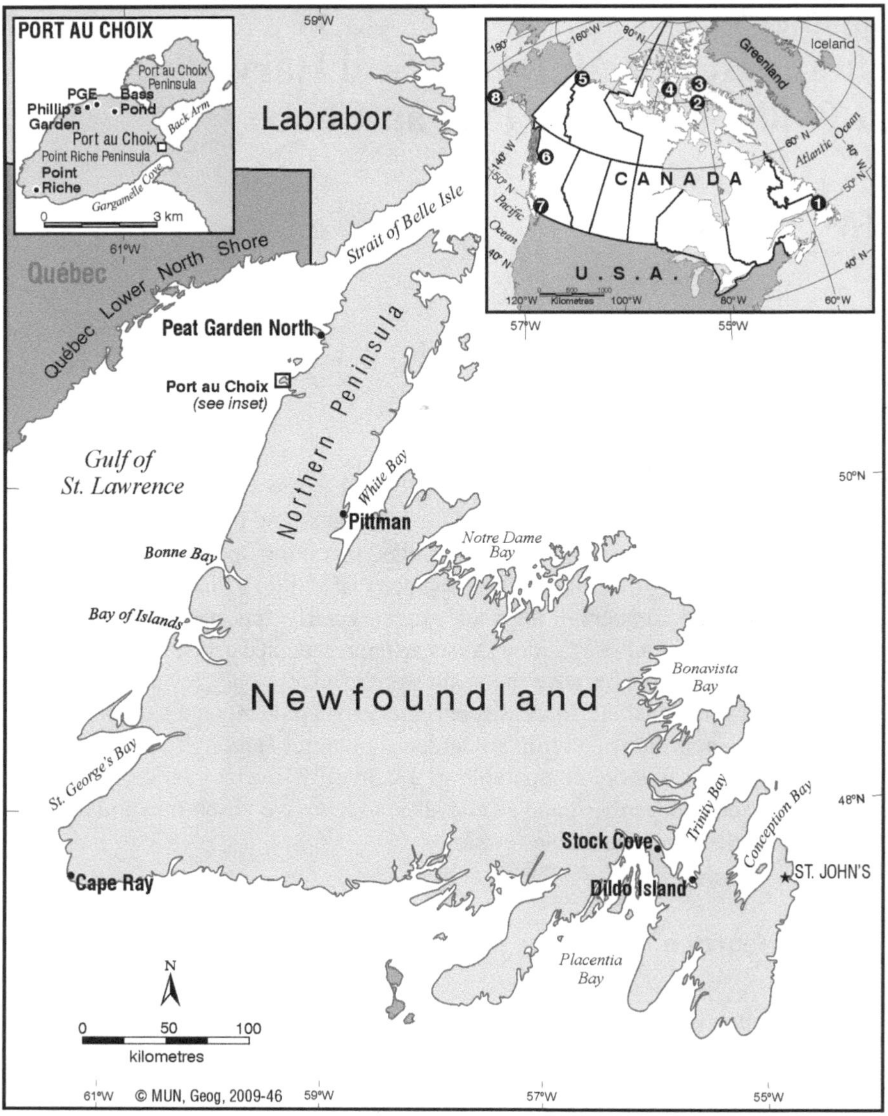


**Fig. 7.1** Place names and site locations mentioned in the text; in the inset map of Port au Choix, *PGE*=Phillip's Garden East. Site names on the inset map of northern North America are (1) Phillip's Garden, Port au Choix, (2) Igloolik sites, (3) Nunguvik, Baffin Island, (4) Qariaraqyuk, Somerset Island, (5) Kuukpak, Mackenzie River, (6) Paul Mason site, Skeena River, (7) Keatley Creek, Fraser River Valley, (8) Agayadan Village, southeast Aleutians

Archaeology Project has identified and mapped 68 dwellings and has noted many more as they become fleetingly visible under changing light conditions. Other dwellings are obscured beneath midden deposits. A 2,600 m magnetometry survey identified four hidden dwellings (Fig. 7.2; see Eastaugh and Taylor, Chap. 9) and we identified

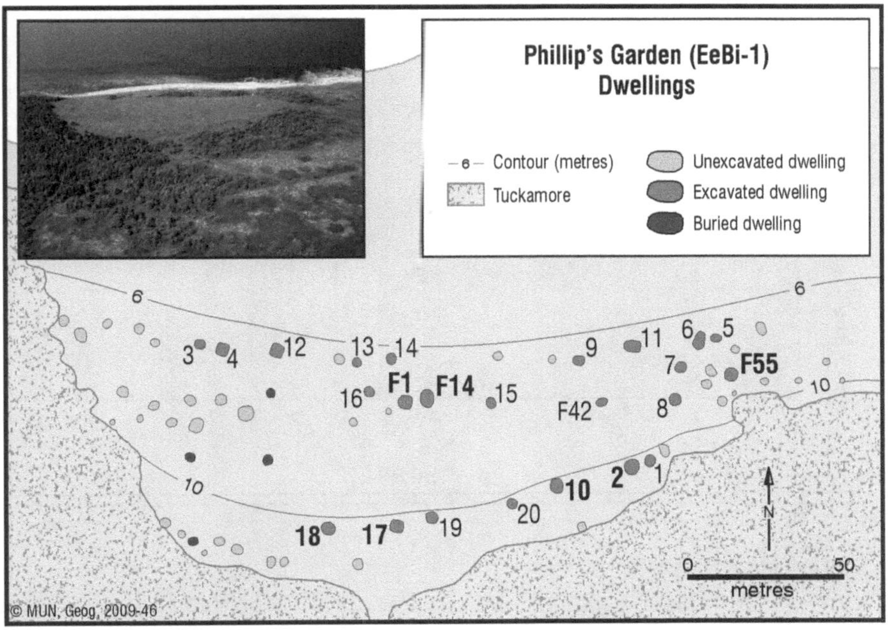


**Fig. 7.2** Phillip's Garden showing house depressions and in inset photograph. Dwellings discussed in the text are bolded

another beneath a midden excavation (Renouf 1986). Intensive test pitting identified thick middens where ground surface was unmarked by depressions, further suggesting that abandoned dwellings are filled in (Renouf 1987). These lines of evidence suggest that the total number of Phillip's Garden dwellings exceeds 68 by a significant margin.

Twenty-four dwellings have been excavated or tested by Harp (1964, 1976) and the Port au Choix Archaeology Project (Renouf 1986, 1987, 1991, 1992, 1993, 1999, 2002). A sample of associated midden deposits was also excavated (Hodgetts 2002; Hodgetts et al. 2003; Murray 1992; Renouf 2000). In addition, we re-investigated four dwellings originally excavated by Harp (Cogswell 2006; Cogswell et al. 2006; Renouf 2006, 2007, Renouf et al. 2005). Most recently, we have focused on outdoor areas contiguous to dwellings (Renouf 2009a, b).

There are >40 charcoal-based dates from the site (see Appendix). On the basis of >30 dates from 15 dwellings, Renouf and Bell (2009) divided site chronology into three arbitrary temporal phases, early (1990–1550 cal BP), middle (1550–1350 cal BP) and late (1350–1180 cal BP). These phases were based on the number of dwellings per decade with overlapping calibrated date ranges at one sigma probability.

This is summarized and updated in the bar diagram in Fig. 7.3a which is based on data in Fig. 7.3b. The middle phase is defined on the basis of the maximum number of dwellings with overlapping dates and the early and late phases are defined relative to the middle phase. In the early phase, the cal BP date ranges (at one sigma)

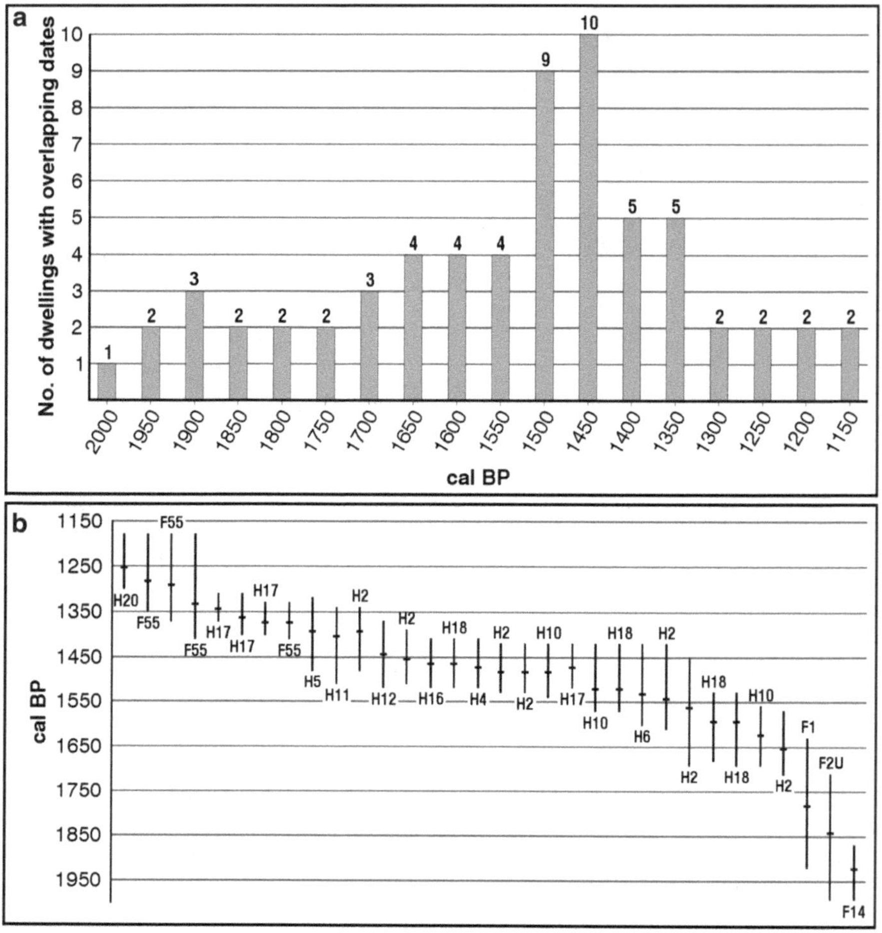


**Fig. 7.3** (**a**) Bar graph showing the number (at the head of each column) of dated dwellings with overlapping charcoal-based radiocarbon dates (cal BP) at one sigma probability. This is the basis of the phase designation and is based on the cal BP dates in (**b**)

of 1-4 dated dwellings overlap, in the middle phase the cal BP ranges of 5–10 dated dwellings overlap and in the late phase the cal BP ranges of 1–3 dated dwellings overlap. This suggests a pattern of occupation from an initial low population to a population maximum followed by a decrease in numbers. Given that the number of dated dwellings is a fraction of the probable total number, it is clear that several dwellings were occupied at the same time throughout the site's history, in particular during the middle phase.

Harp (1976) and Erwin (1995, Chap. 8) arrived at a similar pattern of population establishment, growth and decline based on radiocarbon dates and number of identified dwellings. Unlike Renouf and Bell (2009) they calculated the theoretically possible number of dwellings by extrapolating from the dated dwellings to the total number of identified dwellings. Harp (1976:124) suggested that at its maximum, between 1.8 and 12.6 dwellings could have been occupied at the same time and Erwin (Chap. 8) calculated a potential population maximum of 6–10 dwellings. Meantime, early phase dwellings are likely under-represented since an undetermined number are hidden beneath middens.

Migratory harp seals were the economic basis of Phillip's Garden. Harp seals are available off the Point Riche headland twice yearly, first in mid-December when they are swimming southward in open water from their summer grounds off Greenland to their whelping grounds in the Gulf of St. Lawrence, and second in March–April when they make their return journey northward at the edge of the retreating ice (Sergeant 1991; see Murray, Chap. 11).

The abundance, accessibility and predictability of harp seals differ significantly between mid-December and March–April. In mid-December, harp seals are relatively scarce off Port au Choix and are more commonly found along the Quebec shoreline (LeBlanc 1996:27). Modern Port au Choix sealers tell us that the fall seal hunt was only ever an occasional activity and that seals were sometimes netted off Phillip's Garden. By contrast, in March–April harp seals are concentrated in their thousands off the Point Riche headland and in particular at Phillip's Garden where an ice lead regularly opens up a short distance offshore. According to Port au Choix residents, Phillip's Garden has always been the best location for the spring seal hunt.

However, the abundance of harp seals is available for a short time only. Since harp seals move north following the fast-moving ice they are present off Point Riche sometimes for only a few days or at most a few weeks.

The next section describes faunal, artefactual and architectural data in more detail, to show how the Phillip's Garden Dorset utilized the abundance of harp seals.

## Harp Seals, Architecture and Diversification

### *Seal Hunting*

Faunal material is well preserved and abundant throughout Phillip's Garden and is concentrated in large and small middens. For example, 9 m$^2$ of midden Feature 2 yielded 241,523 bones and bone fragments, or 26,836 bones/m$^2$. Faunal analyses (Hodgetts et al. 2003; Murray 1992; Renouf 2000) show that seal, *Phocidae*, was the focal resource, ranging from 90% of the number of identified specimens in the early and middle phases of site occupation to >70% by the late phase (Hodgetts et al. 2003). Hodgetts et al. (2003:110) noted that seal bone is difficult to identify to species and thus most archaeological seal bones cannot be identified beyond *Phocidae*. However, they also observed that of the *Phocidae* bones that could be identified to species, 92–100% were harp seal, which by extrapolation means that most *Phocidae* remains at Phillip's Garden were harp seal as well (Hodgetts et al. 2003:110).

Harp seal herds have a characteristic age structure at each migration period (Hodgetts 2005a; Sergeant 1991). In mid-December the herd is comprised of adults, immatures (>22 months) and 10 month-old juveniles (born the previous year). In March–April, the herd consists of adults, immatures, 2–3 month-old juveniles and newborns. A small number of bones belonging to newborn seals was found in all identified faunal assemblages, indicating March–April exploitation (Hodgetts et al. 2003:114; Renouf and Murray 1999:127). In addition, four sectioned seal canines indicated a March–June season of death (Hiseler 1997:128). Mid-December exploitation was also identified by Hodgetts (2005a, 2005b) who used long bone measurements to establish the age structure of the faunal assemblage. Her data from three midden assemblages showed the presence of 2–3 month-old and 10 month-old juveniles which indicated a March–April and mid-December hunt, respectively.

Seal hunting is reflected in the tool assemblage (Table 7.1). Of the 29,380 artefacts in the total lithic assemblage, 3,268 (11.1%) are harpoon endblades (Fig. 7.4a). There are many fewer harpoon heads (Fig. 7.4b) ($n=125$, 2.7% of the organic tool assemblage) and foreshafts ($n=35$, 0.01% of the organic tool assemblage). This scarcity compared to the number of endblades suggests that harpoon heads and foreshafts were recycled and reused. Microblades ($n=6{,}624$, 22.5% of the lithic assemblage) are twice as common as endblades. In her study of cut marks on seal bones from a middle phase midden (Feature 2), Wells (1988:15) classified extremely thin, shallow slices as microblade cut marks and wider, deeper cuts as biface cut marks. Since thin shallow marks were by far the most common, Wells (1988:15) concluded that microblades were the main butchering tool. This is supported by Knapp (2008:81) who in her replication studies of hide-processing tools found a microblade to be highly effective in removing the thick layer of blubber from beneath the hide.

## *Seal Butchering*

The relative frequency of seal skeletal elements was determined for two associated midden assemblages from the early phase of site occupation (Murray 1992). Since no body parts were missing that could not be accounted for by taphonomy, Renouf and Murray (1999:128) concluded that whole seals were processed and consumed at the site.

Wells (1988:46) reconstructed the sequence of seal butchery based on cut-mark patterns from a middle phase midden. The seal was laid on its back for skinning and butchering. The hide was cut at the skull near the ears and where the flippers joined the long bones. The head and tongue were removed, the vertebral column was disarticulated into meat packages and the limbs were disarticulated from the flippers. The scapula and pelvis were disarticulated from long bones at the proximal end and the flippers were disarticulated at the distal end. Cut marks showed that meat was removed from the ribs and front limbs. There was some evidence of cutting meat from the flippers.

**Table 7.1** Artefact totals, including fragments, from Phillip's Garden as of November 30, 2009

| Provenience | Date ranges cal BP at 1σ probability | Harpoon endblades | Harpoon heads | Foreshafts | Microblades | Tabular slate tools | Total organic tools | Total lithic tools |
|---|---|---|---|---|---|---|---|---|
| House 1 | Undated | 0 | 0 | 0 | 21 | 3 | 0 | 90 |
| House 2 | 1710–1240 | 451 | 7 | 4 | 926 | 206 | 360 | 4,473 |
| House 3 | Undated | 47 | 1 | 0 | 28 | 0 | 46 | 228 |
| House 4 | 1520–1410 | 268 | 7 | 3 | 408 | 94 | 213 | 2,211 |
| House 5 | 1480–1320 | 26 | 1 | 0 | 78 | 13 | 8 | 216 |
| House 6 | 1600–1420 | 228 | 8 | 4 | 588 | 82 | 487 | 1,840 |
| House 7 | Undated | 19 | 1 | 0 | 17 | 9 | 50 | 112 |
| House 8 | Undated | 12 | 0 | 0 | 14 | 2 | 25 | 62 |
| House 9 | Undated | 4 | 0 | 0 | 0 | 3 | 17 | 30 |
| House 10 | 1690–1420 | 204 | 9 | 3 | 548 | 104 | 257 | 1,975 |
| House 11 | 1510–1340 | 301 | 5 | 1 | 586 | 90 | 199 | 2,248 |
| House 12 | 1520–1370 | 173 | 10 | 6 | 246 | 115 | 289 | 1,501 |
| House 13 | Undated | 67 | 4 | 0 | 83 | 28 | 81 | 432 |
| House 14 | Undated | 9 | 2 | 0 | 34 | 8 | 24 | 89 |
| House 15 | Undated | 20 | 2 | 0 | 27 | 5 | 41 | 115 |
| House 16 | 1520–1410 | 45 | 2 | 0 | 181 | 48 | 45 | 620 |
| House 17 | 1710–1310 | 372 | 25 | 7 | 791 | 252 | 850 | 4,184 |
| House 18 | 1690–1410 | 357 | 16 | 2 | 803 | 168 | 533 | 2,962 |
| House 19 | Undated | 34 | 1 | 0 | 11 | 4 | 53 | 92 |
| House 20 | 1300–1180 | 20 | 2 | 1 | 49 | 11 | 51 | 224 |
| House 17/18 ext. | n/a | 99 | 3 | 1 | 116 | 69 | 306 | 707 |
| House Feature 1 | 1920–1630 | 61 | 1 | 0 | 182 | 31 | 80 | 541 |
| House Feature 1/14 | n/a | 20 | 2 | 0 | 29 | 2 | 23 | 150 |
| House Feature 14 | 1990–1870 | 80 | 0 | 0 | 89 | 19 | 123 | 686 |
| House Feature 55 | 1400–1180 | 47 | 1 | 0 | 164 | 15 | 74 | 594 |
| Midden Feature 2 | 1990–1340 | 133 | 0 | 1 | 205 | 74 | 110 | 1,013 |

(continued)

**Table 7.1** (continued)

| Provenience | Date ranges cal BP at 1σ probability | Harpoon endblades | Harpoon heads | Foreshafts | Microblades | Tabular slate tools | Total organic tools | Total lithic tools |
|---|---|---|---|---|---|---|---|---|
| Midden Feature 42 | Undated | 34 | 0 | 0 | 13 | 1 | 1 | 105 |
| Midden Feature 49 | 1930–1720 | 5 | 0 | 0 | 17 | 3 | 4 | 45 |
| Unknown | | 132 | 15 | 3 | 370 | 104 | 1,940 | 329 |
| Total | | 3,268 | 125 | 35 | 6,624 | 1,563 | 4,714 | 29,380 |

Counts change as excavation and collection analyses continue. For details of radiocarbon dates see Appendix. House Feature 1/14 refers to a line of excavation units where Features 1 and 14 abut. House 17/18 ext. refers to the exterior area between those two structures

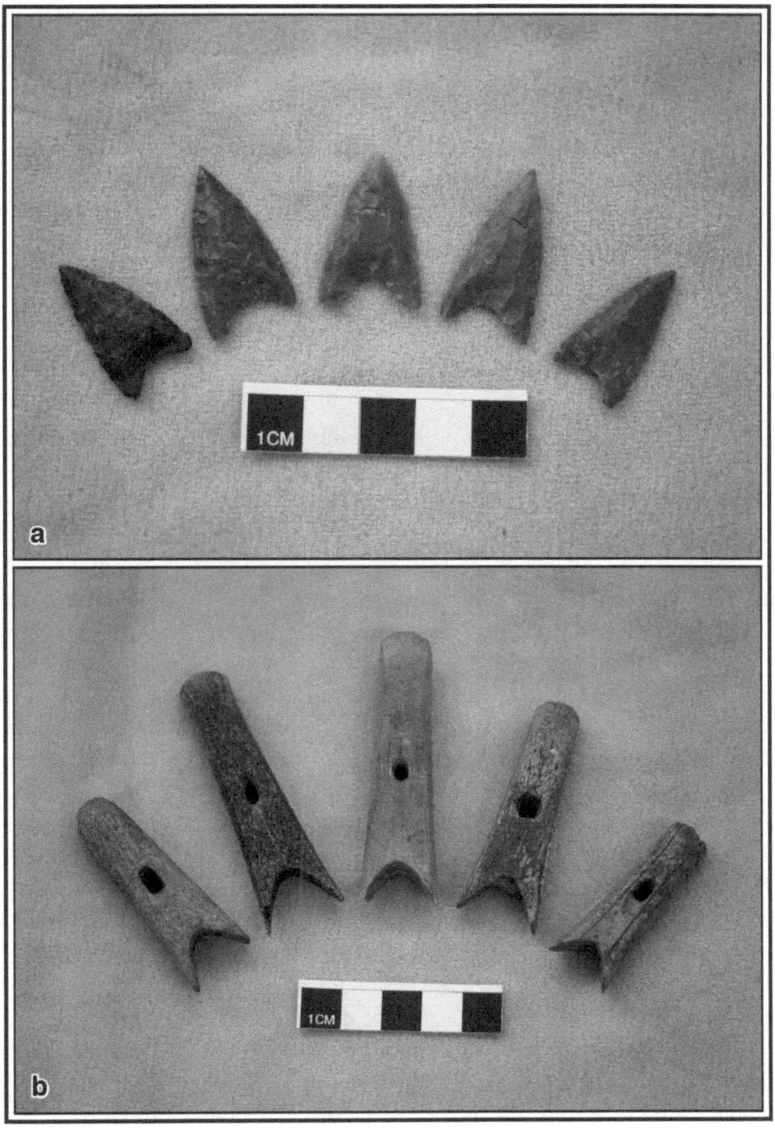


**Fig. 7.4** Seal hunting tools from Phillip's Garden. (**a**) Chert harpoon endblades. (**b**) Bone harpoon heads

## *Importance of Fat*

Seal blubber would have been an important product, used as fuel, food, waterproofing for boots and boats and grease for sled shoes. Phillip's Garden soil is permeated with seal fat, as evidenced by the dark, organically rich soil which has a fatty feel to it. A burned substance coats both sides of the many soapstone vessel fragments from the site, reflecting their function as lamps and cooking pots. We assume this is burned seal fat because the residue is black and crusty and because seal fat is the most readily available fuel supply. However, this is contrary to Deal's (1990) analysis of one burned residue sample for which the fatty acid and stable isotope signatures were similar to fish and bird tissue, respectively. However, as Deal (1990:10) pointed out, there could well be complicating issues of fatty acid degradation and post-depositional contamination.

## *Hide Processing*

Seal hides were another important product, essential for clothing, boots and covers for boats and dwellings. Recent work suggests that hide working was as intensive as the hunt itself (Knapp 2008; Renouf and Bell 2008).

Distinctive tabular slate tools with one or more unifacially bevelled scraping edges comprise 5.3% of the lithic tool assemblage (Table 7.1). Renouf and Bell (2008:37) organized these tools into two broad categories, bevelled-edged (Fig. 7.5a) and rounded-tip (Fig. 7.5b), suggesting the former were scraping tools and the latter were used to fold and pleat leather, for example, to make boot heels and toes. Knapp (2008) tested these suggestions in her use-wear analysis of replica slate tools. She scraped sealskins and creased seal leather with the replica tools and compared use-wear patterns of replica and archaeological specimens. She confirmed that the bevelled-edge slate tools were skin scrapers, but was unable to identify the function of the rounded-tip slate tools which did not display the use-wear of either skin scraping or creasing.

Bell et al. (2005), Renouf and Bell (2008), Bambrick (2009) and Renouf et al. (2009) suggested that sealskins were soaked in Bass Pond, about 500 m east of Phillip's Garden (Fig. 7.1). Modern Northern Peninsula hide workers soak sealskins in shallow warm ponds where bacterial action loosens the hair which then can easily be scraped off (Bock 1991) (Fig. 7.6a, b); the depilated skin is then tanned for colour and preservation and made into waterproof boots. Bell et al. (2005) inferred similar activities for the Dorset based on fossil pollen from Bass Pond sediments.

Analysis of these sediments showed a disturbance across pollen taxa at 2200–1400 cal BP (Bell et al. 2005). In particular, Bell et al. (2005) argued that the appearance of the aquatic herb *Myriophyllum* (2200–1800 cal BP) and an increase in the aquatic alga *Pediastrum* (2000–1400 cal BP) indicated increased bacterial action in the pond. There was also a marked decrease in *Sphagnum* at 2200–1800 cal BP which they interpreted as pond-side trampling associated with hide-soaking activities.

In addition, analysis of chironomids (Rosenberg et al. 2005) indicated an increase in Bass Pond salinity at 2200–1400 cal BP declining to normal levels at 1100 cal BP. Rosenberg et al. (2005:69) suggested that this increase could be related to marine incursions such as storm surges or tsunamis. However, Bell et al. (2005) pointed out

**Fig. 7.5** Sealskin processing tools from Phillip's Garden. (**a**) Bevelled-edged slate tools. (**b**) Rounded-tip slate tools

that at 13 m above sea level and 300 m from the shore, Bass Pond was beyond the reach of storm surges. They also noted that there was no sedimentary evidence in the Bass Pond core of storm surges or tsunamis. They therefore argued that the increase in salinity could be related to tanning activities which are often based on saline solutions of water and bark (Bock 1991).

**Fig. 7.6** Sealskin processing north of Port au Choix in late June 2008. (**a**) Sealskins stretched on frames. (**b**) Framed sealskins soaking in a shallow pond

Bambrick (2009) did a follow-up analysis of a second sediment core from Bass Pond, focusing on *Myriophyllum* and *Pediastrum*. In addition, she looked at values of 15N, an isotopic marker of sea mammal tissue (cf. Douglas et al. 2004). Bambrick's data showed a contemporaneous increase in *Myriophyllum*, *Pediastrum* and 15N values. However, according to the chronology of this core the timing of the disturbance event was earlier than in the first core, occurring at 2400–2200 cal

BP rather than 2200–1400 cal BP. Bambrick's (2009) chronology introduced the possibility that anthropogenic disturbance in Bass Pond could be connected to the Groswater Palaeoeskimo occupation at nearby Phillip's Garden East (Fig. 7.1) dated at 2950–2130 cal BP (Renouf 1994a, 2005). However, since this was a small, intermittently occupied, seal hunting site (Kennett 1990; LeBlanc 1996) it seems a less likely candidate than the larger and more intensively occupied Phillip's Garden. The temporal relationship between the two cores and two sites remains to be resolved.

## *Dwelling Architecture*

Phillip's Garden dwellings were originally defined by Harp (1976:132) based on his extensive excavation of eight and partial excavation of 12 dwellings. Dwellings were oval or rectangular with a rear platform and side walls constructed of stacked limestone shingle and with a northeast entrance facing the sea. There was a central depression bisected by a line of pits which included a pit in the rear platform; these pits were part of the central hearth area. Harp's published plan of the interior of House 2 showed it to be 38.3 m$^2$ (Harp 1976:133; Renouf and Murray 1999:121). Harp (1976:132) suggested that House 2 size and layout reflected two families or a single extended family group.

More recent data have revised these architectural descriptions and size estimates (Renouf 2006, 2009c). We re-excavated four middle phase dwellings previously excavated or tested by Harp (Houses 2, 10, 17, and 18) (Fig. 7.2) and extended our excavations outside the berm that outlined each dwelling. In all except House 10, we dismantled the dwelling and associated features to understand their construction.

In all cases, what originally had been called walls by Harp (1976) were in fact 2–4 m deep platforms constructed of 3–4 layers of fist-sized rocks. There is some variability in platform depth (Table 7.2): in Houses 17 and 18 both side platforms were 2–3 m deep, but Houses 2 and 10 had one deep (4.2 m) and one shallow (1.3 m) lateral platform; it may be that the exceptional width of the wider platform represents some slumping. We redefined the size of these re-excavated dwellings, increasing their footprints to 94–103 m$^2$ (Table 7.2). In light these re-excavations we realized that we likely had underestimated the size of two other dwellings (Features 1 and 14[2]) where we, like Harp, had interpreted the perimeter berms as walls rather than platforms. Since interior space is defined by the footprint, these dwellings are minimally 51.5 and 74.7 m$^2$, respectively (Table 7.2). We had dismantled one other dwelling (Feature 55), revealing a clearly defined ring of post-holes following the outer edge of the perimeter platform and defining an interior space of 28.3 m$^2$. This is the smallest cold-weather dwelling identified at the site (Renouf 2003:409), second in size only to House 5 (16.61 m$^2$) which Harp (1976:130) interpreted as a summer dwelling, and Feature 42 (23.75 m$^2$) which we interpreted as a tent structure (Renouf 2003:394).

[2]Harp used House numbers (e.g. Houses 1–21) as opposed to my use of feature numbers (e.g. dwelling Features 1, 14 and 55).

**Table 7.2** Attributes of Phillip's Garden dwellings excavated or tested by the Port au Choix Archaeology Project

| Dwelling | Shape | E-W x N-S dimensions (m) | Depth of lateral platforms (m) | Depth of rear platform (m) | No. rock layers comprising platforms | Footprint ($m^2$)[a] | Sources additional to PAC Archaeology Project maps |
|---|---|---|---|---|---|---|---|
| House 2 | Subrectangular | 10.5[b] × 9 | *E* = 1.3,[c] *W* = 4.2 | No info | 5 | 94 | Renouf (2006:125) |
| House 10 | Subrectangular | 12.5 × 8.4[b] | *E* = 3.3, *W* = 1.3[c] | No info | 3 | ≥105 | Renouf et al. (2005:16) |
| House 17 | Trilobate | 11.6 × 11.9 | *E* = 2.2, *W* = 2.2 | 3 | 3 | 88 | Renouf (2009c:96) |
| House 18 | Oval | 11.6 × >8.9[b] | *E* = 2.94, *W* = 2.96 | No info | 2 | ≥103 | Cogswell et al. (2006:22–23) |
| Feature 1 | Oval | 9.2 × 7 | *S* = 2.0, *N* = 1.2 | ≥2.1 | No info | 51.5 | Renouf and Murray (1999:134) |
| Feature 14 | Oval | 7.5 × 12 | *E* = 1.7, *W* = 1.3 | ≥2.4 | No info | 74.7 | Renouf (1987:7, 2003:409) |
| Feature 55 | Circular | 6 | Perimeter = 1.1–1.9 | n/a | 3 | 28.3 | Renouf (2003:409, 2006:123) |

[a] Footprint calculation based on shape as follows:
Circlar – $\pi r^2$; oval – $\pi r^2$ based on radius of averaged diameter; subrectangular – length by width; trilobate – the sum of the area of three lobes established using field map; where one dimension is not established – minimum length by minimum width = minimum area (≥)
[b] Estimated from surface contours because not fully excavated
[c] Narrow width suggests this might be a platform similar to Feature 55

The re-excavated dwellings were 10.5–12.5 m wide (Table 7.2) which would have necessitated central support. Upon dismantling the structures we found that what Harp had described as an axial line of pits was in fact a centrally placed pair of post-holes with a third post-hole in the rear of the dwelling. House 2 had the largest and deepest pair of post-holes at 81 and 55 cm deep; each had a rim diameter of 58 cm tapering to 14–23 cm diameter at the base; pit walls were carefully reinforced with rows of small beach cobbles (Renouf 2006:124-125). These were clearly made for large, load-bearing posts in what must have been a substantial dwelling. There were similarly placed post-holes in Houses 10, 17 and 18. Features 1 and 14 each had two central pits and one rear pit which, in light of our recent re-excavations, we now think were post-holes. Feature 55 (Renouf 1992, 1999), the smallest cold-weather house, had no interior pits. The central post-holes were at either end of the burned and fat-stained slab or slab-and-cobble axial cooking area.

At least some, and possibly all, Phillip's Garden dwellings were framed with whale ribs (Renouf 2009c). In those dwellings where we excavated part of the exterior area and dismantled the platforms (Feature 55, Houses 17 and 18) we found deep, narrow, evenly spaced post-holes into which the proximal or distal end of a whale rib could be slotted to stand upright and curve towards the centre of the dwelling (Renouf 2009c:95). In House 10, we found a section of a whalebone post next to a post-hole (Renouf et al. 2005:14). Harp found a number of segments of similar posts, depicted in field notes and photographs and currently in the Phillip's Garden artefact collection. In one case only, House 17, we noted curved indentations surrounding the axial feature, which we interpreted as the imprints of whale ribs laid flat to form a curb (Fig. 7.7) (Renouf 2009c:98).

In the rear of House 18, we found a slab-constructed storage pit. This was a virtual square, measuring 56×60 cm at the top and 49×22 cm at the bottom; it was 60 cm deep. The walls were formed of large rocks and smaller cobbles. Several large boulders formed the floor and several large flat stones, displaced capstones, were slumped into the pit (Cogswell et al. 2006:12). We found a similar but disturbed pit in the rear of House 17 (Renouf 2009c:94). Since we only trenched Houses 2 and 10 we did not uncover the rear of the dwellings; however, Harp's field sketches of Houses 2 and 10 show two rear pits in each dwelling.

According to Harp's field notes and our own excavations, in most dwellings there was a northeast-facing entrance. Although we uncovered the front area of each dwelling we excavated, we did not completely expose the rear area in all cases. Where we did (Features 1, 14 and 55) we identified a second entrance facing southeast (Renouf 2003:393).

With the exception of Feature 55, the smallest cold-weather dwelling, the interior layout was remarkably consistent (Fig. 7.7). Each dwelling had: two central post-holes, a central cooking area, a large rear storage pit, a second pit or post-hole associated with the storage pit, one rear and two side platforms and, where observable, a front and a rear entrance. In contrast to the consistent interior layout, dwelling shape was variable (Table 7.2).

The range of radiocarbon dates associated with these dwellings indicates that they, or their locations, were used over a lengthy period. For example, seven radiocarbon

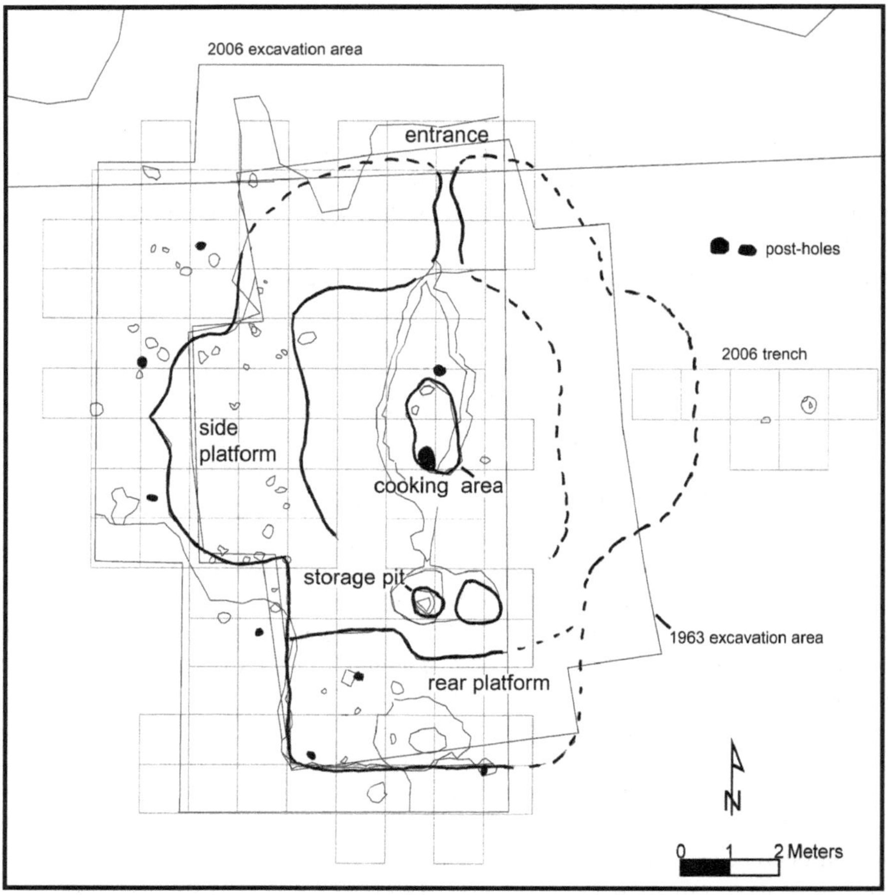


**Fig. 7.7** Interior layout of Phillip's Garden dwellings, illustrated by House 17. The *solid line* traces the dwelling outline based on the Port au Choix Archaeology Project's 2006 excavation and the *dashed line* shows the approximated outline based on Harp's 1963 excavation. The *two dark ovals* in the dwelling centre are post-holes for support posts. The *dark ovals* on the perimeter of the dwelling are post-holes for whale ribs

dates from House 2 and its associated middens span >200 years, from 1650 cal BP[3] (P-692) to 1390 cal BP (Beta 211271). Seven dates from the interior and exterior of House 17 span >300 years, from 1660 cal BP (Beta 238477) to 1340 cal BP (Beta 238479). Four dates from House 18 span >100 years, from 1590 cal BP (P-736) to 1460 cal BP (Beta 211268). While these dates are not precise enough to indicate an exact duration of dwelling use, they clearly represent a very lengthy occupation.

Some dwellings were renovated, re-sized and reused. For example, in House 2 there was evidence of two major phases of construction in the central axial area where the large central post-holes were later filled in and reconstructed to fit narrow

[3]Calibrated median dates are used for these ranges; details of radiocarbon dates from Phillip's Garden are in the Appendix.

posts; in a third sequence of events one of these re-sized post-holes was capped by a slab (Renouf 2006:125). House 18 also underwent phases of re-construction, identified by two distinct perimeters, one large and one small, centred on the axial feature (Cogswell et al. 2006:23).

Recent excavations have focused on the outdoor space between Houses 17 and 18 (Renouf 2009a, b). A complex palimpsest of single and paired post-holes, stake-holes and small indentations suggest that Houses 17 and 18 were associated with a variety of exterior structures. Although most of these data are not yet analysed, we have interpreted one linear and one curved concentration of stake-holes to be the impressions of racks (Fig. 7.8).

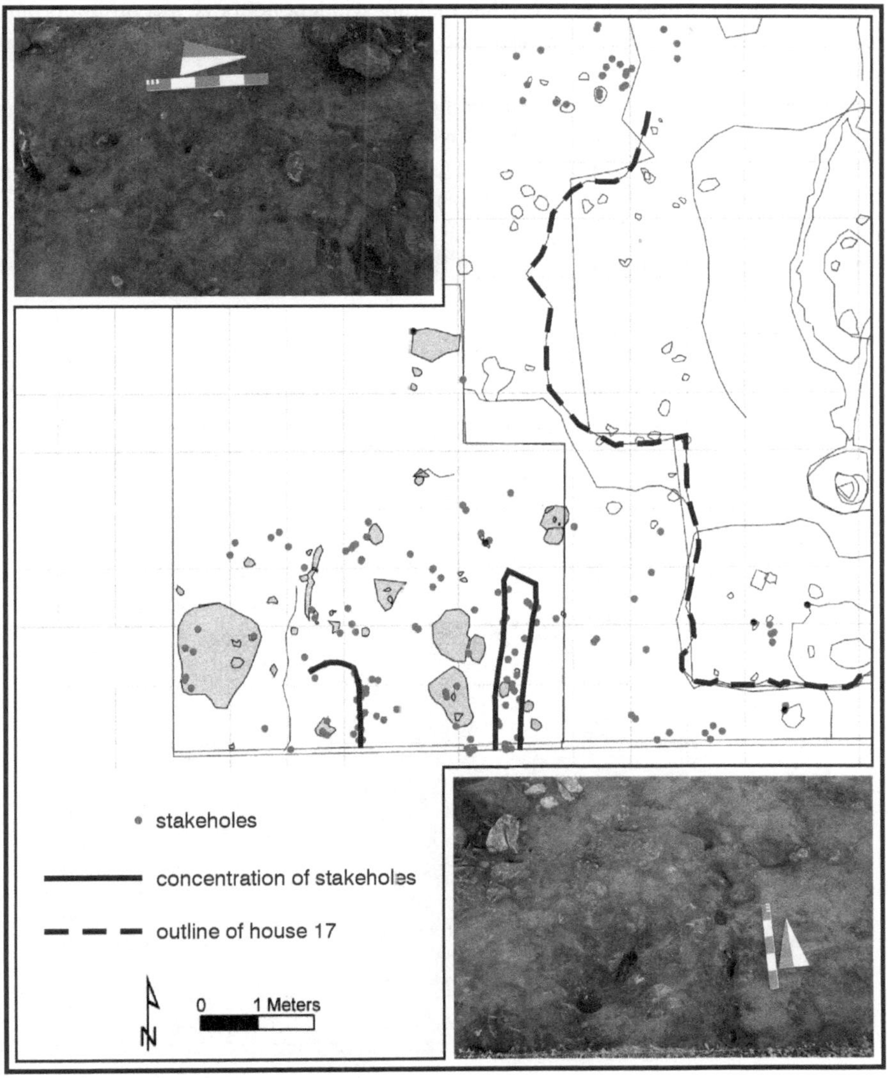


**Fig. 7.8** Stake-holes outside House 17. The two highlighted concentrations, each with an inset photograph, are interpreted as racks

## *Economic, Seasonal and Social Diversification*

Although seal harvesting was the main economic activity, there is also a clear focus on the acquisition and use of whalebone (Renouf 2009c). Not only were whale ribs used to frame dwellings, but whalebone was an important raw material for tools including sled shoes, lances, punches, wedges, planks and posts (Wells, personal communication 2010). A preliminary examination of organic artefacts from Houses 17 and 18 showed that a significant proportion was made of whalebone: 62% (501/807) and 49% (257/520), respectively. A more detailed intra-site description and analysis is underway (Wells 2006). There are many whalebone tool cores (Fig. 7.9) and preforms, indicating that making whalebone tools was an important activity at the site and that Port au Choix must have been near the source of raw material.

The site was also seasonally diversified with evidence of at least occasional occupation at other times of year. Warm weather occupation is suggested by a small (23.75 m$^2$) lightweight tent structure defined on the basis of a ring of post-holes (Renouf 2003:395) and a small (16.61 m$^2$) oval depression with no interior features and few artefacts which Harp (1976:130) interpreted as a summer dwelling. If we are correct that sealskins were soaked in Bass Pond, this would have taken place in

**Fig. 7.9** Whalebone cores from Phillip's Garden

early summer when pond water was sufficiently warm for bacterial action. As already mentioned, Hodgetts (2005a, 2005b) demonstrated that there was some site occupation in early winter.

As suggested elsewhere (Renouf 1994b; Renouf and Bell 2009), it is likely that Phillip's Garden had an important social and ritual function as families gathered together in one place. Population aggregation sites are essential for hunting and gathering populations who live most of the year in dispersed family groups because it gives them the opportunity to engage in social and ritual activities that reinforce cultural identity. These social gatherings occur during times of plenty when a large population can be supported (e.g. Lee 1972) or in times of scarcity when coordinated labour is necessary (e.g. Balikci 1970). In the case of Phillip's Garden, the abundance of harp seals both supported and required a large labour force.

## Interpretation

### *Multi-family Dwellings*

The large size of the Phillip's Garden dwellings, ranging from 74.7 to 105 $m^2$ (excluding Feature 55 which at 28.3 $m^2$ is anomalously small) (Table 7.2) indicates that these were multi-family structures. The rear and side platforms also suggest multiple families. For example, the three demarcated platforms that create the trilobate shape of House 17 suggest a minimum of three family groups. However, the large footprint of 88 $m^2$ suggests this number could be doubled, which would be consistent with historically recorded Mackenzie Inuit cruciform houses where two families occupied each bench (Lee and Reinhardt 2003:178). It is also consistent with Lee and Reinhardt's (2003:173–182) estimates of 2–6 families in Eskimo and Inuit winter houses with a size range of 27–81 $m^2$. Following McGuire and Schiffer (1983) who linked investment in dwelling construction to anticipated duration of use, the substantial nature of Phillip's Garden dwellings suggests they were built with long-term use in mind.

### *Communal Households*

Interior layout reflects household organization. For example, Dawson (2002) argued that in trilobate Thule dwellings of the central Canadian Arctic, demarcated interior space connected by a communal area reflected families integrated into a cohesive household. Similarly, Coupland et al. (2009) analysed the interior layout of multi-family plank houses at several late prehistoric sites along the Pacific north coast. At the northernmost sites, houses had a single central hearth surrounded by sleeping areas which Coupland et al. (2009) argued reflected communal household production and consumption. In multi-family plank houses from sites further south they noted a more complex mix of household communalism and family individualism, reflected in the presence of both central and peripheral cooking areas. Hoffman (1999) made similar arguments for late precontact Aleut dwellings from the eastern Aleutians.

The interior layout of Phillip's Garden dwellings suggests that each household operated as a single economic unit. The single central food preparation area indicates that cooking and eating were communal activities. The family platforms positioned around the central cooking and eating area reinforce the idea of communal household organization, as does the single rear storage pit which indicates a shared food supply. If platform delineation corresponds with family demarcation, House 17 families were delineated more clearly than those in other dwellings where platforms were raised areas contained within the overall shape of the dwelling. Families in Feature 55, the smallest dwelling, would have been the least delineated, since there was a single perimeter platform.

## Communal Hunting

Hunters could reasonably count on the arrival of thousands of harp seals off Phillip's Garden at some point during the end of March and first two weeks of April. However, they could not predict the exact day the seals would arrive nor their length of stay, which could vary from 2–3 days to 2–3 weeks depending on wind and ice conditions. This uncertainty must have been a strong incentive to maximize the number of seals killed during the first few days.

An increase in the number of hunters would have been the surest way to increase the number of seals taken in a short period of time given that harpoon technology involved one hunter killing one seal at a time. The coordination of those hunters would increase hunting efficiency, further increase net returns and, presuming shared distribution of the proceeds, would decrease the risk for each individual hunter. For example, several hunters could surround seals that were on the ice or in the water, preventing their escape. Once a seal was harpooned, other hunters could efficiently dispatch the struggling and wounded animal. Harp seals are heavy, with adult males approaching 130 kg (Lavigne and Kovacs 1988:7), so a number of hunters could help drag carcasses onto the ice or shore. Seals are aggressive when provoked, with strong jaws and sharp teeth, so several hunters could keep snapping animals under control.

## Coordinated Processing

There would have been multiple carcasses requiring processing for food, fat and hides. This processing would have been multi-stage and labour-intensive. For example, animals would have been flensed and the carcasses butchered into meat packages likely for

**Fig. 7.10** A bevelled-edged slate scraper with two sets of reverse bevels

distribution amongst households. Fat would have been set aside for distribution and storage. Each sealskin would have been stretched, either by lacing it into a frame (Fig. 7.6a) or pegging it onto the ground. The stretched skin would have been repeatedly scraped over a period of days or weeks. Those skins destined for leather would have been depilated and tanned.

Specialized skin scraping tools were developed to increase the efficiency of these tasks. A number of bevelled slate scrapers have more than one scraping edge. For example, Fig. 7.10 shows a slate scraper with two sets of reverse bevels so that the hide worker had four sharp edges to use before having to stop and re-sharpen. These are analogous to the ground slate knives that were developed to efficiently process large catches of salmon on the north Pacific coast (Frink et al. 2003; Graesch 2007).

## *Processing for Storage*

If the seal kill was maximized there would have been an inevitable surplus. Cold storage or freezing would not have been practical beyond May or June and we have not found large storage pits on the site. However, as Park (1999) suggested for Thule contexts, seal meat obtained in the spring could have been cut into small packages and strips and dried outdoors in the sun and wind. Park (1999) described ethnographic accounts of Inuit hanging strips of caribou to dry from outdoor racks and suggested that seal meat could have been dried the same way. Betts and Friesen (2004:361) noted that nineteenth-century Mackenzie Inuit dried bird, fish and sea mammal meat which was then stored in pits or on racks. If surplus seal meat was dried at Phillip's Garden, it could have been laid on or suspended from outdoor racks such as those we have identified between Houses 17 and 18 (Fig. 7.8).

### *Diversification*

The intensive seal harvest set the stage for diversification of other activities, for example, obtaining whalebone and making whalebone tools (Renouf 2009c). Modern Port au Choix residents tell us that whales are common off the Point Riche headland throughout the spring when they are often injured by ice, to perish and wash ashore. No doubt the Phillip's Garden Dorset scavenged carcasses such as these.

As a regularly occupied seal hunting location, Phillip's Garden became a permanent place. As a population aggregation site, that permanent place had social and symbolic significance. It is not surprising therefore that Phillip's Garden was occupied on an occasional basis throughout the year. It is also not surprising that a Dorset burial, Crow Head Cave (EeBi-4), lies below a prominent escarpment and overlooks the site (see Brown, Chap. 12).

## Implications

Table 7.3 summarizes data from the known large Dorset sites in Newfoundland and none approach Phillip's Garden in size, number and size of dwellings, and size of artefact assemblage. This is likely linked to the geographic attributes of Phillip's Garden that made it a uniquely productive place for the spring harp seal hunt. Because of this, Phillip's Garden is more comparable to a cross-cultural range of other large and intensively occupied sites in prime localities with access to abundant and predictable resources. However, in none of these cases was the focal resource available for such a short period of time as were the harp seals at Phillip's Garden and in few of these cases were the multi-family houses as large.

For example, the Igloolik region on the Melville Peninsula (Fig. 7.1) is one of the richest archaeological areas of the Canadian Arctic, with many Pre-Dorset, Dorset and Thule sites spread out over a series of raised shorelines. From Dorset times onward, this was a prime walrus hunting area, with extensive shallow waters providing excellent feeding grounds and hauling-out spots during the lengthy period of open water (Murray 1999:476). Based on identified faunal collections from Igloolik Dorset sites, walrus was an important resource (Murray 1996, 1999). Murray (1999) argued that the large size and aggressive behaviour of walrus required a coordinated hunt similar to the requirements of whaling. Some Igloolik

**Table 7.3** Information about the largest Dorset sites so far known in Newfoundland

| Site name | Borden # | Estimated site size (ha) | Number of identified dwellings | Size range of excavated dwellings ($m^2$) | Lithic artefact total[a] | Data source |
|---|---|---|---|---|---|---|
| Phillip's Garden | EeBi-1 | 2 | >67 | 30–110 | 34,094 | PAC Archaeology Project |
| Point Riche | EeBi-20 | 0.33 | 17 | 30–33 | 3,352 | Eastaugh (2002, 2003) |
| Peat Garden North | EgBf-18 | 0.1 | 2 | 20 | 652 | Hartery and Rast (2003) |
| Pittman | DkBe-1 | No info | 0 | n/a | 407 | Linnamae (1975) |
| Cape Ray | CjBt-1 | 0.33 | 1 | 27.5 | 5,086 | Fogt (1998); Linnamae (1975) |
| Dildo Island | CjAj-2 | No info | 2 | 34.4 | 5,562 | LeBlanc (1997, 1998, 2003) |
| Stock Cove | CkAl-3 | 0.2–0.4 | 0 | n/a | 3,096 | Robbins (1985) |

[a] Excluding flakes, retouched and utilized flakes, and tip flute spalls

Dorset sites are particularly large, such as Alarnerk where 208 dwelling remains occurring in clusters of five are spread out along 2.5 km of coastline (Maxwell 1985:182; Meldgaard 1960:588). Although most Dorset dwellings from Igloolik sites are small, averaging 20 $m^2$ (Meldgaard 1960:588; Murray 1996:42), there are also a small number of larger dwellings, up to 98 $m^2$ (Meldgaard 1960:589).

One of the largest Dorset sites in the Canadian Arctic is Nunguvik (Fig. 7.1), north Baffin Island, where 30 dwelling remains are spread out over 7 ha, most occurring singly and a few clustering in groups of 2–3 (Mary-Rousselière 2002:18). Faunal assemblages demonstrated that caribou was the primary resource, supplemented by *Phocidae*, narwhal, birds, fish and small game. Presumably Nunguvik was a particularly good location for caribou, although there are none there today (Mary-Rousselière 2002:10). The five dwelling remains excavated at Nunguvik ranged in size from 29.3 to 52 $m^2$.

Qariaraqyuk, the largest known Thule site in Arctic Canada, is on Somerset Island (Fig. 7.1) in a key summer whaling area. The site extends over several hectares and Whitridge (1999:152) recorded 59 winter dwellings, in addition to dwelling features from other seasons. Whitridge (1999:103) suggested that many of these winter dwellings were contemporaneous and therefore that population at the site was as large as 300 people. Winter dwellings were fairly small, with interior space ranging from 5.4 to 24.0 $m^2$ (Whitridge 1999:150–151). Bones collected from the site surface reflected the importance of bowhead whaling which Whitridge (1999) suggested was carried out by whaling crews led by a captain.

The late precontact Mackenzie Inuit site of Kuukpak on the Mackenzie River (Fig. 7.1) is near shallow waters that were excellent for drive hunting beluga during the summer months (Betts and Friesen 2004; Friesen and Arnold 1995). The site is large, extending along the Mackenzie River for about 800 m, and 21 dwelling remains have been identified (Arnold 1994); the identified faunal assemblage was dominated by beluga (Friesen and Arnold 1995:28). Based on the plan drawing of House 1 (Arnold 1994:90) interior dwelling space was about 45 $m^2$. Betts and Friesen (2004:363) suggested that similar-sized Mackenzie Inuit dwellings housed up to 30 people.

There are many large sites with multi-family dwellings on the British Columbia coast and interior waterways. For example, in northern British Columbia the early prehistoric Paul Mason site (Fig. 7.1) is situated on a constriction of the Skeena River which made it an ideal spot for salmon fishing (Coupland 1996), which would have taken place over at least a couple of months (cf. Schalk 1977:220). Consistent with this, a high proportion of salmon bones was found in cooking hearths. Ten multi-family dwellings, 45.7–72.6 $m^2$, were spread over about 2,000 $m^2$. Coupland (1996:124) noted that salmon fishing was likely carried out individually at this site, using dip nets and basket traps, and that processing and preservation would have required cooperative labour.

Further south, in the Middle Fraser River Valley, Keatley Creek (Fig. 7.1) is a large early-to-late prehistoric site situated on a major salmon fishing river where at least 50 dwelling remains are spread out over 4 ha (Hayden 2000:10). Salmon was available for a number of months in this area of the river, and abundance of salmon bones indicated that they were the main resource of the site's inhabitants (Hayden 1997; Kusmer 2000). Although some Keatley Creek dwellings were small (38.5 $m^2$) medium-sized dwellings were 79–177 $m^2$ and there were a few large dwellings >177 $m^2$ (Hayden 2005:173).

The late precontact Aleut site of Agayadan Village, southwestern Alaska, is situated at the entrance to a large lagoon known to be an ecological hotspot in general and a good location for salmon in particular. Sockeye and chum peaked in July and early August, and were followed by coho salmon which were available as late as October and November (Hoffman 1999:42). Salmon remains in faunal assemblages from the site showed it to be a major resource (Hoffman 1999:151, 2002:32). Twenty multi-family dwellings were spread over 2.5 ha and dwelling size ranged from 48.4 to 177.5 $m^2$ (Hoffman 1999:153).

These few examples of large and intensively occupied sites are all situated in choice locations where particular resources were abundant, accessible and predictable. Like the March–April harp seals at Phillip's Garden, these resources provided the opportunity for intensive exploitation by a large and coordinated labour force. In most cases, this labour force was organized through multi-family households.

It is interesting that the Phillip's Garden multi-family dwellings are amongst the largest. This may be linked to the fact that, unlike the examples reviewed above, the focal resource at Phillip's Garden was available for a brief period only, a few days to a few weeks. This temporal compression of otherwise unlimited abundance necessitated intense and coordinated economic activity which was mobilized through singularly large multi-family households.

## Conclusions

Phillip's Garden is one of the largest and richest Dorset sites in the Canadian North. It was a seasonally occupied, permanent location from which harp seals were intensively hunted during their northward migration when they appeared off the Point Riche headland in late March–early April. Although the general timing was predictable, the particular timing was not. Which day the seals turned up and for how long depended on wind and ice conditions, which were variable. Phillip's Garden was a prime seal hunting spot because seals appeared in their thousands a short distance from shore where they were concentrated in a regularly occurring ice lead. From the hunters' perspective, harp seals were an infinite resource limited only by technology, person-power and period of availability. Timing was critical. When the seals appeared, hunters had to mobilize and take as many seals as possible before the winds shifted and the ice moved out, taking the seals with them. Seal carcasses had to be skinned and butchered. If the abundance of meat was to be spread out over the year, it had to be dried and stored. Similarly, fat had to be collected and stored. Skins had to be turned into usable hides. Like any agricultural harvest, all hands were required for the short burst of critically timed work. At Phillip's Garden this labour force was coordinated through large multi-family households. The regularity of the enterprise made it worthwhile to invest time and effort into constructing substantial dwellings that were used year after year. This regularity was the basis for diversification, such as acquiring whalebone from which tools were made and dwellings framed. The aggregation of families was the ideal opportunity to engage in critical social and ritual activities. Phillip's Garden must have been an important place to which families had strong ties of memory and cultural identity. Appropriately, it was overlooked by the dead.

## *Epilogue*

However, Phillip's Garden did not go on forever and by 1180 cal BP it was abandoned. Bell and Renouf (2008) and Renouf and Bell (2009) argued that this was linked to an increase in temperature which disrupted the pattern of ice and associated seals. According to modern Port au Choix sealers, spring ice conditions are linked to winds and tides which can be unpredictable and therefore there are years when seals are inaccessible or absent from Phillip's Garden. This is contrary to the archaeological evidence of construction and reconstruction of substantial multi-family dwellings which were a significant investment in infrastructure that would not have been made had there been a risk of seals not turning up in any given year. On the other hand, some Phillip's Garden dwellings were eventually abandoned, to become the garbage dumps of subsequent structures. Perhaps these contrasting data suggest variability in ice conditions over the seven centuries of site occupation, with periods of greater and lesser predictability of sea ice and harp seals and consequent variability in hunting patterns. Such variability, if it did exist, would be key to understanding the development, success and eventual abandonment of this remarkable site.

**Acknowledgements** The Port au Choix Archaeology Project is generously funded by the Social Sciences and Humanities Research Council of Canada, the Canada Research Chairs Program, Parks Canada, Memorial University and the Provincial Archaeology Office. I thank each member of the many Phillip's Garden excavation crews over the years and the many graduate students who analysed pieces of the Phillip's Garden puzzle. I thank the staff of the Port au Choix National Historic Site for their continued support. Thanks to marine harvester, Dwight Spence, for his insight on harp seal hunting from Port au Choix. Patty Wells and Dominique Lavers updated the bone artefact catalogue, Dominique Lavers took the artefact photographs, and Charles Conway, Memorial University Geography Department, drafted all but one figure. Thanks to Gary Coupland, Max Friesen, Bjarne Grønnow and Patty Wells for their careful comments on an earlier version of this paper. A particular acknowledgement to the late Elmer Harp for generously allowing us to use his field notes, drawings and photographs which have allowed us to build on his pioneer work.

## References

Arnold, C.D. 1994. Archaeological investigations on Richards Island. In *Bridges across time: The NOGAP archaeological project*, ed. J.-L. Pilon, 85–93. Canadian Archaeological Association Occasional Paper No. 2, Ottawa.

Balikci, A. 1970. *The Netsilik Eskimos*. New York: Garden City Press.

Bambrick, J. 2009. Palaeoeskimo Sealskin Processing in Port au Choix, Northwestern Newfoundland: A Paleoenvironmental Analysis. Master's thesis, Department of Archaeology, Memorial University of Newfoundland, St. John's.

Bell, T., J.B. Macpherson, and M.A.P. Renouf. 2005. Late prehistoric impact on Bass Pond, Port au Choix. *Newfoundland and Labrador Studies* 20(1): 107–129.

Bell, T., and M.A.P. Renouf. 2008. The domino effect: culture change and environmental change in Newfoundland, 1500-1100 cal BP. *The Northern Review* 28: 72–94.

Betts, M.W., and T.M. Friesen. 2004. Quantifying hunter-gatherer intensification: A zooarchaeological case study from Arctic Canada. *Journal of Anthropological Archaeology* 23: 257–384.

Bock, A. 1991. *Out of necessity: The story of sealskin boots in the Strait of Belle Isle*. Peninsula Typesetting and Design, St. Anthony, NL.

Coupland, G. 1996. The evolution of multi-family households on the northwest coast of North America. In *People who lived in big houses: Archaeological perspectives on large domestic structures*, ed. G. Coupland and E.B. Banning, 121–130. Monographs in World Archaeology, 27. Madison: Prehistory Press.

Coupland, G., T. Clark, and A. Palmer. 2009. Hierarchy, communalism, and the spatial order of northwest coast plank houses. *American Antiquity* 74(1): 77–106.

Cogswell, A.E. 2006. House 18 and the Middle Phase of Dorset Occupation at Phillip's Garden (EeBi-1). Master's thesis, Department of Anthropology, Memorial University of Newfoundland, St. John's.

Cogswell, A.E., M.A.P. Renouf, and P. Wells. 2006. 2005 Excavations at the Port au Choix National Historic Site. On file, Parks Canada, Archaeology, Atlantic Region, Halifax.

Dawson, P.C. 2002. Space syntax analysis of Central Inuit snow houses. *Journal of Anthropological Archaeology* 21(4): 464–480.

Deal, M. 1990. Exploratory analyses of food residues from prehistoric pottery and other artifacts from eastern Canada. *Society for Archaeological Sciences Bulletin* 13(1): 6–12.

Douglas, M.S.V., J.P. Smol, J.M. Savelle, and J.M. Blais. 2004. Prehistoric Inuit whalers affected Arctic freshwater ecosystems. *Proceedings of the National Academy of Science* 101(6): 1613–1617.

Eastaugh, E.J.H. 2002. The Dorset Palaeoeskimo Site at Point Riche: An Intra-site Analysis. Master's thesis, Department of Anthropology, Memorial University of Newfoundland, St. John's.

Eastaugh, E.J.H. 2003. A Middle Dorset semi-subterranean dwelling at Point Riche, Newfoundland. *Études/Inuit/Studies* 27(1–2): 451–471.

Erwin, J.C. 1995 An Intra-Site Analysis of Phillip's Garden: A Middle Dorset Palaeo-Eskimo Site at Port au Choix, Newfoundland. Master's thesis, Department of Anthropology, Memorial University of Newfoundland, St. John's.

Fogt, L.M. 1998. The Excavation and Analysis of a Dorset Palaeoeskimo Dwelling at Cape Ray, Newfoundland. Master's thesis, Department of Anthropology, Memorial University of Newfoundland, St. John's.

Friesen, T.M., and C.D. Arnold. 1995. Zooarchaeology of a focal resource: Dietary importance of beluga whales to the precontact Mackenzie Inuit. *Arctic* 48(1): 22–30.

Frink, L., B.W. Hoffman, and R.D. Shaw. 2003. Ulu knife use in western Alaska: A comparative ethnoarchaeological study. *Current Anthropology* 44(1): 116–123.

Graesch, A.P. 2007. Modeling ground slate knife production and implications for the study of household labor contributions to salmon fishing on the Pacific northwest coast. *Journal of Anthropological Archaeology* 26(4): 576–606.

Harp, E. 1964. *The cultural affinities of the Newfoundland Dorset Eskimo*. National Museums of Canada Bulletin 200, Ottawa.

Harp, E. 1976. Dorset settlement patterns in Newfoundland and southeastern Hudson Bay. In *Eastern Arctic prehistory: Paleoeskimo problems*, ed. M.S. Maxwell, 119–138. Washington, D.C.: Society for American Archaeology.

Hartery, L.J., and T.L. Rast. 2003. A Middle Dorset Palaeoeskimo structure at Peat Garden North, northwest Newfoundland. *Études/Inuit/Studies* 27(1–2): 473–492.

Hayden, B. 1997. *The pithouses of Keatley Creek: Complex hunter-gatherers of the northwest Plateau*. New York: Harcourt-Brace College Publishers.

Hayden, B. 2000. Socioeconomic factors influencing housepit assemblages at Keatley Creek. In *The ancient past of Keatley Creek, volume 2*, ed. B. Hayden, 3–27. Burnaby, B.C.: Simon Fraser University Archaeology Press.

Hayden, B. 2005. The emergence of large villages and large residential corporate group structures among complex hunter-gatherers at Keatley Creek. *American Antiquity* 70(1): 169–180.

Hiseler, G. 1997. Harp Seal (*Phoca groenlandica*) Canine Dentine Incremental Annuli as Indicators of Age and/or Season of Death. Master's thesis, Department of Anthropology, Memorial University of Newfoundland, St. John's.

Hodgetts, L.M. 2002. Report of the 2001 Excavations at Phillip's Garden, Port au Choix National Historic Site. On file, Parks Canada, Archaeology, Atlantic Region, Halifax.

Hodgetts, L.M. 2005a. Using bone measurements to determine the season of harp seal hunting at Phillip's Garden. *Newfoundland and Labrador Studies* 20(1): 91–106.

Hodgetts, L.M. 2005b. Dorset Palaeoeskimo harp seal exploitation at Phillip's Garden (EeBi-1), northwestern Newfoundland. In *The exploitation and cultural importance of sea mammals*, Proceedings of the 9th Conference of the International Council of Archaeozoology, Durham, August 2002, ed. G. Monks, 62–76. Oxford: Oxbow Books.

Hodgetts, L.M., M.A.P. Renouf, M.S. Murray, D. McCuaig-Balkwill, and L. Howse. 2003. Changing subsistence practices at the Dorset Paleoeskimo site of Phillip's Garden, Newfoundland. *Arctic Anthropology* 40(1): 106–120.

Hoffman, B.W. 1999. Agayadan village: household archaeology on Unimak Island, Alaska. *Journal of Field Archaeology* 26(2): 147–161.

Hoffman, B.W. 2002. The Organization of Complexity: A Study of Late Prehistoric Village Organization in the Eastern Aleutian Region. PhD thesis, Department of Anthropology, University of Wisconsin, Madison.

Kennett B.L. 1990. Phillip's Garden East: An Examination of the Groswater Palaeo-Eskimo Phase. Master's thesis, Department of Anthropology, Memorial University of Newfoundland, St. John's.

Knapp, R.E. 2008. An Analysis of Tabular Slate Tools from Phillip's Garden (EeBi-1), A Dorset Palaeoeskimo Site in Northwestern Newfoundland. Master's thesis, Department of Anthropology, Memorial University of Newfoundland, St. John's.

Kusmer, K. 2000. Zooarchaeological analysis at Keatley Creek: Socioeconomy. In *The ancient past of Keatley Creek, volume 2*, ed. B. Hayden, 119–133. Burnaby, B.C.: Simon Fraser University Archaeology Press.

Lavigne, D.M., and K.M. Kovacs. 1988. *Harps and hoods: Ice-breeding seals of the North Atlantic*. Waterloo: University of Waterloo Press.

LeBlanc, S. 1996. A Place with a View: Groswater Subsistence-Settlement Patterns in the Gulf of St. Lawrence. Master's thesis, Department of Anthropology, Memorial University of Newfoundland, St. John's.

LeBlanc, S. 1997. Dildo Island Archaeological Project: The Dorset Occupation of Dildo Island Preliminary Field Report, 1996. On file, Provincial Archaeology Office, Government of Newfoundland and Labrador, St. John's.

LeBlanc, S. 1998. Dildo Island 1997 Field Season: Interim Report. On file, Provincial Archaeology Office, Government of Newfoundland and Labrador, St. John's.

LeBlanc, S. 2003. A Middle Dorset dwelling in Trinity Bay, Newfoundland. *Études/Inuit/Studies* 27(1–2): 493–513.

Lee, M., and G.A. Reinhardt. 2003. *Eskimo architecture: Dwelling and structure in the early historic period*. Fairbanks: University of Alaska Press.

Lee, R.J. 1972. Intensification of social life among the !Kung Bushmen. In *Population growth: Anthropological implications*, ed. B. Spooner, 343–350. Cambridge: MIT Press.

Linnamae, U. 1975. *The Dorset culture: A comparative study in Newfoundland and the Arctic*. Technical Papers of the Newfoundland Museum 1, St. John's.

Mary-Rousselière, G. 2002. *Nunguvik et Saatut: Site Paléoeskimaux de Navy Board Inlet, Île de Baffin*. Mercury Series Archaeology Paper 162, Canadian Museum of Civilization, Ottawa.

Maxwell, M.S. 1985. *Prehistory of the eastern Arctic*. New York: Academic Press.

McGuire, R.H., and M.B. Schiffer. 1983. A theory of architectural design. *Journal of Anthropological Archaeology* 2: 227–303.

Meldgaard, J. 1960. Prehistoric sequences in the eastern Arctic as elucidated by stratified sites at Igloolik. In *Men and cultures: Selected papers of the fifth international congress of anthropological and ethnological sciences*, ed. A.F.C. Wallace, 588–595. Philadelphia: University of Pennsylvania Press.

Murray, M.S. 1992. Beyond the Laundry List: The Analysis of Faunal Remains from a Dorset Dwelling at Phillip's Garden (EeBi 1), Port au Choix, Newfoundland. Master's thesis, Department of Anthropology, Memorial University of Newfoundland, St. John's.

Murray, M.S. 1996. Economic Change in Palaeoeskimo Prehistory of Foxe Basin, N.W.T. PhD thesis, Department of Anthropology, McMaster University, Hamilton.

Murray, M.S. 1999. Local heroes: The long-term effects of short-term prosperity – An example from the Canadian Arctic. *World Archaeology* 30(3): 466–483.

Park, R.W. 1999. Seal use and storage in the Thule culture of Arctic North America. *Revista de Arqueologia Americana/Journal of American Archaeology* 16: 77–97.

Renouf, M.A.P. 1986. Archaeological Investigations at Phillip's Garden and Point Riche, Port au Choix National Historic Park Report of 1985 Field Activities. On file, Parks: Canada, Archaeology, Atlantic Region, Halifax.

Renouf, M.A.P. 1987. Archaeological Excavations at the Port au Choix National Historic Park: Report of 1986 Field Activities. On file, Parks Canada, Archaeology, Atlantic Region, Halifax.

Renouf, M.A.P. 1991. Archaeological Investigations at the Port au Choix National Historic Park: Report of the 1990 Field Season. On file, Parks Canada, Archaeology, Atlantic Region, Halifax.

Renouf, M.A.P. 1992. The 1991 Field Season, Port au Choix National Historic Park: Report of Archaeological Excavations. On file, Parks Canada, Archaeology, Atlantic Region, Halifax.

Renouf, M.A.P. 1993. The 1992 Field Season, Port au Choix National Historic Park: Report of Archaeological Excavations. On file, Parks Canada, Archaeology, Atlantic Region, Halifax.

Renouf, M.A.P. 1994a. Two transitional sites at Port au Choix, northwestern Newfoundland. In *Threads of Arctic prehistory: Papers in honour of William E. Taylor, Jr.*, ed. D. Morrison and J.-L. Pilon, 165–196. Mercury Series Archaeology Paper 149, Canadian Museum of Civilization, Ottawa.

Renouf, M.A.P. 1994b. Hunter-gatherer population aggregations in prehistory: A case study from Port au Choix, northwestern Newfoundland. In *Hunters and gatherers in the modern context*, Proceedings of the Seventh International Conference on Hunting and Gathering Societies, ed. L. Ellanna, 529–539. Department of Anthropology, University of Alaska at Fairbanks, Fairbanks

Renouf, M.A.P. 1999. *Ancient cultures, bountiful seas: The story of Port au Choix*. Historic Sites Association of Newfoundland and Labrador, St. John's.

Renouf, M.A.P. 2000. Symbolism and subsistence: seals and caribou at Port au Choix, northwestern Newfoundland. In *Animal bones, human societies*, ed. P. Rowley-Conwy, 65–73. Oxford: Oxbow Books.

Renouf, M.A.P. 2002. *Archaeology at Port au Choix, northwestern Newfoundland 1990-1992*. Occasional Papers in Northeastern Archaeology 12. St. John's: Copetown Press.

Renouf, M.A.P. 2003. A review of Palaeoeskimo dwelling structures in Newfoundland and Labrador. *Études/Inuit/Studies* 27(1–2): 375–416.

Renouf, M.A.P. 2005. Phillip's Garden West: A Newfoundland Groswater variant. In *Contributions to the study of the Dorset Palaeo-Eskimos*, ed. P.D. Sutherland, 57–80. Mercury Series Archaeology Paper 167, Canadian Museum of Civilization, Gatineau.

Renouf, M.A.P. 2006. Re-investigating a middle phase Dorset dwelling at Phillip's Garden, Port au Choix, Newfoundland. In *Dynamics of northern societies*, Proceedings of the SILA/NABO Conference on Arctic and North Atlantic Archaeology, Copenhagen, May 10th–14th, 2004, ed. J. Arneborg and B. Grønnow, 119–128. Publications from the National Museum Studies in Archaeology and History Vol. 10, Copenhagen.

Renouf, M.A.P. 2007. Re-excavating House 17 at Phillip's Garden, Port au Choix: Report of the 2006 Field Season. On file, Parks Canada, Archaeology, Atlantic Region, Halifax.
Renouf, M.A.P. 2009a. 2008 Field Season at Phillip's Garden, Port au Choix National Historic Site. *Provincial Archaeology Office Archaeology Review* 7: 133–135.
Renouf, M.A.P. 2009b. Outside House 17: 2008 Field Season at Phillip's Garden, Port au Choix National Historic Site. On file, Parks Canada, Archaeology, Atlantic Region, Halifax.
Renouf, M.A.P. 2009c. Dorset Palaeoeskimo whalebone use at Phillip's Garden, Port au Choix. In *On the track of the Thule culture, from Bering Strait to east Greenland*, Proceedings of the SILA Conference "The Thule Culture – New Perspectives in Inuit Prehistory", ed. B. Grønnow, 91–104. Publications from the National Museum Studies in Archaeology and History Vol. 15, Copenhagen.
Renouf, M.A.P., and M.S. Murray. 1999. Two winter dwellings at Phillip's Garden, a Dorset site in northwestern Newfoundland. *Arctic Anthropology* 36(1–2): 118–132.
Renouf, M.A.P., P. Wells, and J.R. Pickavance. 2005. The 2004 Field Season at the Port au Choix National Historic Site: Phillip's Garden (EeBi-1) and Barbace Cove (EeBi-12). On file, Parks Canada, Archaeology, Parks Canada, Atlantic Region, Halifax.
Renouf, M.A.P., and T. Bell. 2008. Dorset Palaeoeskimo sealskin processing at Phillip's Garden, Port au Choix, northwestern Newfoundland. *Arctic* 61(1): 35–47.
Renouf, M.A.P., and T. Bell. 2009. Contraction and expansion in Newfoundland prehistory, AD 900–1500. In *The northern world AD 900-1400*, ed. H. Maschner, O. Mason, and R.J. McGhee, 263–278. Salt Lake City: University of Utah Press.
Renouf, M.A.P., T. Bell, and J. Macpherson. 2009. Hunter-gatherer impact on Subarctic vegetation: Amerindian and Palaeoeskimo occupations of Port au Choix, northwestern Newfoundland. *Arctic Anthropology* 46(1–2): 176–191.
Robbins, D.R. 1985. Stock Cove Trinity Bay: The Dorset Eskimo Occupation of Newfoundland from a Southeastern Perspective. Master's thesis, Department of Anthropology, Memorial University of Newfoundland, St. John's.
Rosenberg, S., I. Walker, and J. Macpherson. 2005. Environmental changes at Port au Choix as reconstructed from fossil midges. *Newfoundland and Labrador Studies* 20(1): 57–73.
Schalk, R.F. 1977. The structure of an anadromous fish resource. In *For theory building in archaeology*, ed. L.R. Binford, 207–249. New York: Academic Press.
Sergeant, D.E. 1991. *Harp seals, man, and ice.* Canadian Special Publications of Fisheries and Aquatic Sciences No. 114. Department of Fisheries and Oceans, Ottawa.
Stuiver, M., and P.J. Reimer. 1993. Extended 14C database and revised CALIB radiocarbon calibration program. *Radiocarbon* 35: 215–230.
Wells, P.J. 1988. An Analysis of the Butchery Practices of Two Palaeo-Eskimo Groups. Honour's dissertation, Department of Anthropology, Memorial University of Newfoundland, St. John's.
Wells, P.J. 2006. Analysis of Dorset Palaeoeskimo Bone, Antler and Ivory Industry: A Regional and Site-Specific Exploration of Cultural Adaptation. PhD thesis proposal, Department of Anthropology, Memorial University of Newfoundland, St. John's.
Whitridge, P.J. 1999. The Construction of Social Difference in a Prehistoric Inuit Whaling Community. PhD thesis, Department of Anthropology, Arizona State University, Tempe.

# Chapter 8
# The Changing Nature and Function of Phillip's Garden: A Diachronic Perspective

**John C. Erwin**

An understanding of the function of Phillip's Garden (EeBi-1) has long been impeded by the complexity and the overlap of the many temporal and physical components of this large archaeological site (Fig. 8.1). Consequently, many questions relating to settlement patterning, site development and house reoccupation have remained unanswered. As there is strong evidence that archaeological remains at Phillip's Garden represent different occupational events, site functions and various seasons of use (Renouf 1991a:60–62, 1993a:59, Chap. 7), it is apparent that establishing contemporaneity amongst the archaeological features is crucial to understanding the function of the site (Renouf 1991a:62). Hence, the basic methodological question is how to determine which of the houses were occupied simultaneously at any given time (Harp 1976:120). In response to this problem, it was the aim of my initial research (Erwin 1995) to establish the function(s) of Phillip's Garden by determining how different occupational sequences contributed to the overall archaeological assemblage. As such, it can be argued that a definition of site function can only be derived by separating out the different components of residency and determining what the archaeological patterns should look like for different occupational events.

## Residency Models and Archaeological Patterning

The function of Phillip's Garden has generally been interpreted in relation to an annual spring hunt of migratory harp seals, since the recovered faunal material is dominated by seal remains, with only trace amounts of caribou, fox, beaver, birds and fish (Harp 1976:128; Hodgetts et al. 2003; Murray, Chap. 11; Renouf 1991a:60; Renouf and Murray 1999:127). This model can be described as a *regular seasonal occupation* which is distinguished by a narrow range of seasonality, similar house types, similar tool assemblages, and simultaneously occupied houses. Alternatively, if

J.C. Erwin (✉)
Department of Archaeology, Memorial University, St. John's, NL, Canada, A1C 5S7
e-mail: jerwin@mun.ca

M.A.P. Renouf (ed.), *The Cultural Landscapes of Port au Choix*, Interdisciplinary Contributions to Archaeology, DOI 10.1007/978-1-4419-8324-4_8,

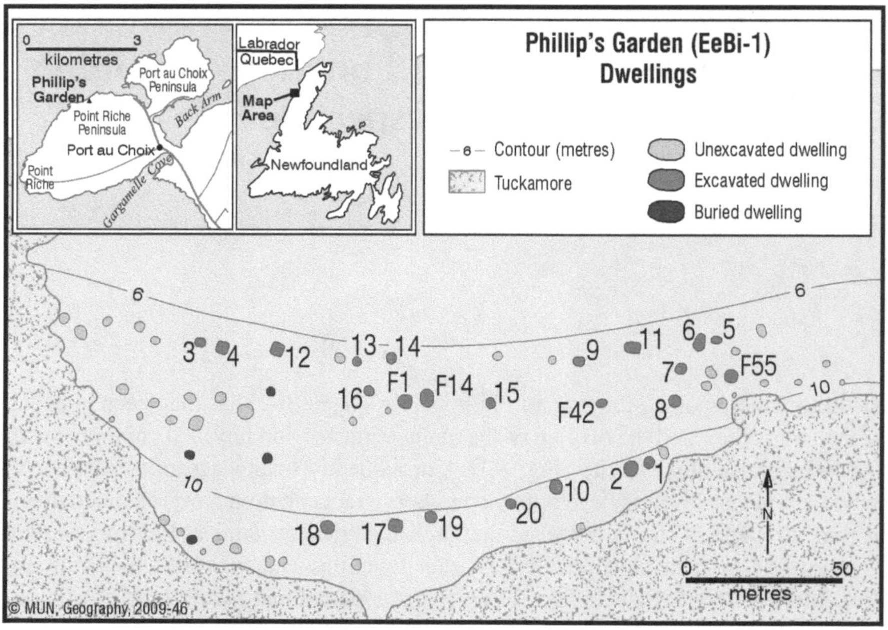


**Fig. 8.1** Phillip's Garden dwellings with inset map showing locations of sites mentioned in the text

the occupation and use of Phillips Garden was more variable, a model for *varied seasonal occupation* might be more applicable. If, for example, Phillip's Garden served as a residential base during the fall (Murray 1992), or during both spring and winter seal hunts (Hodgetts 2005), or even as a staging area from which small groups could have organized for inland caribou hunting and salmon fishing (Harp 1976:132; Renouf 1991a:61), these activities should be distinguished by different house types, a wide range of seasonality, dissimilar tool assemblages, and fewer simultaneously occupied houses. A third possible model, described as *shifting seasonal occupation*, could explain the economic potential of Phillip's Garden relative to its place within an overall sequence of hunter-base camp moves (e.g., Binford 1982:12). If, for example, the function of Phillip's Garden shifted according to the requirements prescribed by overriding settlement and subsistence strategies, this should produce wide diversity in site function and seasonality. The consequence of such site use should include fewer houses occupied concurrently, a wider range of house forms, and increased diversity in tool assemblages. As a *meeting place*, Phillip's Garden may have served to further social relationships through population aggregation and communal activities (Renouf 1991a:61, 1999:36, Chap. 7). Since population aggregation could have only occurred when it was "economically feasible, or when cooperation in hunting and gathering [was] economically necessary" (Renouf 1991a:60), social functions are likely to have cross-cut all other settlement types. In this model, social functions at Phillip's Garden should produce larger structures with communal purpose (Renouf 2006, 2009), uniformity in house styles reflecting similar utilization of the site, and the simultaneous occupation of a larger number of houses. Lastly, the occupation of Phillip's Garden may be described as a *composite occupational pattern*, which includes combinations of the foregoing occupational scenarios. Accordingly, Phillip's Garden's residency patterning could have developed from many different functional events and varying seasons of use over time. This intermixing of residency types would produce less recognizable archaeological patterning (Janes 1983:27) and be responsible for the creation of substantial noise in the archaeological data sets.

The reason for the initial settlement of Phillip's Garden is likely related to the availability of large herds of migrating harp seals in the vicinity of the Strait of Belle Isle (Sergeant 1991). These same seal herds are also what facilitated the concentrated and persistent occupation of Phillip's Garden between 1990 and 1180 cal BP[1] (Renouf 1993b, 2006, Chap. 7). Based on the premise that one might expect redundancy in settlement locations where there is access to reliable resources (Binford 1980, 1990), the subsistence and settlement patterns of Dorset peoples at Phillip's Garden can be interpreted as anchored as a result of their seasonal dependence upon the reliable and abundant resources of the sea. Although it has been demonstrated that Phillip's Garden was an important access window to maritime resources (Harp 1964, 1976; Renouf 1991a, 1993b) what is less certain is whether the Dorset were simply seasonally tethered to it or occupied it in some semi-sedentary fashion (Renouf 1991b:94). In view of the recognized variability in the use of this site over long periods of time (Erwin 1995; Renouf 1991a), it is apparent that a further understanding of Phillip's Garden requires a temporal perspective that considers potential variability of site function. Therefore, the issue of contemporaneity is not only crucial to understanding the function(s) of the site over time and at any point in time (Erwin 1995; Harp 1976; Renouf 1991a), but one that is practical if viewed as phases of occupation.

## Temporal Evidence

At large complex archaeological sites, such as Phillip's Garden, the lack of fine temporal control makes it difficult to determine how different occupational events, site functions, and various seasons of use contributed to the overall archaeological assemblage. Since it is rare that absolute contemporaneity between archaeological units can ever be established, we are left to consider this problem within a temporal framework in which broader units of time are assessed. As such, we are left to question how many houses could have been occupied during any given phase at Phillip's Garden. A number of techniques based upon radiocarbon dating have been applied to this problem (e.g., Shott 1992; Thomas 1986). Most of these techniques are based on the comparison of statistical error inherent within radiocarbon results to determine

[1] Except where indicated, all calendar dates in this chapter were calibrated using Calib 3.0 (Stuiver and Reimer 1993). The one sigma probability range is used here for consistency with Renouf (Chap. 7). Elsewhere in this chapter, the two sigma probability range is used.

whether differences between results are real or whether they are simply a factor of statistical error. Where differences are found to be a result of statistical error, a case can then be made for potential contemporaneity.

In attempting to establish which houses at Phillip's Garden were occupied simultaneously, the problem is further complicated by the likelihood that some of the structures were re-occupied (Renouf 2006:126) and that the radiocarbon results are representative of only a point in the continuum of the occupation. The number of times a structure is re-occupied may be a factor of structure durability and/or function, the life span of owner-occupant, the season(s) in which the structure was used, and the availability of building materials for the construction of a new dwelling vs. repairing the old. Facing this problem in Younger Stone Age settlements in Norway, Helskog (1984) and Helskog and Schweder (1989) demonstrated that it is possible to calculate the statistical probability of contemporaneous occupation through the comparison of radiocarbon results to hypothetical reoccupation intervals which account for house longevity. More specifically, their technique estimates the number of contemporaneous houses by comparing radiocarbon dates to fixed ages with assumed reoccupation intervals. Since house longevity is an unknown variable at Phillip's Garden and is not easily interpreted from the archaeological remains, three different reoccupation intervals (25, 50 and 75 years) were employed for comparative purposes.

As with any comparison of statistical probability, this method relies upon a test of significance. However, in addition to testing the significance of the difference between a radiocarbon date and a fixed age, house longevity is added as a factor in these calculations. As such, this method requires that a fixed time ($t$) and a length of reoccupation ($L$) be chosen and that they be compared to a radiocarbon result (x). The probable number of houses ($N$) which were occupied at any given time ($t$) is calculated by summing the resulting probabilities as shown:

$$N(t) = \sum_{i=1}^{n} p(t|\mathrm{x}_i)$$

(Helskog and Schweder 1989:166).

The summation of the resultant probabilities provides the potential contemporaneity for each house. Unlike other tests in which potential contemporaneity can only be assumed whenever it is demonstrated that the difference between two dates is a result of statistical error, the Helskog and Schweder method calculates probabilities for actual house contemporaneity as they are generated from the radiocarbon data for each dwelling.

The use of this technique requires radiocarbon dates for each of the features which are to be compared. At Phillip's Garden, we were afforded such an opportunity from the long history of research that includes over 30 radiocarbon results (see Appendix) derived from samples obtained from 16 different dwellings. In all but two instances, seal fat-based dates were rejected in favour of the more reliable wood charcoal-based dates for dwellings that were considered in this study. Although seal fat samples are considered unreliable due to marine reservoir effects

**Table 8.1** Radiocarbon ages used in this study

| Context | Lab No. | Material | C14 years BP | Cal BP 1 σ |
|---|---|---|---|---|
| H20 | P-737 | Charcoal | 1321±49 | 1290–1189 |
| F55 | B-66435 | Charcoal | 1410±100 | 1390–1270 |
| H17 | P-734 | Charcoal | 1465±51 | 1400–1300 |
| H5 | P-676 | Charcoal | 1502±49 | 1410–1320 |
| H11 | P-696 | Charcoal | 1509±47 | 1410–1330 |
| H13 | P-731 | Fat | 1891±56 | 1500–1350 |
| H12 | P-729 | Charcoal | 1538±55 | 1510–1350 |
| H16 | P-733 | Charcoal | 1565±53 | 1520-1360 |
| H4 | P-727 | Charcoal | 1580±54 | 1530–1400 |
| H10 | P-694 | Charcoal | 1602±49 | 1540–1410 |
| H6 | P-679 | Charcoal | 1623±47 | 1550–1420 |
| H18 | P-736 | Charcoal | 1683±49 | 1680–1530 |
| H2 | P-AVE | Charcoal | 1698±35 | 1690–1540 |
| F1 | B-15379 | Charcoal | 1850±110 | 1890–1620 |
| H15 | P-732 | Fat | 2294±51 | 1940–1830 |
| F14 | B-AVE | Charcoal | 2016±52 | 2000–1890 |

*H* house; *F* feature; *B-AVE* the average of samples B-23976 and B-23977 and *P-AVE* the average of samples P-692 and P-693
All radiocarbon dates are calibrated using the intercept method at one sigma through Calib 3 (Stuiver and Reimer 1993). Dates from Houses 13 and 15 are on seal fat, corrected using Calib 3. Additional dates from Phillip's Garden, available since this study, are in the Appendix

(Bowman 1990:25), exceptions were made for Houses 13 and 15, since no other dates were available. Corrections for these results were made with the Calib 3 radiocarbon calibration programme (Stuiver and Reimer 1993). The radiocarbon ages used in this analysis are given in Table 8.1.

Following Helskog and Schweder (1989), the number of houses which could have been occupied at any point in time (Fig. 8.2) was calculated. It is noted that only 16 of the total of identified houses at Phillip's Garden are included in this analysis, since the majority have not yet been excavated. It is also noted that 13 of the 16 houses which are included in this analysis were initially chosen for excavation by Harp on the basis of a random sampling (Harp 1976:120) and that the three remaining houses excavated by Renouf (1993a:19) were chosen on the basis of assumed differences to those excavated by Harp, as a method of assessing the range of house variation at the site.

The sum of these calculations, as shown in Fig. 8.3, demonstrates that the highest potential house contemporaneity at Phillip's Garden was reached at about 1400 years cal BP after a long and slow increase, and an early and brief peak near 1850 cal BP. After 1300 cal BP, the use of the site declined rapidly until it was abandoned at about 1100 cal BP. Based upon house longevity, and an estimated 50[2] possible dwellings, it

[2]When this study was originally carried out (Erwin 1995) the number of identified dwellings at Phillip's Garden was only 50; with further site mapping this number increased, most recently to 68 (Renouf 2006:121).

**Comparison to a Fixed Age with a Reoccupation Interval of 25 Years as a Test for Contemporaneity**

Years Before Present

| House | 1050 | 1100 | 1150 | 1200 | 1250 | 1300 | 1350 | 1400 | 1450 | 1500 | 1550 | 1600 | 1650 | 1700 | 1750 | 1800 | 1850 | 1900 | 1950 | 2000 | 2050 | 2100 |
|---|---|---|---|---|---|---|---|---|---|---|---|---|---|---|---|---|---|---|---|---|---|---|
| H20 | | 0.04 | 0.18 | 0.43 | 0.12 | 0.09 | 0.02 | | | | | | | | | | | | | | | |
| F55 | | | 0.01 | 0.07 | 0.23 | 0.43 | 0.15 | 0.08 | 0.02 | | | | | | | | | | | | | |
| H17 | | | | 0.01 | 0.09 | 0.38 | 0.14 | 0.12 | 0.02 | | | | | | | | | | | | | |
| H5 | | | | | 0.03 | 0.23 | 0.36 | 0.17 | 0.04 | | | | | | | | | | | | | |
| H11 | | | | | 0.01 | 0.18 | 0.49 | 0.19 | 0.04 | | | | | | | | | | | | | |
| H12 | | | | 0.01 | 0.04 | 0.11 | 0.24 | 0.34 | 0.12 | 0.08 | 0.04 | 0.01 | | | | | | | | | | |
| H13 | | | | 0.01 | 0.03 | 0.11 | 0.25 | 0.34 | 0.13 | 0.08 | 0.03 | 0.01 | | | | | | | | | | |
| H16 | | | | 0.01 | 0.03 | 0.09 | 0.20 | 0.32 | 0.02 | 0.09 | 0.05 | 0.02 | 0.01 | | | | | | | | | |
| H4 | | | | | | 0.02 | 0.10 | 0.28 | 0.24 | 0.13 | 0.07 | 0.02 | | | | | | | | | | |
| H10 | | | | | | 0.01 | 0.08 | 0.25 | 0.38 | 0.14 | 0.08 | 0.02 | | | | | | | | | | |
| H2 | | | | | | | | 0.01 | 0.04 | 0.13 | 0.29 | 0.19 | 0.12 | 0.07 | 0.02 | 0.01 | | | | | | |
| H6 | | | | | | 0.01 | 0.06 | 0.21 | 0.40 | 0.12 | 0.09 | 0.03 | 0.01 | | | | | | | | | |
| H18 | | | | | | | | 0.01 | 0.06 | 0.15 | 0.29 | 0.14 | 0.11 | 0.06 | 0.02 | 0.01 | | | | | | |
| F1 | | | | | | | | 0.01 | 0.02 | 0.04 | 0.07 | 0.12 | 0.17 | 0.21 | 0.08 | 0.07 | 0.06 | 0.04 | 0.03 | 0.01 | 0.01 | |
| H15 | | | | | | | | | | | | | | | 0.04 | 0.17 | 0.42 | 0.09 | 0.10 | 0.03 | | |
| F14 | | | | | | | | | | | | | | | | 0.03 | 0.15 | 0.40 | 0.01 | 0.11 | 0.03 | |
| | 0.00 | 0.04 | 0.19 | 0.53 | 0.57 | 1.65 | 2.08 | 2.32 | 1.52 | 0.97 | 1.00 | 0.55 | 0.41 | 0.34 | 0.16 | 0.28 | 0.62 | 0.53 | 0.13 | 0.14 | 0.04 | 0.00 |

Probability

**Comparison to a Fixed Age with a Reoccupation Interval of 50 Years as a Test for Contemporaneity**

Years Before Present

| House | 1050 | 1100 | 1150 | 1200 | 1250 | 1300 | 1350 | 1400 | 1450 | 1500 | 1550 | 1600 | 1650 | 1700 | 1750 | 1800 | 1850 | 1900 | 1950 | 2000 | 2050 | 2100 |
|---|---|---|---|---|---|---|---|---|---|---|---|---|---|---|---|---|---|---|---|---|---|---|
| H20 | 0.01 | 0.11 | 0.40 | 0.43 | 0.01 | 0.18 | 0.05 | 0.01 | | | | | | | | | | | | | | |
| F55 | | | 0.03 | 0.16 | 0.47 | 0.35 | 0.11 | 0.16 | 0.05 | 0.01 | | | | | | | | | | | | |
| H17 | | | | 0.03 | 0.25 | 0.68 | 0.27 | 0.23 | 0.06 | 0.01 | | | | | | | | | | | | |
| H5 | | | | 0.01 | 0.11 | 0.53 | 0.30 | 0.22 | 0.10 | 0.01 | | | | | | | | | | | | |
| H11 | | | | | 0.07 | 0.46 | 0.31 | 0.18 | 0.10 | 0.01 | | | | | | | | | | | | |
| H12 | | | | 0.02 | 0.09 | 0.24 | 0.47 | 0.31 | 0.09 | 0.16 | 0.08 | 0.03 | 0.01 | | | | | | | | | |
| H13 | | | | 0.02 | 0.07 | 0.23 | 0.49 | 0.30 | 0.13 | 0.16 | 0.07 | 0.02 | | | | | | | | | | |
| H16 | | | | 0.02 | 0.06 | 0.19 | 0.40 | 0.45 | 0.08 | 0.19 | 0.10 | 0.04 | 0.01 | | | | | | | | | |
| H4 | | | | | 0.01 | 0.06 | 0.24 | 0.56 | 0.27 | 0.20 | 0.14 | 0.04 | 0.01 | | | | | | | | | |
| H10 | | | | | 0.01 | 0.04 | 0.19 | 0.50 | 0.32 | 0.15 | 0.15 | 0.05 | 0.01 | | | | | | | | | |
| H2 | | | | | | | | 0.02 | 0.09 | 0.28 | 0.56 | 0.25 | 0.20 | 0.13 | 0.05 | 0.01 | | | | | | |
| H6 | | | | | | 0.03 | 0.16 | 0.44 | 0.39 | 0.04 | 0.18 | 0.07 | 0.01 | | | | | | | | | |
| H18 | | | | | | | 0.01 | 0.04 | 0.13 | 0.32 | 0.57 | 0.23 | 0.21 | 0.13 | 0.05 | 0.01 | | | | | | |
| F1 | | | | | | | 0.01 | 0.02 | 0.04 | 0.08 | 0.14 | 0.23 | 0.33 | 0.41 | 0.15 | 0.14 | 0.12 | 0.08 | 0.05 | 0.03 | 0.01 | 0.01 |
| H15 | | | | | | | | | | | | | | 0.01 | 0.10 | 0.38 | 0.47 | 0.04 | 0.19 | 0.06 | 0.01 | |
| F14 | | | | | | | | | | | | | | | 0.01 | 0.08 | 0.34 | 0.54 | 0.13 | 0.21 | 0.07 | 0.01 |
| | 0.01 | 0.11 | 0.44 | 0.69 | 1.14 | 3.00 | 3.00 | 3.43 | 1.85 | 1.61 | 1.99 | 0.95 | 0.79 | 0.68 | 0.36 | 0.63 | 0.92 | 0.66 | 0.37 | 0.29 | 0.09 | 0.02 |

Probability

**Comparison to a Fixed Age with a Reoccupation Interval of 75 Years as a Test for Contemporaneity**

Years Before Present

| House | 1050 | 1100 | 1150 | 1200 | 1250 | 1300 | 1350 | 1400 | 1450 | 1500 | 1550 | 1600 | 1650 | 1700 | 1750 | 1800 | 1850 | 1900 | 1950 | 2000 | 2050 | 2100 |
|---|---|---|---|---|---|---|---|---|---|---|---|---|---|---|---|---|---|---|---|---|---|---|
| H20 | 0.04 | 0.24 | 0.67 | 0.30 | 0.10 | 0.26 | 0.09 | 0.02 | | | | | | | | | | | | | | |
| F55 | | 0.01 | 0.08 | 0.32 | 0.73 | 0.30 | 0.01 | 0.25 | 0.09 | 0.02 | | | | | | | | | | | | |
| H17 | | | 0.01 | 0.11 | 0.51 | 0.35 | 0.33 | 0.23 | 0.11 | 0.02 | | | | | | | | | | | | |
| H5 | | | | 0.03 | 0.29 | 0.61 | 0.25 | 0.09 | 0.17 | 0.03 | | | | | | | | | | | | |
| H11 | | | | 0.02 | 0.22 | 0.72 | 0.22 | 0.03 | 0.19 | 0.03 | | | | | | | | | | | | |
| H12 | | | 0.01 | 0.05 | 0.16 | 0.39 | 0.69 | 0.30 | 0.00 | 0.24 | 0.13 | 0.05 | 0.01 | | | | | | | | | |
| H13 | | | 0.01 | 0.04 | 0.15 | 0.39 | 0.69 | 0.30 | 0.04 | 0.24 | 0.11 | 0.03 | 0.01 | | | | | | | | | |
| H16 | | | 0.01 | 0.04 | 0.12 | 0.32 | 0.60 | 0.36 | 0.17 | 0.24 | 0.16 | 0.07 | 0.02 | | | | | | | | | |
| H4 | | | | | 0.03 | 0.14 | 0.42 | 0.59 | 0.30 | 0.12 | 0.21 | 0.07 | 0.02 | | | | | | | | | |
| H10 | | | | | 0.02 | 0.10 | 0.36 | 0.71 | 0.30 | 0.04 | 0.24 | 0.09 | 0.02 | | | | | | | | | |
| H2 | | | | | | | 0.01 | 0.05 | 0.19 | 0.46 | 0.54 | 0.30 | 0.15 | 0.20 | 0.08 | 0.02 | | | | | | |
| H6 | | | | | 0.01 | 0.08 | 0.30 | 0.68 | 0.31 | 0.07 | 0.26 | 0.11 | 0.03 | | | | | | | | | |
| H18 | | | | | | | 0.02 | 0.08 | 0.23 | 0.51 | 0.48 | 0.30 | 0.18 | 0.19 | 0.08 | 0.02 | 0.01 | | | | | |
| F1 | | | | | | | 0.01 | 0.03 | 0.06 | 0.13 | 0.23 | 0.35 | 0.49 | 0.37 | 0.21 | 0.15 | 0.17 | 0.12 | 0.08 | 0.04 | 0.02 | 0.01 |
| H15 | | | | | | | | | | | | | | 0.04 | 0.23 | 0.64 | 0.31 | 0.15 | 0.26 | 0.10 | 0.02 | |
| F14 | | | | | | | | | | | | | | | 0.03 | 0.19 | 0.59 | 0.34 | 0.22 | 0.26 | 0.11 | 0.02 |
| | 0.04 | 0.25 | 0.79 | 0.90 | 2.34 | 3.67 | 3.98 | 3.71 | 2.16 | 2.13 | 2.34 | 1.36 | 0.92 | 0.81 | 0.63 | 1.02 | 1.08 | 0.61 | 0.56 | 0.40 | 0.15 | 0.03 |

Probability

**Fig. 8.2** Comparison of houses to a fixed age at 25, 50 and 75 year re-occupation intervals

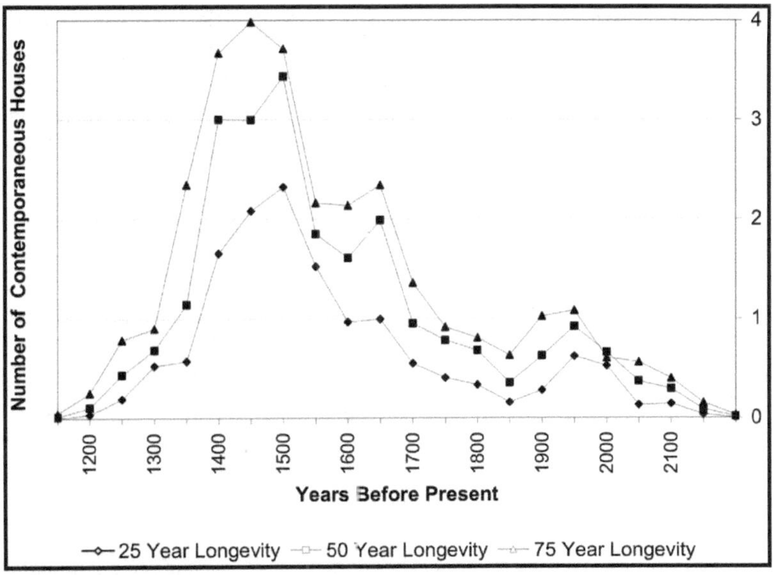


**Fig. 8.3** Potential house contemporaneity at 25, 50 and 75 year re-occupation intervals

can also be concluded that 6–10 house features were occupied simultaneously during the height of the occupation at about 1400 cal BP (Erwin 1995:42) and that such a level of use of the site persisted for a period of about 300 years. These figures are not dissimilar to Harp's slightly more generous calculations, in which he estimated a hypothetical peak of 12 simultaneously occupied households (Harp 1976:124) from a total of only 36 identified house pits (Harp 1976:120).

With the establishment of this temporal framework, observations relating to intra-site spatial patterning, settlement permanence, and site function over time are tested as means of untangling the various strands of archaeological evidence which have combined to produce the palimpsest of archaeological remains that are known as Phillip's Garden.

## Intra-Site Spatial Analysis

The interpretation of the spatial patterning of archaeological house features at large complex sites can be based upon the assumption that site development is not a random process, since people tend to organize their space in relatively patterned ways (Binford 1978; Ferring 1984: Oetelaar 1993; Portnoy 1981). In an effort to understand the organization of living space and the archaeological patterns which they produce, ethnoarchaeology has demonstrated that there are both functional and symbolic reasons for such organization (Binford 1978; Hodder 1987; Rossignol and Wandsnider 1992). From a functional perspective, the location of houses at Phillip's Garden may have been influenced by an availability of building materials and a proximity to other natural resources. For example, a conveniently located quantity of paving stones, or a supply of sod, may have been important considerations in the building or rebuilding of a Dorset semi-subterranean house. Likewise, the positioning of structures may also have been according to the relationship between residential and/or non-residential tasks and the seasonal availability of open water.

Another factor which may have influenced the positioning of houses is the expected duration of residence, and the avoidance of refuse from previous occupations, as suggested by Hayden and Cannon (1983). They demonstrated that there is a likely correlation between occupation length and the distance between residential areas and refuse heaps. Alternatively, less permanent structures may have been quickly abandoned when amounts of refuse within close proximity became intolerable. Finally, the abandonment of more permanent houses may have occurred at a time when structures were no longer serviceable and when building a new dwelling was more practical.

The location of houses may also have been influenced by rules of social behaviour and the relationships amongst the site occupants. The relatedness of the occupants, for example, could have been a factor which determined the proximity of dwellings. Other matters such as kinship patterns, the sexual division of space, and conventions regarding the use, or even ownership, of existing dwellings and their subsequent reoccupation could also have influenced the spatial development of the site.

On the basis of these possibilities, the dwelling locations were compared using a Nearest Neighbour Analysis (e.g., Carr 1984; Whallon 1974) and then compared to the temporal framework previously established. While an appreciation for spatial relationships might simply be gained through the casual inspection of a site map, this "eyeball method" lacks a means for making standardized observations and comparisons. Consequently, a method of analysis was employed which defined pairs and clusters of archaeological features as a measure of their spatial relationship. Operationally, this involved the measurement of the nearest distance between every known house at Phillip's Garden.

Assuming a normal distribution of measurements, the mean ($\mu$) and the standard deviation ($\sigma$) of the nearest neighbour (NN) distances were calculated. These figures are then used to determine the "cut-off" distance ($D$) at 1.65 standard deviations above the mean (Carr 1984:181).

$$D_{1.65} = 1.65 \times \sigma + \text{NN}\mu.$$

The cut-off distance is expected to be greater than 95% of all the measured nearest neighbour distances. The limits of clustering are then identified by drawing circles (with radii equal to this value) around every house. A cluster is identified where the circles meet or overlap. From these calculations, it was established that the mean nearest neighbour distance for dwelling units at Phillip's Garden was

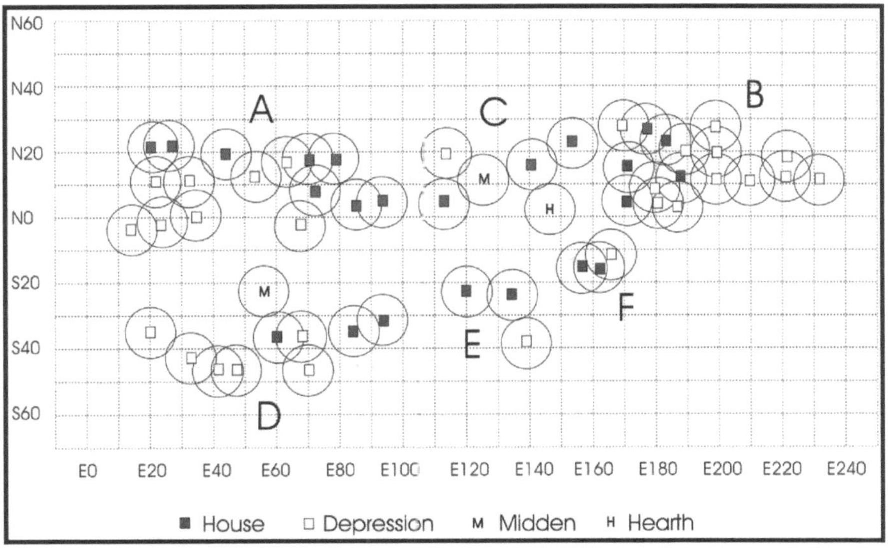


**Fig. 8.4** Nearest Neighbour Analysis and house position on a 10 m grid

approximately 2 m. At 1.65 standard deviations above the mean, the cut-off distance was thus calculated as 7.5 m, which was used to draw radii around every feature (Fig. 8.4). This analysis resulted in the identification of six clusters of dwellings (A–F) as summarized in Table 8.2, which also indicates the geographic orientation of each pair of houses as either east/west (EW) or north/south (NS). By comparing the age of each dwelling to its relative orientation, and spatial position as nearest neighbours, it is concluded that: (1) house occupation was not solely based upon an E/W linear arrangement dictated by beach ridge locations, (2) the entire site was accessible to occupation during the whole period of its Dorset occupation and (3) houses which are spatially related are more likely to be temporally related (Table 8.3).

Based upon these observations, it can also be argued that the spatial patterning of houses at Phillip's Garden is somewhat unique for a Palaeoeskimo site. As the shoreline has slowly been emerging through the process of post-glacial isostatic rebound (Bell et al. 2005), it would generally be expected that the earliest occupations should be located along the higher beach ridges, and that the later occupations should be situated along the lower ridges. However, such is not the case at Phillip's Garden, where there is little apparent temporal association along the beach ridges and a number of cases where there is greater potential contemporaneity across them (Erwin 1995:61–65). As such, it is argued that these temporal and spatial relationships may be evidence for some level of site planning associated with long-term reuse of the site.

**Table 8.2** Comparison of nearest neighbour analysis and house position

| Cluster | Houses compared | Nearest neighbour | Relative position |
|---|---|---|---|
| A | H3 and H4 | Yes | E/W |
| | H3 and H12 | No | E/W |
| | H4 and H12 | No | E/W |
| | H13 and H14 | Yes | E/W |
| | H13 and H16 | Yes | N/S |
| | H13 and F1 | No | N/S |
| | H13 and F14 | No | E/W |
| | H14 and H16 | Yes | N/S |
| | H14 and F1 | No | N/S |
| | H14 and F14 | No | N/S |
| | H16 and F1 | Yes | E/W |
| | H16 and F14 | No | E/W |
| | F14 and F1 | Yes | E/W |
| B | H5 and H6 | Yes | E/W |
| | H5 and H7 | Yes | E/W |
| | H5 and H8 | No | N/S |
| | H5 and F55 | Yes | N/S |
| | H6 and H7 | Yes | N/S |
| | H6 and H8 | No | N/S |
| | H6 and F55 | No | N/S |
| | H7 and H8 | Yes | N/S |
| | H7 and F55 | No | E/W |
| | H8 and F55 | No | E/W |
| C | H9 and H11 | Yes | E/W |
| | H9 and H15 | No | E/W |
| | H9 and F2 | No | E/W |
| | H11 and H15 | No | E/W |
| | H11 and F2 | No | E/W |
| | H15 and F2 | Yes | E/W |
| D | H17 and H18 | No | E/W |
| | H17 and H19 | Yes | E/W |
| | H18 and H19 | No | E/W |
| E | H10 and H20 | Yes | E/W |
| F | H1 and H2 | Yes | E/W |

## Settlement Permanence

Settlement permanence can generally be interpreted from a variety of lines of evidence, including the presence of substantial architectural remains, as well as midden features, storage pits (Ames 1991; Chatters 1987; Kelley 1982; Rafferty 1985) and the presence of communal structures and cemeteries (Charles and Buikstra 1983; Chatters 1987; Price and Brown 1985). Each of these indicators is considered in turn.

**Table 8.3** Comparison of radiocarbon dated houses and house clusters

| Cluster | Description of feature | C14 date |
|---|---|---|
| A | House 3 | n/a |
| | House 4 | 1465±65 |
| | House 12 | 1427±81 |
| | House 13 | 1423±74 |
| | House 14 | n/a |
| | House 16 | 1441±82 |
| | House Feature 1 | 1753±32 |
| | House Feature 14 | 1942±56 |
| B | House 5 | 1367±44 |
| | House 6 | 1482±65 |
| | House 7 | n/a |
| | House 8 | n/a |
| | Feature 55 | 1327±62 |
| C | House 9 | n/a |
| | House 11 | 1371±41 |
| | House 15 | 1887±57 |
| | Feature 2 | 1830±102 |
| | Feature 42 | n/a |

Additional radiocarbon dates are available from Phillip's Garden since this study was done; details are in the Appendix. Dates from Houses 13 and 15 are on seal fat

## *Architecture and Occupation*

Assuming that architectural variation can be attributed to changing style, seasonality, social organization, and human idiosyncrasies, it can be argued that there should be a tendency toward a greater degree of house permanence during periods of planned re-occupation compared to periods of sporadic occupation. In this regard, Harp initially suggested that the construction of the substantial and culturally standardized semi-subterranean dwellings at Phillip's Garden indicated that the site was likely a "seasonal base of some permanence from which they could easily sortie inland for caribou, as well as along the coast to the main salmon rivers during the summer" (Harp 1976:132). Harp's initial descriptions of the Phillip's Garden dwellings included suggested season of occupation. Larger, deeper and more substantial house remains were generally interpreted as cold-weather occupations while less prominent remains were equated with warm-weather occupations (Harp 1976:130).

More recent investigations have shown a greater diversity in house form, and such a simple summer/winter dichotomy has since been questioned (Jensen 1993; Murray 1992; Renouf 1987). A comparison of house permanence relative to the temporal evidence for occupation can be used to further identify patterning in the occupation of Phillip's Garden. Based upon the assumption that increased permanence in house construction indicates planned re-occupation, it is suggested that the less organized and lightly built structures should reflect a less regularly planned occupation. To date, at least four basic forms of shelter have been identified at Phillip's Garden. Winter houses have been recognized as rectangular semi-subterranean stone structures with internal dimensions ranging in size from 5×6 to 5×7 m (Harp 1964; Renouf 1993a:24). A shallow 3×5 m oval arrangement of rocks was identified by Harp (1976:130) as a summer house. Originally identified as an external hearth, Feature 42 is also believed to represent "a warm-weather or short-term structure" (Renouf 2003c:394–395). Dwelling Feature 1 was interpreted by Renouf and Murray (1999) as an ephemeral structure built for short-term cold-weather use. A more recent re-examination of House 2 (Renouf 2006), previously described by Harp (1976:130–132) as a winter house, suggested that this structure may have been more or less permanently occupied as a place of seasonal social gathering during the site's middle phase, which is also the phase for greatest potential house contemporaneity. Substantial structural elements, evidence for major renovation, and a lengthy occupation are cited as evidence for such use (Renouf 2006:129). Renouf showed that House 2 and three other middle phase houses excavated by Harp were not only substantially built but were very large, ranging from 88 to >100 $m^2$ (Renouf 2006, 2009, Chap. 7).

Based upon a comparison of generalized house types and the depth of cultural deposits of 14 excavated house structures and their relative ages (Table 8.4), there is some evidence to suggest that a number of houses were poorly defined and lightly occupied during the earliest period of occupation, as evidenced by the relatively thin deposits and lighter construction of Feature 1, House 15 and Feature 14. During the

**Table 8.4** Comparison on depth of house deposits with degree of house permanence

| House | Median age cal BP | Depth of deposits (cm) | House type |
|---|---|---|---|
| H20 | 1234±56 | 43 | Insufficient data to determine |
| F55 | 1327±62 | 32 | Well-defined semi-subterranean |
| H17 | 1348±48 | 39 | Well-defined semi-subterranean |
| H5 | 1367±44 | 22 | Poorly defined lacking internal structure |
| H11 | 1372±41 | 23 | Poorly defined lacking internal structure |
| H12 | 1427±81 | 28 | Well-defined semi-subterranean |
| H16 | 1441±82 | 20 | Insufficient data to determine |
| H4 | 1465±65 | 33 | Well-defined semi-subterranean |
| H10 | 1473±64 | 20 | Well-defined semi-subterranean |
| H6 | 1482±65 | 36 | Poorly defined lacking internal structure |
| H18 | 1606±78 | 28 | Insufficient data to determine |
| H2 | 1612±73 | 23 | Well-defined semi-subterranean |
| F1 | 1753±132 | 15 | Moderately defined semi-subterranean |
| F14 | 1942±56 | 14 | Well-defined semi-subterranean |

period of greatest potential house contemporaneity, both substantial and ephemeral constructions are noted. Likewise, the depths of the cultural deposits in the various houses are correspondingly variable, suggesting that the use of Phillip's Garden appears to shift between regular and sporadic periods of occupation, which is likely a reflection of the site's changing use, permanence, and seasonality over time.

## *Midden and Storage Features*

The presence of midden features and storage pits suggest site permanence and potential for reoccupation (Bailey and Parkington 1988; Hayden and Cannon 1983; Henry 1985; Testart 1982, 1988). In the case of Phillip's Garden, the presence of substantial midden material across virtually the entire site is evidence of a heavy use and subsequent long-term reoccupation. In addition to materials which fill many of the abandoned house pits (Eastaugh and Taylor, Chap. 9), Renouf has identified a number of relatively shallow and often poorly defined pit features within house structures she has excavated. These features are approximately 15–30 cm deep (Renouf 1986, 1987, 1993a) and are often filled with faunal material; they cannot be identified as storage features and she has recently suggested they are post-holes filled with secondary material (Renouf, Chap. 7). In the two middle phase houses where her excavations extended to the rear platform she found well-formed pits, approximately 60×60×60 cm which she identified as household storage pits (Renouf, Chap. 7). However, there is no evidence of large and substantial storage facilities at Phillip's Garden that would indicate large-scale storage of the large amounts of available seal meat. Therefore, it might be argued that resources were utilized immediately and discarded on site, or else taken elsewhere. Considering the hundreds of thousands of faunal specimens that have been recovered from this site, it would appear that much of these resources remained and were utilized at the site. The implication of a lack of large-scale storage is that the site's season of occupation corresponded to the times of the year when resources were readily available. While seal herds are predictable, their availability to hunters can often be limited by environmental factors such as wind and ocean currents. As such, I would argue that the site's use on a seasonal basis should be considered as variable. More specifically, recent investigations of midden materials by Hodgetts (2005) provide direct evidence for an early winter use of the site during the early and terminal periods of the occupation of Phillip's Garden, with a greater variability in season of use during the middle of the occupation.

## *Burial Sites*

The presence of human burials in the vicinity of Port au Choix (Brown 1988, Chap. 12) is also relevant to the discussion of site permanence and planned reoccupation, since it has been argued elsewhere that the construction and maintenance of cemeteries requires an investment of time and labour that correlates with more sedentary mobility strategies, and one which also serves as a cultural marker for increasing territoriality (Charles and Buikstra 1983:119). Although the cave and rock shelter burials in Port au Choix are not necessarily comparable to cemeteries, they nevertheless represent an investment in time, and might, as Brown has suggested, be "markers for later visitations" (Brown 1988:114).

Due to the lack of datable materials and disturbed contexts, little can be said about the temporal variation in these burials. The fact, however, that burials exist in Port au Choix, and almost nowhere else in Newfoundland, aside from Englee on the other side of the Northern Peninsula, might be attributed to the significance of Phillip's Garden as a meeting place. It also may be indicative of the distinctiveness of the west coast variant of Dorset culture in Newfoundland (LeBlanc 2000, 2008; Robbins 1986). Like the earlier Maritime Archaic cemetery in Port au Choix (see Renouf and Bell, Chap. 3), it can be speculated that the Dorset burials may have symbolically or physically acted as territorial markers of Dorset land use over this area and the vast resources of the sea during their tenure.

## Discussion

The occupation of Phillip's Garden initially was interpreted as relatively homogeneous and stable. In recent years, Renouf has determined that much of the variation at this large and complex site is due to overlapping functions, seasons and mobility patterns. If we accept, as Lee (1976:95) suggested, that there is no "typical year" for hunter-gatherers, and that subsistence and settlement strategies vary significantly according to time, place and circumstance, then the varying use of Phillip's Garden should be detectable. However, at sites like Phillip's Garden which were heavily reoccupied for long periods of time, such archaeological patterning must be observed through a diachronic perspective if changing functions are to be understood.

Although there is little doubt that harp seal herds played a central role in the repeated occupation of Phillip's Garden, temporal evidence combined with changing house structures, patterns of seasonality and site permanence demonstrates broad trends in the site's use. For example, there is strong evidence for a *regular seasonal occupation* during the period of highest potential house contemporaneity. During this period, many of the houses are structurally and functionally similar, and are generally situated along the lower terrace, nearest open water. This is consistent with Renouf's initial interpretation of the site as a late winter/early spring seal hunting location (Renouf 1993a:59). Beyond this main function, Renouf had initially interpreted other activities at Phillip's Garden only as "noise". However, put into chronological perspective, the "noise" becomes more understandable when varying occupations of the site are interpreted over time.

Recently confirmed variations in seasonality also suggest that a *varied seasonal occupation* was likely at the beginning and the end of the site's occupation (Hodgetts 2005; Hodgetts et al. 2003; Renouf and Murray 1999). It is during these periods that Phillip's Garden seems to have been occupied much less regularly, possibly for different functions over a wider range of seasons. The combination of houses along both the upper and lower terraces also suggests a wider range of seasonality which is consistent with this type of occupation. Evidence for a *shifting seasonal occupation* during periods when occupation of the site was less regular is tenuous at best, as the high proportion of harp seal remains strongly attests. As a *meeting place*, Phillip's Garden seems to have been well suited during a *regular seasonal occupation* when the greatest number of concurrently occupied houses is suggested. In view of Renouf's (2006, 2009) recent evidence for multi-family houses at Phillip's Garden and the burial evidence in the immediate area of the Port au Choix and Point Riche Peninsulas, it can be concluded that social functions were an important part of Phillip's Garden occupation.

## Conclusions

Though the function of Phillip's Garden has long been explained, the complexity of this extensive and well-preserved site is only now being fully recognized. In retrospect, it can be argued that the initial interpretations of Phillip's Garden tended to oversimplify the nature and function of the site as a consequence of the abundance and overwhelming complexity of the archaeological evidence. Despite over 75 years of research, the importance of continuing investigations of Phillip's Garden continues to hold an important key to our further understanding of Palaeoeskimo cultures. However, if we are to explain the complexity of the archaeological patterning of Phillip's Garden, we must adopt a diachronic approach to future analyses.

**Acknowledgements** Thanks to Charles Conway, Memorial University Geography Department, for drafting Figs. 11.1 and 11.2 and also to two anonymous reviewers for their helpful comments on an earlier draft of this manuscript.

## References

Ames, K.M. 1991. Sedentism: A temporal shift or a transitional change in hunter-gatherer mobility patterns. In *Between bands and states*, ed. S.A. Gregg, 108–134. Center for Archaeological Investigations Occasional Paper No. 9. Southern Illinois University at Carbondale, Carbondale.

Bailey, G., and J. Parkington. 1988. The archaeology of prehistoric coastlines: An introduction. In *The archaeology of prehistoric coastlines*, ed. G. Bailey and J. Parkington, 1–10. Cambridge: Cambridge University Press.

Bell, T.B., I.R. Smith, and M.A.P. Renouf. 2005. Postglacial sea-level history and coastline change at Port au Choix, Great Northern Peninsula, Newfoundland. *Newfoundland and Labrador Studies* 20(1): 9–31.

Binford, L.R. 1978. *Nunamiut ethnoarchaeology*. New York: Academic Press.

Binford, L.R. 1980. Willow smoke and dogs' tails: hunter-gatherer settlement systems and archaeological site formation. *American Antiquity* 41(1): 4–20.

Binford, L.R. 1982. The archaeology of place. *Journal of Anthropological Archaeology* 1: 5–31.

Binford, L.R. 1990. Mobility, housing, and environment: a comparative study. *Journal of Anthropological Research* 46(2): 119–152.

Bowman, S. 1990. *Radiocarbon dating: interpreting the past.* Berkeley: University of California Press.

Brown, S.C. 1988. Archaeological investigations at Crow Head Cave and the Gargamelle Rockshelter in the Port au Choix National Historic Park, Newfoundland: Report of the 1986 Field Activities. On file, Parks Canada, Archaeology, Atlantic Region, Halifax.

Carr, C. 1984. The nature of organization of intrasite archaeological records and spatial analytic approaches to their investigation. In *Advances in archaeological method and theory*, vol. 7, ed. M.B. Schiffer, 103–222. New York: Academic Press.

Charles, D.K., and J.E. Buikstra. 1983. Archaic mortuary sites in the central Mississippi drainage: Distribution, structure, and behavioral implications. In *Archaic hunters and gatherers in the American Midwest*, ed. J.L. Phillips and J.A. Brown, 117–146. New York: Academic Press.

Chatters, J.C. 1987. Hunter-gatherer adaptations and assemblage structure. *Journal of Anthropological Archaeology* 6: 336–375.

Erwin, J.C. 1995. An Intrasite Analysis of Phillip's Garden: A Middle Dorset Palaeo-Eskimo Site at Port au Choix, Newfoundland. Master's thesis, Department of Anthropology, Memorial University of Newfoundland, St. John's.

Ferring, C.R. 1984. Intrasite spatial patterning: Its role in settlement-subsistence systems analysis. In *Intrasite spatial analysis in archaeology*, ed. H.J. Hietala, 116–126. Cambridge: Cambridge University Press.

Harp, E. 1964. *The cultural affinities of the Newfoundland Dorset Eskimo.* National Museums of Canada Bulletin 200, Ottawa.

Harp, E. 1976. Dorset settlement patterns in Newfoundland and southeastern Hudson Bay. In *Eastern Arctic prehistory: Paleoeskimo problems*, ed. M.S. Maxwell, 119–138. Washington, D.C.: Society for American Archaeology.

Hayden, B., and A. Cannon. 1983. Where the garbage goes: Refuse disposal in the Maya Highlands. *Journal of Anthropological Archaeology* 2: 117–163.

Helskog, K. 1984. The younger stone age settlements in Varanger, north Norway: Settlement and population size. *Acta Borealia* 1: 39–69.

Helskog, K., and T. Schweder. 1989. Estimating the number of contemporaneous houses from $^{14}$C dates. *Antiquity* 63: 166–172.

Henry, D.O. 1985. Preagricultural sedentism: The Natufian example. In *Prehistoric hunter-gatherers: The emergence of cultural complexity*, ed. T.D. Price and J.A. Brown, 365–384. Orlando: Academic Press.

Hodder, I. 1987. The meaning of discard: Ash and domestic space in Baringo. In *Method and theory for activity area research: An ethnoarchaeological approach*, ed. S. Kent, 424–448. New York: Columbia University Press.

Hodgetts, L.M. 2005. Using bone measurements to determine the season of harp seal hunting at the Dorset Palaeoeskimo site of Phillip's Garden. *Newfoundland and Labrador Studies* 20(1): 91–106.

Hodgetts, L.M., M.A.P. Renouf, M.S. Murray, D. McCuaig-Balwill, and L. Howse. 2003. Changing subsistence practices at the Dorset Paleoeskimo site of Phillip's Garden, Newfoundland. *Arctic Anthropology* 40(1): 106–120.

Janes, R.R. 1983. *Archaeological ethnography among the Mackenzie Basin Dene, Canada.* The Arctic Institute of North America, Technical Paper 28. University of Calgary, Alberta.

Jensen, J.F. 1993. Intrasite Analysis of Palaeo-Eskimo Campsites. Master's thesis, Institut for Forhistorisk og Klassisk Arkaeologi, University of Copenhagen.

Kelley, K.B. 1982. Ethnoarchaeology of the Black Hat Navajos: Historical and ahistorical determinants of site features. *Journal of Anthropological Research* 38: 45–74.

LeBlanc, S. 2000. Middle Dorset (1900-1100 BP) regional variability on the island of Newfoundland and in Saint-Pierre et Miquelon. In *Identities and cultural contacts in the Arctic*, ed. M. Appelt, J. Berglund, and H.C. Gulløv, 97–105. Copenhagen: The Danish National Museum and Danish Polar Center.

LeBlanc, S. 2008. Middle Dorset Variability and Regional Cultural Traditions: A Case Study from Newfoundland and Saint-Pierre and Miquelon. PhD dissertation, Department of Anthropology, University of Alberta, Edmonton.

Lee, R.B. 1976. Kung spatial organization: An ecological and historical perspective. In *Kalahari hunter-gatherers*, ed. R.B. Lee and I. DeVore, 73–97. Cambridge: Harvard University Press.

Murray, M.S. 1992. Beyond the Laundry List: The Analysis of Faunal Remains from a Dorset Dwelling at Phillip's Garden (EeBi 1), Port au Choix. Master's thesis, Department of Anthropology, Memorial University of Newfoundland, St. John's.

Oetelaar, G.A. 1993. Identifying site structure in the archaeological record: An Illinois Mississippian example. *American Antiquity* 58(4): 662–687.

Portnoy, A. 1981. A microarchaeological view of human settlement space and function. In *Modern material culture: The archaeology of us*, ed. R.A. Gould and M.B. Schiffer, 213–224. New York: Academic Press.

Price, T.D., and J.A. Brown. 1985. Aspects of hunter-gatherer complexity. In *Prehistoric hunter-gatherers: The emergence of cultural complexity*, ed. T.D. Price and J.A. Brown, 3–20. Orlando: Academic Press.

Rafferty, J.E. 1985. The archaeological record on sedentariness: Recognition, development, and implications. In *Advances in archaeological method and theory*, vol. 8, ed. M.B. Schiffer, 113–156. Orlando: Academic Press.

Renouf, M.A.P. 1986. Archaeological Investigations at Phillip's Garden and Point Riche, Port au Choix National Historic Park: Report of 1985 Field Activities. On file, Parks Canada, Archaeology, Atlantic Region, Halifax.

Renouf, M.A.P. 1987. Archaeological Investigations at the Port au Choix National Historic Park; Report of 1986 Field Activities. On file, Parks Canada, Archaeology, Atlantic Region, Halifax.

Renouf, M.A.P. 1991a. Archaeological Investigations at the Port au Choix National Historic Park: Report of the 1990 Field Season. On file, Parks Canada, Archaeology, Atlantic Region, Halifax.

Renouf, M.A.P. 1991b. Sedentary hunter-gatherers: A case for northern coasts. In *Between bands and states*, ed. S.A. Gregg, 89–107. Center for Archaeological Investigations Occasional Paper No. 9. Southern Illinois University at Carbondale, Carbondale.

Renouf, M.A.P. 1993a. The 1992 Field Season, Port au Choix National Historic Park: Report of Archaeological Excavations. On file, Parks Canada, Archaeology, Atlantic Region, Halifax.

Renouf, M.A.P. 1993b. Palaeoeskimo seal hunters at Port au Choix, northwestern Newfoundland. *Newfoundland Studies* 9(2): 188–210.

Renouf, M.A.P. 1999. *Ancient cultures, bountiful seas: The story of Port au Choix*. Historic Sites Association of Newfoundland and Labrador, St. John's.

Renouf, M.A.P. 2003c. A review of Palaeoeskimo dwelling structures in Newfoundland and Labrador. *Études/Inuit/Studies*, 27(1–2) 375–416.

Renouf, M.A.P. 2006. Re-investigating a middle phase Dorset dwelling at Phillip's Garden, Port au Choix, Newfoundland. In *Dynamics of northern societies*, Proceedings of the SILA/NABO Conference on Arctic and North Atlantic Archaeology, Copenhagen, May 10th-14th, 2004, ed. J. Arneborg and B. Grønnow, 119–128. Publications from the National Museum Studies in Archaeology and History Vol. 10, Copenhagen.

Renouf, M.A.P. 2009. Dorset Palaeoeskimo whalebone use at Phillip's Garden, Port au Choix. In *On the track of the Thule culture, from Bering Strait to east Greenland*, Proceedings of the SILA Conference "The Thule Culture – New Perspectives in Inuit Prehistory", ed. B. Grønnow, 91–104. Publications from the National Museum Studies in Archaeology and History Vol. 15, Copenhagen.

Renouf, M.A.P., and M.S. Murray. 1999. Two winter dwellings at Phillip's Garden, a Dorset site in northwestern Newfoundland. *Arctic Anthropology* 36(1–2): 118–132.

Robbins, D.T. 1986. "Newfoundland Dorset" culture? In *Palaeo-Eskimo cultures in Newfoundland, Labrador and Ungava*, 119–124. Reports in Archaeology No. 1, Memorial University of Newfoundland, St. John's.

Rossignol, J., and L. Wandsnider (eds.). 1992. *Space, time, and archaeological landscapes*. New York: Plenum Press.

Sergeant, D.E. 1991. *Harp seals, man, and ice.* Canadian Special Publications of Fisheries and Aquatic Sciences No. 114. Department of Fisheries and Oceans, Ottawa.

Shott, M.J. 1992. Radiocarbon dating as a probabilistic technique: The Childers site and late Woodland occupation in the Ohio Valley. *American Antiquity* 57(2): 202–230.

Stuiver, M., and P.J. Reimer. 1993. Extended 14C database and revised CALIB radiocarbon calibration program. *Radiocarbon* 35: 215–230.

Testart, A. 1982. The significance of food storage among hunter-gatherers: Residence patterns, population densities and social inequalities. *Current Anthropology* 23(5): 523–537.

Testart, A. 1988. Some major problems in the social anthropology of hunter-gatherers. *Current Anthropology* 29(1): 1–31.

Thomas, D.H. 1986. *Refiguring anthropology: First principles of probability & statistics*. Prospect Heights, Illinois: Waveland Press.

Whallon, R. 1974. Spatial analysis of occupation floors II: The application of nearest neighbour analysis. *American Antiquity* 39: 16–34.

# Chapter 9
# Settlement Size and Structural Complexity: A Case Study in Geophysical Survey at Phillip's Garden, Port au Choix

**Edward J.H. Eastaugh and Jeremy Taylor**

## Introduction

Phillip's Garden (EeBi-1) is the largest, most complex and most extensively investigated Dorset Palaeoeskimo site in Newfoundland (Renouf, Chap. 7). Despite this, the precise number of dwellings at the site is unknown. To date, estimates have been based on visible house depressions that cover the upper two terraces at the site (Harp 1976) and to a lesser extent through test pitting (Renouf 1985). However, as extensive midden deposits are known to have buried many dwellings at the site (Renouf and Murray 1999:119) and the number of these buried depressions has never been systematically assessed, calculations of total number of dwellings at the site have tended to be best-guess estimates. The magnetometer survey conducted at Phillip's Garden in 2001 provides the first systematic approach towards a meaningful calculation.

## Phillip's Garden

Phillip's Garden is located on the Point Riche Peninsula approximately four kilometres west of the town of Port au Choix (Fig. 9.1). It is situated in a 1.8-ha meadow bordered on three sides by thick stunted-spruce forest (Renouf 2006:121, 2009:91; Renouf and Murray 1999:119). The meadow encompasses three raised beach terraces ranging from 6 to 11 m above present sea level. Cultural material and numerous house depressions cover the upper two terraces that range between 8 and 11 m above sea level (Harp 1976:120). Radiocarbon dates on charcoal from the site range

E.J.H. Eastaugh (✉)
Department of Anthropology, University of Western Ontario,
London, ON, Canada, N6A 5C2
e-mail: eeastaug@uwo.ca

M.A.P. Renouf (ed.), *The Cultural Landscapes of Port au Choix*, Interdisciplinary Contributions to Archaeology, DOI 10.1007/978-1-4419-8324-4_9,

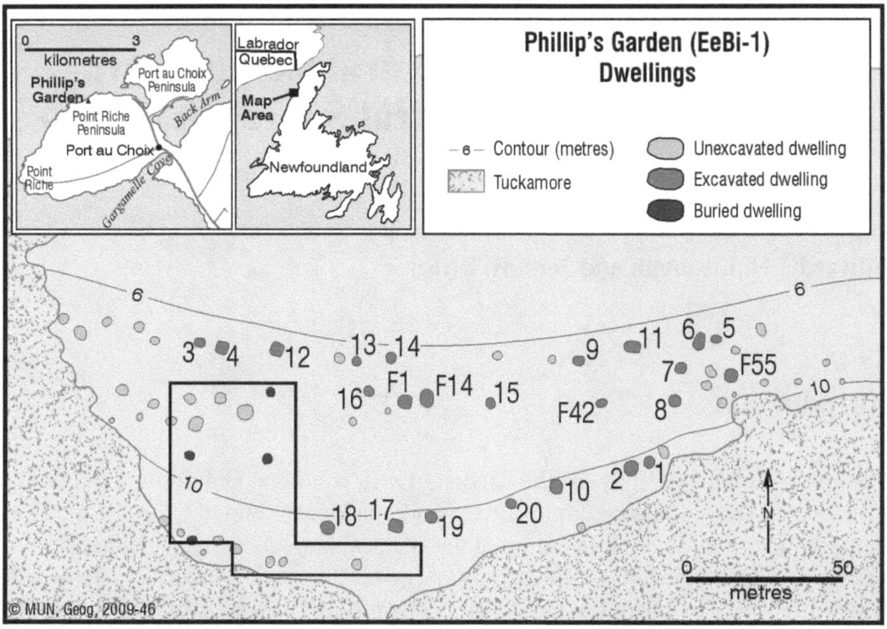


**Fig. 9.1** Phillip's Garden showing location of geophysics grid. The *inset* map shows the location of sites mentioned in the text

between 1990 cal BP[1] and 1180 cal BP (see Appendix), indicating that it was occupied for approximately 800 years.

Phillip's Garden has been interpreted as a fixed hub in an otherwise mobile seasonal round where families gathered every spring to take advantage of the predictable and abundant harp seal populations that migrated past the site between the end of March and early April each year (Renouf, Chap. 7). Analysis of the faunal material has demonstrated the dominance of seal, particularly harp seal, in all assemblages from the site (Hodgetts et al. 2003). Erwin (1995, Chap. 8) recognized three broad phases in the site's history. During the first phase, dating between 1950 and 1400 cal BP, the site was generally small and only intermittently used on a variable seasonal basis. The second phase, dating from 1400 to 1300 cal BP, saw a more regular and denser occupation which focused on the seal hunt in the spring. The third phase, in the 200 or so years following 1300 cal BP, saw a return to a more varied, sporadic seasonal occupation and smaller population levels. The faunal material confirms this general pattern, particularly the changes in the subsistence practices towards the end of the site's history (Hodgetts et al. 2003). Evidence from midden Feature 73, which dates to the third phase of site occupation (1365 cal BP, Beta 160976; 1285 cal BP, Beta 66436; 1277 cal BP, Beta 160977) indicates a marked decrease in the reliance on harp seal and a move towards increased dietary breadth including more fish and bird (Hodgetts et al. 2003:110).

[1]Except where indicated, all calendar dates in this chapter were calibrated using Calib 6.0html (Stuiver and Reimer 1993) and are represented by the one sigma probability range.

The majority of excavated dwellings at Phillip's Garden have been identified as winter structures (Harp 1976; Renouf 1985, 1986, 1987, 1991, 1992, 1993, 2006, 2009; Renouf and Murray 1999). Harp (1976:130) presented House 2 as the model of this type of dwelling. Although the exact outline of the dwelling was never determined, he described it as a sub-rectangular dwelling measuring approximately 5 by 8 m and constructed on the limestone beach. According to Harp (1976) a 5 m depression in the centre of the dwelling was cleared of rocks to form a semi-subterranean area which he believed was used as the main living space. A line of stone-lined pits ran through the centre of this depression, forming an axis through the dwelling and functioning as a general hearth area. A sleeping platform, consisting of a semi-circular area approximately 30 cm higher than the main floor, was situated at the rear of the house. A low wall 30–32 cm high, constructed from limestone rocks cleared from the central area, was constructed around the outside of the depression (Fig. 9.2). Harp (1976) also noted from one of the excavation profiles that turf blocks had been banked against the outside of the wall.

A second house type identified by Harp (1976:130) was a summer dwelling (House 5). It was initially observed as a shallow depression and unlike the winter dwellings it had no peripheral ring of stones, internal hearth or any other noteworthy features. It consisted of an oval area, $3 \times 6$ m, which contained noticeably fewer rocks than the surrounding area.

Subsequent fieldwork by Renouf (1985, 1986, 1987, 1991, 1992, 1993, 2002, 2006, 2009) demonstrated considerably more variation in dwelling type at Phillip's

**Fig. 9.2** Photograph of Harp's excavation of the winter dwelling, House 2, looking south; grid stakes are placed 5 ft (1.5 m) apart. Note the central depression, the perimeter of stones and the central trough. Photo: E. Harp, Jr. by permission of the Centre for Newfoundland Studies, Memorial University of Newfoundland

Garden than presented by Harp. Renouf excavated three dwellings (Features 1, 14, 55) (Fig. 9.1) and noted that, although similar to Harp's "standardized" winter structure, these dwellings had considerably less well-defined walls, axial features, central depressions and rear platforms. She subsequently trenched Harp's House 2 and based on her results she suggested that the perimeter of rocks surrounding the dwelling, originally interpreted by Harp as house walls, were instead internal structures to the dwelling, representing platforms and benches. This meant that House 2 was considerably larger than originally thought, with an exterior footprint of 94 $m^2$ (Renouf 2006:123). She suggested that rather than representing the "standard" house type at Phillip's Garden, House 2 was in fact unusually large and well-constructed. Subsequent re-excavation of Harp's House 17 (Renouf 2009) and House 18 (Cogswell 2006) confirmed the large size and sturdy construction of two more Phillip's Garden dwellings and Renouf (2009:101, Chap. 7) suggested that this was the new standard, at least during the middle phase of site occupation.

Renouf was also the first to conduct limited excavation in areas away from the visible house depressions. In one of these areas she identified a ring of post-holes around a central flagstone feature (Feature 42) (Fig. 9.1). No pits, platforms, central depression or other features were found associated with this structure. She interpreted Feature 42 as a summer tent (Renouf 2003:394).

The number of suggested dwellings at Phillip's Garden has steadily grown over the years as our knowledge of the site and its features has increased. Harp (1976:120), who excavated nine and tested a further 11 house depressions between 1949–1959 and 1961–1963, estimated that there was a total of 36 dwellings. This number was subsequently raised to over 50 following Renouf's fieldwork between 1983 and 1992 (Renouf and Murray 1999:119). The most recent survey of Phillip's Garden in 2001 raised the number again to 62, including 39 unexcavated and 23 excavated house depressions (PAC Project 2001) and Renouf (Chap. 7) has since increased this to 68.

## The Magnetometer Survey

A 2,600 m area in the southwest corner of Phillip's Garden was covered in a one-day magnetometry survey (Fig. 9.1). This area was chosen because it was the only part of the site that had not undergone substantial excavation in previous years. The geophysics grid was aligned to the site grid set up by Renouf in 1984 (1985) and was established to an accuracy of ±5 cm with a total station theodolite. All visible topographic features, including house depressions, old excavation trenches and back-dirt piles were also surveyed. This was to facilitate the comparison of the magnetometer results to known surface features. The geophysics grid consisted of five 20×20 m and three 20×10 m blocks. Readings were logged at 1 m intervals along parallel traverses spaced 1 m apart.

The raw data were processed in Geoplot 2 where they were converted into images in the grey-scale format. This format divides a given range of readings into a set of number classes, each with a predefined shade of grey (Ovenden-Wilson 1997).

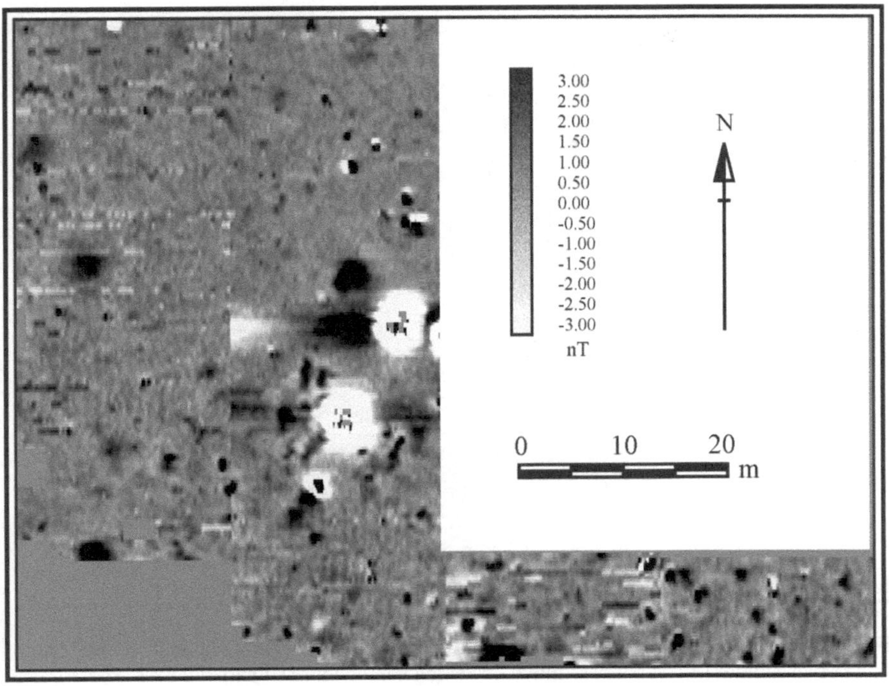


**Fig. 9.3** Magnetometer results

An increase in tone corresponds with an increase in number class. Processing included: (1) de-spiking to remove excessively strong magnetic anomalies that most likely resulted from modern metal objects in the ground; (2) zero mean traversing to remove slight striping effects in the graphics that often occur in fluxgate gradiometer data; and (3) interpolating *X* and *Y* to smooth the graphics data between sample points. The magnetometry results are presented in Fig. 9.3 and the interpreted plot is presented in Fig. 9.4.

## Discussion

The interpretation of the magnetometer survey results is limited by the fact that time constraints did not allow us to ground-truth any of the identified anomalies. In ground-truthing, anomalies identified in the survey are subsequently excavated to reveal the underlying features, making possible a direct comparison between the characteristics of the anomaly (e.g. shape, size and strength) and the feature that created it. In appropriate circumstances, similar unexcavated anomalies can then be identified from magnetometry characteristics similar to the ground-truthed anomalies. Some comparative data exist for similar ground-truthed anomalies at the nearby Palaeoeskimo site at Point Riche (EeBi-20) (Fig. 9.1) which had similar, but

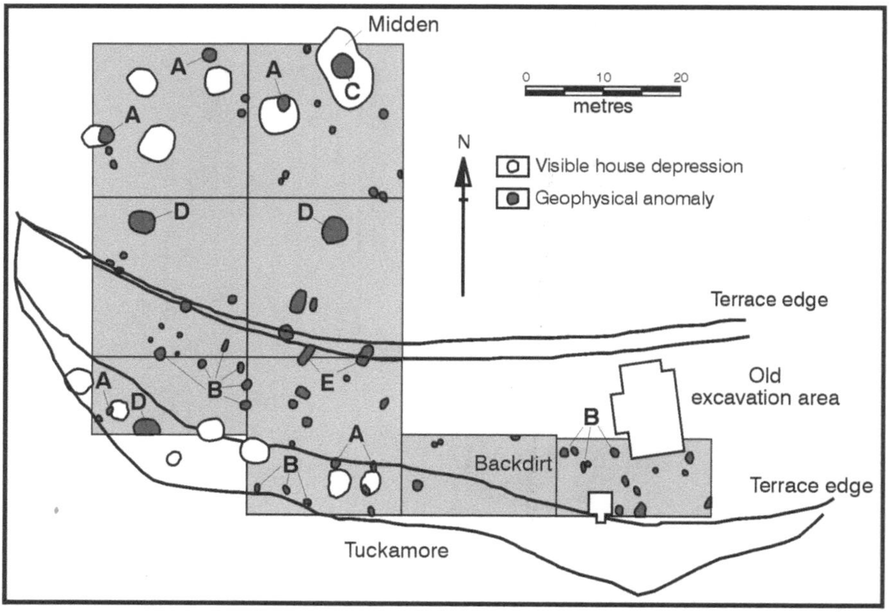


**Fig. 9.4** Interpreted plot of geophysics results: *A* – dwelling midden; *B* – unidentified features (summer dwellings?); *C* – midden filling a dwelling depression; *D* – buried dwelling depression; *E* – unidentified oblong features

smaller, dwelling depressions (Eastaugh 2003; Eastaugh and Taylor 2005). In most instances, however, the interpretation of anomalies at Phillip's Garden is based on comparisons between the shape and size of the anomalies to known feature types identified during excavation and to identifiable topographic features at the site.

The results of the magnetometer survey (Fig. 9.3) show two round negative (white) anomalies in the centre of the survey area. These were produced by two metal grid pegs and therefore should be ignored. The archaeological features all show as positive (dark grey) anomalies of various shape and size across the whole survey area, and more particularly, concentrated towards the southern half.

One striking aspect of the magnetometer survey is that despite the large number of anomalies, there are only two instances where an anomaly corresponds to the location of a house depression identified in the topographic survey (Fig. 9.4). This indicates that the dwelling structures are not detected by the magnetometer. This is probably due to the type of limestone substrate at Phillip's Garden, which lacks the iron minerals necessary for detection. As the main architectural features of the dwelling were all constructed from this limestone, there is no magnetic contrast between them and so they are essentially invisible to the magnetometer. Although this was expected, Harp (1976:132) had indicated that there was some evidence for the use of turf in the construction of house walls. These were anticipated to be detectable, as the earth platform of House Feature 30 at Point Riche had showed as a clear positive anomaly surrounding the central depression

(Eastaugh and Taylor 2005). The absence of similar anomalies at Phillip's Garden may indicate that turf was not commonly used as a building material at the site. This is supported by the absence of turf in all the dwellings excavated by Renouf (1985, 1986, 1987, 1991, 1992, 1993, 2002, 2006, 2009).

Although the dwellings themselves are not detectable, the magnetometer survey does show small positive anomalies immediately adjacent to six of the ten house depressions identified in the topographic survey (Fig. 9.4: anomaly A). These anomalies are almost certainly middens associated with the dwellings. Midden deposits are commonly found at the front and sides of dwellings of many Arctic cultures (Fogt 1998:16; Morrison 1983:53; Newell 1981:203; Simonsen 1961) and have been identified outside dwelling entrance ways at Phillip's Garden, including midden Feature 52 located immediately outside House Feature 1 (Renouf and Murray 1999:124) and midden Feature 77 identified in front of House 2 by Harp (Harp 1976; Hodgetts 2005; Renouf and Murray 1999:122). Midden deposits are extremely susceptible to magnetometer survey as they usually contain large quantities of magnetically enhanced burnt material and decayed organic matter (Clark 1990:101). The middens at Phillip's Garden are commonly thick, greasy, black deposits containing abundant bone, burnt seal fat and charcoal (Hodgetts 2001; Renouf 1987). We would therefore expect midden material at Phillip's Garden to be highly detectable. This is confirmed by a large positive anomaly in the magnetometer survey which coincides with the location of an exposed midden in the topographic survey (Fig. 9.4: anomaly C).

The numerous other small anomalies that cover the survey area, particularly in the southern half, are more difficult to interpret (Fig. 9.4: anomalies B and E). Most are located away from the identified house depressions, in areas where no surface features were identified. As most excavation at Phillip's Garden has concentrated on the house depressions, there is little information on archaeological features that occur away from these structures. The only significant investigation to take place away from the dwellings was by Renouf (1993, 2003) who excavated a 5×4 m trench in the centre of the 8 m terrace and discovered a summer tent structure (Feature 42). It is possible that some of the small round anomalies represent central hearths similar to that identified within Feature 42 (Renouf 1993). However, if some of these anomalies relate to summer structures it is more likely that they represent midden deposits associated with the summer dwellings, given that the more substantial architectural features of the semi-subterranean dwellings were undetectable. Alternatively, these as well as the oblong anomalies may represent different feature types that we would expect on large residential sites, including lean-tos, rubbish pits and external activity areas and caches.

The spatial separation of different feature types, including summer and winter dwellings, has been noted on proto-historic Inuit sites (Friesen and Stewart 1994:349; McGhee 1972:72) and Norwegian Younger Stone Age sites (Renouf 1989:228) in the Arctic. Phillip's Garden may have a similar pattern but since a thick midden deposit covers much of the site these less substantial features may be buried and therefore hidden from view. Given that only 8 m$^2$ (0.05%) of the site has been investigated away from the semi-subterranean dwellings, it is possible that previous investigations have missed these less substantial feature types, including any areas containing summer occupation. Our current understanding of site seasonality, which stresses late winter/early spring occupancy, may merely reflect the tendency of past researchers to concentrate their efforts on the large semi-subterranean winter/spring dwellings. The possibility of significant numbers of warm-weather dwellings at Phillip's Garden is certainly tantalizing and if demonstrated it would require a reinterpretation of the seasonal use of the site to incorporate greater summer occupancy.

The largest anomalies identified at Phillip's Garden measure 3.5–4 m in diameter (Fig. 9.4: anomalies C and D). The only excavated features that conform to this shape and size are the central depressions of the semi-subterranean dwellings (Harp 1976; Renouf 1985, 1986, 1987, 1991, 1992, 1993). Although we have noted that dwelling architecture (e.g. depressions, axial features, built-up stone perimeters) was undetectable, it has been demonstrated that midden deposits were clearly detectable. Excavation and test pitting have demonstrated that dwelling depressions at Phillip's Garden are often filled with midden material after their abandonment (Hodgetts 2001:2; Renouf 1985:39; Renouf and Murray 1999:119). In one instance a depression, inferred to be a dwelling, was identified beneath and totally obscured by midden Feature 2 (Renouf 1987:27). It is extremely likely therefore that the four large round anomalies identified by magnetometer survey are house depressions filled with midden material.

If we are correct in our identification of four buried dwellings in the survey area the total number of dwellings at Phillip's Garden is likely to be significantly higher than previously estimated. Assuming that the distribution of dwellings is evenly spaced throughout the site, we would expect a total of between 20 and 25 buried depressions across the site as a whole. When added to the 68 house depressions already identified, this would give us a figure of approximately 88 semi-subterranean dwellings at Phillip's Garden, a far higher number than any previous estimate.

## Conclusions

The magnetometer survey at Phillip's Garden demonstrates a higher number and more complex arrangement of archaeological features at Phillip's Garden than previously understood. The survey identified numerous small features, particularly in areas away from the dwelling depressions. Although it is not yet clear what these anomalies are, they represent a wider range of features and activities at the site than previously appreciated.

The possibility of four buried dwelling depressions suggests that the number of winter-type dwellings is probably greater than previously estimated. Although the presence of these buried depressions has been known for some time, there has been no reliable way to establish their number until now. It is estimated that the total number of semi-subterranean dwellings at Phillip's Garden is likely nearer to 88 rather than 68 as previously suggested. If correct, this increase in site size will have substantial implications for our understanding of dwelling contemporaneity and occupational history at Phillip's Garden.

Clearly, many of the suggestions outlined in this paper are speculative; without excavation our interpretations of the anomalies identified during the magnetometer survey cannot be confirmed. However, whether some, none, or all of our interpretations are correct, the identification of numerous features away from the dwellings raises a number of tantalizing and exciting possibilities for our understanding of Phillip's Garden. If nothing else, the magnetometer survey surely encourages a return to the site to investigate those areas that have not been the focus of major excavation.

**Acknowledgements** We would like to thank the following agencies for their financial support: Parks Canada, the Institute of Social and Economic Research, the J.R. Smallwood Foundation for Newfoundland and Labrador studies, the Government of Newfoundland and Labrador Culture and Heritage Division, and the Social Sciences and Humanities Research Council of Canada (grant awarded to Renouf). The Newfoundland Archaeological Heritage Outreach Program provided a graduate fellowship for Eastaugh. We would also like to thank the School of Archaeology and Ancient History, University of Leicester, England who provided the magnetometer and resistivity meter. We are also indebted to Priscilla Renouf for allowing us to use facilities and equipment belonging to the Port au Choix Archaeology Project throughout the field season and for providing maps and photographs from the Port au Choix project archive. Thanks to Charles Conway, Memorial University Geography Department, who drafted Figs. 9.1 and 9.4. The authors are also grateful to John Erwin, Priscilla Renouf and two anonymous reviewers for their comments on an earlier draft of this paper.

## References

Clark, A. 1990. *Seeing beneath the soil.* London: Batsford Press.

Cogswell, A.E. 2006. House 18 and the Middle Phase of Occupation at Phillip's Garden (EeBi-1). Master's thesis, Department of Anthropology, Memorial University of Newfoundland, St. John's.

Eastaugh, E.J.H. 2003. A Middle Dorset semi-subterranean dwelling at Point Riche, Newfoundland. *Études/Inuit/Studies* 27(1–2): 451–471.

Eastaugh, E.J.H., and J. Taylor. 2005. Geophysical survey of the Dorset Palaeoeskimo site of Point Riche. *Newfoundland and Labrador Studies* 20(1): 157–173.

Erwin, J.C. 1995. An Intrasite Analysis of Phillip's Garden: A Middle Dorset Palaeo-Eskimo Site at Port au Choix, Newfoundland. Master's thesis, Department of Anthropology, Memorial University of Newfoundland, St. John's.

Harp, E. 1976. Dorset settlement patterns in Newfoundland and southeastern Hudson Bay. In *Eastern Arctic prehistory: Paleoeskimo problems*, ed. M.S. Maxwell, 119–138. Washington, D.C.: Society for American Archaeology.

Hodgetts, L. 2001. Report on the 2001 Excavations at Phillip's Garden, Port au Choix National Historic Site. On file, Parks Canada, Archaeology, Atlantic Region, Halifax

Hodgetts, L. 2005. Using bone measurements to determine the season of harp seal hunting at the Dorset Palaeoeskimo site of Phillip's Garden. *Newfoundland and Labrador Studies* 20(1): 91–106.

Hodgetts, L.M., M.A.P. Renouf, M.S. Murray, D. McCuaig-Balkwill, and L. Howse. 2003. Changing subsistence practices at the Paleoeskimo site of Phillip's Garden, Newfoundland. *Arctic Anthropology* 40(1): 106–120.

Fogt, L. 1998. The Excavation and Analysis of a Dorset Palaeoeskimo Dwelling at Cape Ray, Newfoundland. Master's thesis, Department of Anthropology, Memorial University of Newfoundland, St. John's.

Friesen M., and A. Stewart. 1994. Protohistoric settlement patterns in the interior district of Keewatin: Implications for Caribou Inuit social organisation. In *Threads of Arctic prehistory: Papers in honour of William E. Taylor, Jr.*, ed. D. Morrison and J.-L. Pilon, 341–360. Mercury Series Archaeology Paper 149, Canadian Museum of Civilization, Ottawa.

McGhee, R. 1972. *Copper Eskimo Prehistory*. National Museums of Canada, Ottawa.

Morrison, D.A. 1983. *Thule culture in western Coronation Gulf, N.W.T.* Mercury Series Archaeology Paper 149, National Museums of Canada, Ottawa.

Newell, R.R. 1981. The intermound and extramound tests. In *The 1981 excavations at the Utqiagvik archaeological site, Barrow, Alaska*, ed. E.S. Hall and L. Fullerton, 174–262. North Slope Borough Commission on Inupiat History, Barrow.

Ovenden-Wilson, S. 1997. Grange Park, Northamptonshire: Report on the Geophysical Survey. On file, Geophysical Surveys of Bradford, Bradford, England.

PAC Project. 2001. Map on file with the Port au Choix Archaeology Project, Department of Archaeology, Memorial University of Newfoundland, St. John's.

Renouf, M.A.P. 1985. Archaeology of the Port au Choix National Historic Park: Report of 1984 Field Activities. On file, Parks Canada, Archaeology, Atlantic Division, Halifax.

Renouf, M.A.P. 1986. Archaeological Investigations at Phillip's Garden and Point Riche, Port au Choix National Historic Park: Report of 1985 Field Activities. On file, Parks Canada, Archaeology, Atlantic Region, Halifax.

Renouf, M.A.P. 1987. Archaeological Excavations at the Port au Choix National Historic Park: Report of 1986 Field Activities. On file, Parks Canada, Archaeology, Atlantic Region, Halifax.

Renouf, M.A.P. 1989. *Prehistoric hunter-fishers of Varangerfjord, northeastern Norway: Reconstruction of settlement and subsistence during the Younger Stone Age*. British Archaeological Reports International Series 487. British Archaeological Reports, Oxford.

Renouf, M.A.P. 1991. Archaeological Investigations at the Port au Choix National Historic Park: Report of the 1990 Field Season. On file, Parks Canada, Archaeology, Atlantic Region, Halifax.

Renouf, M.A.P. 1992. The 1991 Field Season, Port au Choix National Historic Park: Report of Archaeological Excavations. On file, Parks Canada, Archaeology, Atlantic Region, Halifax.

Renouf, M.A.P. 1993. The 1992 Field Season, Port au Choix National Historic Park: Report of Archaeological Excavations. On file, Parks Canada, Archaeology, Atlantic Region, Halifax.

Renouf, M.A.P. 2002. *Archaeology at Port au Choix, northwestern Newfoundland 1990-1992*. Occasional Papers in Northeastern Archaeology 12. St. John's: Copetown Press.

Renouf, M.A.P. 2003. A review of Palaeoeskimo dwelling structures in Newfoundland and Labrador. *Études/Inuit/Studies* 27(1–2): 375–416.

Renouf, M.A.P. 2006. Re-investigating a middle phase Dorset dwelling at Phillip's Garden, Port au Choix, Newfoundland. In *Dynamics of northern societies*, Proceedings of the SILA/NABO Conference on Arctic and North Atlantic Archaeology, Copenhagen, May 10th-14th, 2004, ed. J. Arneborg and B. Grønnow, 119–128. Publications from the National Museum Studies in Archaeology and History Vol. 10, Copenhagen.

Renouf, M.A.P. 2009. Dorset Palaeoeskimo whalebone use at Phillip's Garden, Port au Choix. In *On the track of the Thule culture, from Bering Strait to east Greenland*, Proceedings of the SILA Conference "The Thule Culture – New Perspectives in Inuit Prehistory", ed. B. Grønnow, 91–104. Publications from the National Museum Studies in Archaeology and History Vol. 15, Copenhagen.

Renouf, M.A.P., and M.S. Murray. 1999. Two winter dwellings at Phillip's Garden, a Dorset site in northwestern Newfoundland. *Arctic Anthropology* 36(1–2): 118–132.

Simonsen, P. 1961. Varangerfunnene II: Fund og udgravninger på fjordens sydkyst. *Tromsø Museums Skrifter* 7(2): 1–524.

Stuiver, M., and P.J. Reimer. 1993. Extended 14C database and revised CALIB radiocarbon calibration program. *Radiocarbon* 35: 215–230.

# Chapter 10
# Down the Labrador: Ramah Chert Use at Phillip's Garden, Port au Choix*

**Robert J. Anstey and M.A.P. Renouf**

## Introduction

This chapter presents an analysis of Ramah chert tools and debitage from the Dorset Palaeoeskimo site of Phillip's Garden (EeBi-1). Although the presence of Ramah chert at this site has long been acknowledged (Anstey 2008; Bell and Renouf 2008; Kennett 1985; Renouf 1986), details of temporal, technological and social dynamics of Ramah chert use at the site have not been examined. These dynamics of Ramah chert use are addressed in three ways. First, based on a sample of six dwelling assemblages we describe the temporal patterns of Ramah chert use at Phillip's Garden. Second, we describe Ramah chert reduction practices at the site. Third, we discuss the implications of our results for understanding the social dynamics at Phillip's Garden and between Dorset populations in Newfoundland and Labrador.

Ramah chert can only be acquired from a narrow sedimentary formation that occurs between Nachvak Fiord and Saglek Fiord, northern Labrador (Fitzhugh 1972; Gramly 1978; Lazenby 1980; Loring 2002; Morgan 1975) (Fig. 10.1). Yet, it occurs on archaeological sites as far south as Pennsylvania and Maryland, over 3,000 km from the source (Loring 2002). Extended social and lithic exchange networks must have been in place for this material to have reached areas so far from its source. Therefore, Ramah chert can be used as a proxy indicator for the intensity of this exchange and social interaction. In the context of this chapter, high quantities of Ramah chert at Phillip's Garden would indicate greater exchange and social communication with Labrador and low quantities would indicate lesser exchange and social communication.

* "Down the Labrador" refers to the northward journey to Labrador by fishermen from Newfoundland (DNE 1990). The use of "down" when referring to north is based on sailing against or "down" the prevailing southwesterly winds.

R.J. Anstey (✉)
Department of Archaeology, Memorial University, St. John's, NL, Canada, A1C 5S7
e-mail: ranstey@mun.ca

M.A.P. Renouf (ed.), *The Cultural Landscapes of Port au Choix*, Interdisciplinary Contributions to Archaeology, DOI 10.1007/978-1-4419-8324-4_10,

**Fig. 10.1** Map showing location of places mentioned in the text

Our results indicate that Ramah chert was used throughout the occupation of Phillip's Garden although in small amounts. Relative proportions of artefacts and flakes were <2.5% of the total dwelling assemblage in the case of artefacts and of a sample of debitage in the case of flakes. Ramah chert was represented primarily by late-stage reduction practices, namely retouch and rejuvenation of preforms and finished tools. However, the intensity of use – represented by the relative frequency of Ramah chert – varies throughout the period of site use, that is, between 1990 and 1180 cal BP.[1] During the early phase (1990–1550 cal BP) of occupation Ramah chert use is relatively low. There appears to have been steady use throughout the middle phase (1550–1350 cal BP). Throughout the late phase (1350–1180 cal BP), Ramah chert use increased which we argue is connected to increased mobility which in turn is linked to increased social communication and exchange with contemporaneous populations in Labrador.

## Context: Phillip's Garden

Phillip's Garden was occupied by Dorset Palaeoeskimo groups for a nearly 800-year period, which lasted from about 1990 to 1180 cal BP. This occupational span has been divided into three chronological phases which represent an initial low population, followed by a population maximum and a subsequent decline: early (1990–1550 cal BP), middle (1550–1350 cal BP) and late (1350–1180 cal BP), respectively (Erwin 1995, Chap. 8; Harp 1976; Renouf 2006:122; Renouf and Bell 2008:36) (Table 10.1; see also Appendix).

Well-preserved and abundant faunal assemblages from the site demonstrate that the late-winter harp seal hunt was almost exclusively the subsistence base (Harp 1976; Hodgetts et al. 2003; Murray 1992, Chap. 11; Renouf 1993). Harp seals were valued not only for their meat but also for their hides which would have been used to make clothing, boots, and other items (Knapp 2008; Renouf and Bell 2008). Today the entire Point Riche peninsula is known as a particularly important locale for the harp seal hunt. Phillip's Garden is particularly important because harp seals appear not far offshore in association with a regularly occurring ice lead. This exceptional resource niche is likely the primary reason why the site location was chosen by Dorset populations (Hodgetts et al. 2003; Murray 1992; Renouf and Murray 1999).

By virtue of both its lengthy occupation and large size (2 ha), Phillip's Garden also has a very large number of dwellings: at least 68 identified dwelling features (Cogswell 2006; Harp 1976; Renouf 2002, 2003a, 2006; Renouf and Murray 1999). Of these dwellings, six examples were chosen for the present analysis: two from the early phase, three from the middle phase, and one from the late phase. Each of the dwellings in this analysis, except for Feature 1 and Feature 55, are consistent in their interior architecture and artefact assemblage variability and are typical of other Dorset dwellings found at the site (Cogswell 2006; Renouf 2003a, 2009).

[1] All dates in this paper are expressed in calibrated calendar years before present (cal BP) at a one-sigma probability range and were calibrated using Calib 6.0html (Stuiver and Reimer 1993).

**Table 10.1** Charcoal-based radiocarbon dates for dwellings sampled at Phillip's Garden

| Dwelling | C14 years BP | Lab No. | Calibrated median age BP | Cal BP range 1σ | Cal BP range 2σ |
|---|---|---|---|---|---|
| F14 | 1970±60 | Beta 23977 | 1920 | 1990–1870 | 2110–1740 |
| F14 (midden) | 1770±120 | Beta 42968 | 1690 | 1820–1550 | 1950–1410 |
| F1 | 1850±110 | Beta 15379 | 1780 | 1920–1630 | 2040–1530 |
| H18 | 1683±49 | P-736 | 1590 | 1690–1530 | 1710–1420 |
| H18 | 1680±40 | Beta 211266 | 1590 | 1690–1530 | 1700–1450 |
| H18 (midden) | 1630±40 | Beta 211267 | 1520 | 1570–1420 | 1680–1410 |
| H2 (midden) | 1736±48 | P-692 | 1650 | 1710–1570 | 1810–1540 |
| H2 (midden) | 1659±48 | P-693 | 1560 | 1690–1450 | 1690–1420 |
| H2 (midden) | 1640±70 | Beta 160975 | 1540 | 1610–1420 | 1710–1290 |
| H2 | 1593±49 | P-683 | 1480 | 1530–1420 | 1600–1370 |
| H17 | 1580±40 | Beta 238480 | 1470 | 1520–1420 | 1550–1380 |
| H17 (midden) | 1480±40 | Beta 238481 | 1370 | 1400–1330 | 1510–1300 |
| H17 | 1465±51 | P-734 | 1360 | 1390–1310 | 1510–1290 |
| H17 (midden) | 1450±40 | Beta 238479 | 1340 | 1370–1310 | 1400–1300 |
| F55 (midden) | 1480±40 | Beta 160976 | 1370 | 1400–1330 | 1510–1300 |
| F55 | 1410±100 | Beta 66435 | 1330 | 1410–1180 | 1530–1090 |
| F55 (midden) | 1370±90 | Beta 66436 | 1290 | 1370–1180 | 1510–1070 |
| F55 (midden) | 1360±80 | Beta 160977 | 1280 | 1350–1180 | 1410–1080 |

Radiocarbon dates were calibrated using Calib 6.0html (Stuiver and Reimer 1993). Dates are from Harp (1976) and Renouf (2006)

## *Dwellings Examined*

The oldest of the early phase dwellings is Feature 14 which dates to 1990–1870 cal BP (Renouf 2006:122). This dwelling was a large (74.7 $m^2$) oval depression defined by a 1 m wide perimeter of raised limestone shingle (Renouf 2003a:394, 409). A narrow linear depression indicated a 3 m long cold-trap entrance passage facing southeast and a slight break in the northeast wall suggested a secondary entrance (Renouf 1987:17, 2003a:394). The presence of a cold-trap passageway is unusual when compared with other Dorset dwellings on the island of Newfoundland; however, similar structural elements have been found in Dorset dwellings in Labrador (Cox 1978, 2003; Renouf 2003a). There was a central pit feature and a cleared raised area at the back of the dwelling interpreted as a rear sleeping platform (Renouf 2003a:394). Based on the presence of a cold-trap entrance, Renouf (1987:17) interpreted the structure as a winter dwelling.

Feature 1 is the other early phase dwelling in our analysis and dates to 1920–1630 cal BP (Renouf 2006:122). Feature 1 was a relatively small (51.5 $m^2$) oval-shaped and less formalized dwelling with an east–west axial feature consisting of two stone-lined and bone-filled pits and two upright rocks that were possibly lamp supports (Renouf 2003a:392). The perimeter was defined by a 1 m wide area of stacked limestone shingle; a break in the northeast and southeast of this perimeter was interpreted as a primary and secondary entrance, respectively (Renouf 2003a:392; Renouf and Murray 1999:123–124). There were three raised areas within the dwelling which were interpreted as sleeping platforms (Renouf 2003a:392). Based on faunal data, this dwelling was interpreted by Renouf and Murray (1999) as a winter structure.

The earliest dwellings from the middle phase of site occupation are House 18 and House 2. Both are roughly contemporaneous, dating to 1590–1460 cal BP and 1650–1390 cal BP, respectively (Harp 1976:137; Renouf 2006:122). House 18 was a large (103.2 $m^2$) rectangular dwelling which contained a north-facing entrance, a north–south axial feature, elevated platforms, and one large and several small storage pits (Cogswell 2006:43, 60–63). Cogswell (2006) interpreted House 18 as a winter occupation based on its size and evidence of substantial construction. A smaller tent-like structure (21.3 $m^2$) was superimposed on the larger dwelling, which indicates either subsequent downsizing or else re-occupation of the original footprint during a warmer time of the year (Cogswell 2006:63, 65).

House 2 was a large (94.1 $m^2$) square structure with a north-facing entrance and a raised platform at the rear and limestone berms to the side which Harp (1976:132–133) interpreted as walls but Renouf (2006:124) re-interpreted as one wall and one sitting platform. A north–south axial feature consisting of a row of deep stone-lined pits ran centrally through the dwelling. Renouf (2003a:392, 2006:125) interpreted these pits as post-holes for central supports. Her re-excavations showed that these post-holes had been subsequently re-modelled to accommodate smaller post-holes which she interpreted as evidence of downsizing (Renouf 2006:126). House 2 was interpreted by Harp (1976) as a winter occupation based on the abundance of associated seal bone; this was confirmed in later analyses (Hodgetts et al. 2003; Renouf and Murray 1999).

Furthermore, due to the unusual robustness suggested by the central posts and the lengthy time range of radiocarbon dates, Renouf (2006:127) suggested that House 2 was built as a permanent structure for repeated seasonal use.

The third middle phase dwelling in our analysis is House 17, dating to 1660–1340 cal BP (Renouf 2006:122). House 17 was a somewhat smaller (88.0 $m^2$) trilobate dwelling which contained a centrally located entrance in the northern wall, well-defined perimeter platforms, a north–south axial feature with two central post-holes, two rear storage pits, and several large perimeter post-holes. These were of appropriate size and shape to accommodate whale ribs which would have likely formed the basic framework of the dwelling (Renouf 2007:57–60, 2009:94). Given the large size and substantial construction of House 17, this was likely a permanent structure meant for repeated seasonal use.

The late phase dwelling in our analysis is Feature 55, dating to 1410–1180 cal BP (Renouf 2006:122). Feature 55 is the smallest cold weather dwelling identified on the site (28.3 $m^2$). It was a circular structure defined by a perimeter of limestone shingle and bisected by an east–west axial feature. Unlike the axial features of the other dwellings described which were defined by pits, the Feature 55 axial feature consisted of a cobble pavement without pits (Renouf 2002:97, 2003a:394, 409). The dwelling contained two entrances, a bench on the outer rim, and 12 stone-lined post-holes which probably held whalebone rib wall supports similar to House 17 (Renouf 2002:95–99, 2003a:394, 2009:93). Based on this structural evidence, Renouf (2002:99) suggests that Feature 55 was a dome-shaped or yurt-like dwelling. Based on faunal data recovered from an associated midden, Hodgetts et al. (2003:116) suggested that Feature 55 was likely occupied from the winter to late spring, and possibly in the early summer.

## *Dorset Palaeoeskimo Mobility, Social Networks and the Environment*

The generally larger size of Dorset Palaeoeskimo dwellings on the island of Newfoundland and, in particular Phillip's Garden, led Renouf (2003a:410) to postulate that this was the result of increased group size and a general decrease in residential mobility. This is consistent with studies of Dorset regionalization based on localized differences in Palaeoeskimo raw material use and endblade style (LeBlanc 2000, 2008; Robbins 1986). This reduced mobility is most clearly seen at Phillip's Garden, based on the large size and substantial construction of the dwellings, in particular during the middle phase.

Bell and Renouf (2008) suggested that Phillip's Garden was a population aggregation site, characteristic of hunter–gatherers, where family groups would gather for social and economic reasons. On the basis of this important social function together with the location of Phillip's Garden in the Gulf of St. Lawrence just south of the Strait of Belle Isle (Fig. 10.1), Bell and Renouf (2008) argued that Phillip's Garden may have served as a gateway to Labrador through which trade, exchange and social relations transpired. Bell and Renouf (2008) further argued that the increased summer temperatures seen in chironomid data from Bass Pond near Phillip's Garden at 1500–1100 cal BP (Rosenberg et al. 2005) might have undermined the predictability of ice conditions and consequently the timing and location of the harp seal herds. They linked this to the abandonment of the site at 1180 cal BP.

This is consistent with Hodgetts et al. (2003) who demonstrated that the reliance on harp seal, although consistently significant, decreased over time. Based on faunal remains from five chronologically separated midden features, they showed that throughout the early and most of the middle phase of site occupation there was a high reliance on harp seal at 97.6–99.4% (Hodgetts et al. 2003:110, 113). However, at the tail end of the middle and throughout the late phase of site occupation, reliance on harp seal decreased to 70.8% with an increasing reliance on bird and fish species (Hodgetts et al. 2003:110, 113). Hodgetts et al. (2003:116) linked this pattern to increasing temperatures generally seen in the western North Atlantic during 2700–1500 cal BP (see also O'Brien et al. 1995).

## Methods

The lithic artefact and debitage assemblages from six dwellings were examined; two from the early phase of site occupation (Features 14 and 1), three from the middle phase (House 18, 2, and 17), and one from the late phase (Feature 55). Preliminary examination of the artefact assemblages of each dwelling involved separating and tallying those artefacts made of Ramah chert. Ramah chert was visually identified on the basis of colour, texture, lustre and opacity. Artefacts that were originally misidentified in the Phillip's Garden database as Ramah chert were removed from the sample. According to Fitzhugh (1972:41) and Tuck (1976:52), Ramah chert has a granular but lustrous appearance, unlike most cherts and chalcedonies, and resembles wet sugar, or sleet on a windshield. Its colour ranges from greyish to blue–grey to dark grey and occasionally will exhibit tints of yellow or green (Lazenby 1980:635–636; Nagle 1984:100). It can have specks, swirls and bands of dark grey or black on a pale background (Lazenby 1980:635–636). It can also occasionally have iron inclusions and staining (Fitzhugh 1972:41; Lazenby 1980:637). Although the visual identification of lithic raw material is subjective and therefore imprecise (Andrefsky 2005:43–46; LeBlanc et al. 2010; Odell 2003:24–41; Rutherford and Stephens 1991:35–36), Ramah chert is sufficiently distinctive in appearance that misidentifications are likely few and the relative quantities in our samples are reasonably accurate.

Examination of the debitage from each dwelling required a multi-staged approach. Following current lithic analysis standards (Ahler 1989; Andrefsky 2001, 2005; Kooyman 2000; Odell 2003; Shott 1994) particular attributes (frequency, weight, size) of the debitage were examined and used as a basis of analysis. A sample of 2,500 flakes was randomly taken from the debitage assemblages of each dwelling for a total of 15,000 flakes; these samples consisted of debitage from different unit bags throughout the dwelling. Each individual debitage assemblage sample was then sorted by lithic type, and any Ramah chert debitage was separated. The number of Ramah chert flakes for each individual sample was totalled. Each individual flake was given a size grade based on length × width: small (<20 $mm^2$), medium (20–40 $mm^2$) or large (>40 $mm^2$). As well, the weight of both the total individual sample and the Ramah chert debitage portion was taken. Since the electronic scale did not register anything under 2 g it was necessary, in order for a more accurate result, that all of the individual flakes that weighed less than 2 g be weighed as a whole. The sum weight was then divided by the number of flakes. The result was a more accurate, albeit average, weight for each individual flake.

## Data Description

To date, there have been approximately 36,000 artefacts recovered from Phillip's Garden. This number comprises a large variety of artefact types, both organic and lithic. The size of lithic assemblages for each dwelling in our sample varied considerably, with notably larger lithic assemblages from the middle phase dwellings House 18, House 2 and House 17 (Table 10.2). The size of the lithic assemblages for the early and late phase dwellings is considerably smaller than those of the middle phase assemblages. A chi-square test ($\chi^2=12.497$, df=5, $p=0.029$) was used to compare the relative frequencies of Ramah chert artefacts and Ramah chert debitage from all dwelling assemblage samples and indicated that the observed differences were very unlikely to have been caused by random sampling; that is, the differences are not likely a function of assemblage size. Also, the results of a Student's $t$-test ($t=0.662$, df=10, $p=0.523$) indicated that the sample means were very similar and could have come from populations with similar means.

**Table 10.2** Quantitative attributes of Ramah chert artefacts and debitage from dwellings sampled at Phillip's Garden

| | F14 | F1 | H18 | H2 | H17 | F55 |
|---|---|---|---|---|---|---|
| Total lithic artefacts | 646 | 661 | 2,961 | 4,474 | 4,147 | 486 |
| Ramah chert artefacts ($n$) | 4 | 6 | 10 | 36 | 25 | 12 |
| Endblade | 1 | 1 | 3 | 9 | 7 | 3 |
| Biface | 3 | 1 | 5 | 5 | 4 | 5 |
| Microblade | 0 | 3 | 0 | 5 | 3 | 1 |
| Endscraper | 0 | 0 | 0 | 9 | 3 | 0 |
| Preform | 0 | 1 | 1 | 7 | 7 | 2 |
| Core | 0 | 0 | 1 | 1 | 1 | 1 |
| Ramah chert debitage ($n$) | 18 | 10 | 20 | 28 | 32 | 10 |
| % Ramah chert artefacts ($n$/total) | 0.6 | 0.9 | 0.3 | 0.8 | 0.6 | 2.5 |
| % Ramah chert debitage ($n$/2,500) | 0.7 | 0.4 | 0.8 | 1.1 | 1.3 | 0.4 |
| % Mean Ramah chert artefacts:debitage | 0.65 | 0.65 | 1.1 | 0.95 | 0.95 | 1.45 |

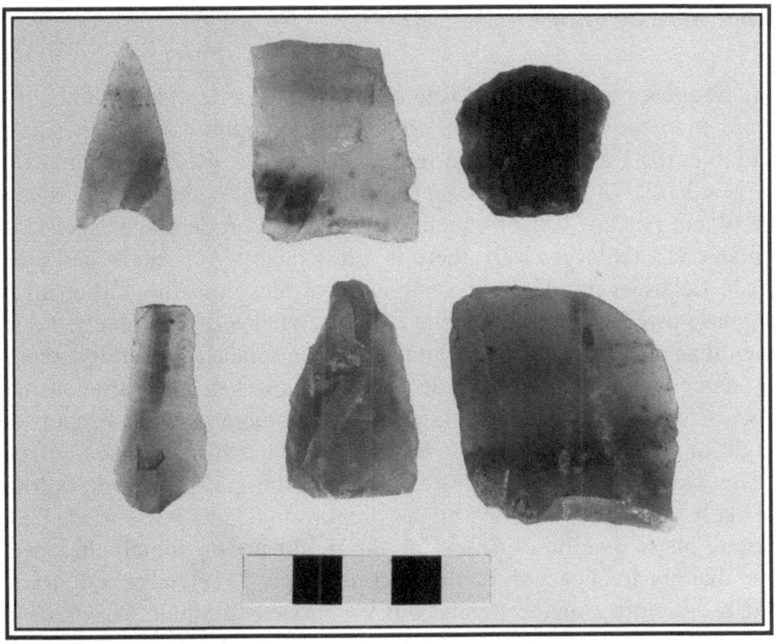

**Fig. 10.2** Typical Ramah chert tools recovered from Phillip's Garden. *First row*: endblade, biface, endscraper; *second row*: microblade, preform, core fragment

With regard to completeness of lithic items, many of these recorded are fragmentary, but since all tools are equally subject to fragmentation, fragments still contribute to the relative frequencies.

The majority of flaked lithic items recovered from the site are made of Cambro-Ordovician Cow Head chert (Coniglio 1987) which has excellent flaking properties because of its fine grain size; this chert is available about 75 km to the south of Port au Choix. There are lesser quantities of quartz crystal, rhyolite and Ramah chert. One hundred and seventy-eight items (0.5%) have been identified as Ramah chert. One hundred and fifty-seven (0.4%) of these are formal tools: endblades, bifaces, microblades, endscrapers, preforms and cores (Fig. 10.2). Ramah chert specimens categorized as tip-flute spalls ($n=6$), unidentified tool fragments ($n=11$), or retouched/utilized flakes ($n=2$) are not included in the analysis.

## *Ramah Chert Artefacts*

The frequency of Ramah chert, compared with the frequency of other lithic materials, is consistently low (Table 10.2). All Ramah chert artefact frequencies, as proportions of each total lithic assemblage of each dwelling, are less than 1% with the exception of Feature 55.

## *Ramah Chert Debitage*

Both the frequency and weight of Ramah chert in each debitage sample of 2,500 flakes are low; the debitage is predominantly composed of small-size grade flakes (Tables 10.2 and 10.3). The frequencies are <1.4%. The weight of each sample is <0.10% of the total sample weight. For the early phase dwellings, Feature 14 and Feature 1, the frequency of Ramah chert debitage exceeds that of the Ramah chert tools. As well, there is a low quantity of small- and medium-size grade flakes in the debitage samples from these dwellings. For all of the middle phase dwellings except House 2, the frequency of Ramah chert debitage is higher than the frequency of Ramah chert artefacts. Generally, there is a greater frequency of small-size grade debitage; however, there are considerable frequencies of medium- and large-size grade debitage as well. Proportionally, the weight of Ramah chert debitage for House 18 and House 2 is essentially the same. The weight of Ramah chert debitage for House 17, however, is low; this is most likely due to the relatively higher proportion of small-size grade debitage. For the late phase dwelling Feature 55, the frequency of Ramah chert artefacts is greater than the frequency of Ramah chert debitage. The frequency of debitage is low and the majority of the debitage is comprised of small-size grade flakes which overall weight very little as a proportion of the total weight of the debitage sample of 2,500 flakes.

To summarize the patterns described above, the frequency of Ramah chert artefacts and debitage across all assemblages is relatively low. The frequency of Ramah chert debitage is greater than the frequency of Ramah chert artefacts in four of the six house assemblages. The frequencies are all <1.4%. The weight of each sample is <0.10% of the total sample weight. The size of flakes in each individual debitage sample of 2,500 varied considerably, with relatively higher frequencies of small-size grade debitage in all samples. Both the Ramah chert artefacts and debitage indicate a predominance of late-stage lithic reduction of preforms and finished tools.

**Table 10.3** Weight and size of Ramah chert debitage for dwellings sampled at Phillip's Garden

| Dwelling | Total sample weight (g) | Ramah chert weight (% total sample) | Small (<20 mm$^2$) | Medium (20–40 mm$^2$) | Large (>40 mm$^2$) |
|---|---|---|---|---|---|
| F14 | 708 | 2 (0.3) | 18 | 0 | 0 |
| F1 | 658 | 2 (0.3) | 7 | 3 | 0 |
| H18 | 1,204 | 6 (0.5) | 12 | 7 | 1 |
| H2 | 1,290 | 12 (0.9) | 19 | 6 | 3 |
| H17 | 2,198 | 12 (0.5) | 25 | 5 | 2 |
| F55 | 914 | 2 (0.2) | 9 | 1 | 0 |

## Observations

During the occupation of the early phase dwellings Feature 14 and Feature 1, Ramah chert artefact and debitage frequencies were the lowest. Erwin (1995, Chap. 8), Renouf and Murray (1999) and Renouf (2006) have suggested that this was a time when occupation of Phillip's Garden was in an ephemeral and exploratory stage. If so, the first groups to occupy the site may have brought with them small numbers of Ramah chert tools from Labrador (LeBlanc 2008:162). The low frequency, low weight and small size of the debitage from Feature 14 and Feature 1 indicate late-stage reduction practices such as retouch and rejuvenation. Throughout the middle phase, Ramah chert artefact and debitage frequencies were relatively higher. The higher relative frequency of Ramah chert artefacts suggests that exchange networks with Labrador continued to be maintained. Although both medium- and large-size grade debitage are present, the larger percentage of small-size grade debitage indicates the predominance of late-stage reduction practices such as retouch and rejuvenation. The medium-size grade debitage may have resulted from thinning preforms. The large-size grade flakes may have served as flake blanks (Pecora 2001:175) or have resulted from core reduction. The late phase Ramah chert frequencies and debitage frequencies are exceptionally high relative to the other house assemblages. In the next section, we address the implications of this increased frequency.

## Discussion

From these data, a number of inferences can be made about the temporal, technological and social dynamics of Ramah chert use at Phillip's Garden. By temporal dynamics we mean the relative frequency of Ramah chert use over time. Technological dynamics refer to the sequence of Ramah chert lithic reduction. Social dynamics refer to communication and exchange networks embedded in Ramah chert use.

### *Temporal Dynamics of Ramah Chert Use*

The relative frequency of Ramah chert use at Phillip's Garden differs through time, from the early phase through to the late phase of occupation. Based on the Ramah chert artefact and debitage frequencies for Feature 14 and Feature 1, the early phase of occupation is characterized by the use of low quantities of Ramah chert artefacts along with relatively higher quantities of late-stage reduction debitage. As previously suggested, these tools may have been brought to the site by its pioneer occupants from Labrador. Based on the Ramah chert artefact and debitage frequencies for

House 18, House 2 and House 17, the middle phase can be characterized by an increased use of Ramah chert. During the occupation of House 17, the use of Ramah chert is less pronounced compared to House 2. The late phase, represented by Feature 55, is characterized by an increase in Ramah chert use.

## *Technological Dynamics of Ramah Chert Use*

There is a generally similar lithic reduction pattern throughout site occupation. The higher frequency of finished tools and the higher frequency of preforms compared to cores suggest that Dorset Palaeoeskimos were receiving Ramah chert preforms and finished tools at Phillip's Garden. The attributes of the Ramah chert debitage indicate that there was principally middle- and late-stage reduction occurring. This, in turn, suggests that the Phillip's Garden Dorset were not manufacturing Ramah chert tools which would have involved early-stage reduction. Rather, all that was occurring was likely preform thinning and tool retouch and resharpening, all essentially middle- and late-stage reduction activity.

There are three logical possibilities to explain how Ramah chert reached Phillip's Garden: trade from the quarry in Labrador direct to Newfoundland; direct procurement by the Dorset Palaeoeskimo occupants of Phillip's Garden; or down-the-line exchange (Nagle 1984, 1986; Renfrew 1977; Renouf 1999:413). Down-the-line exchange involves raw material being passed through a network of related groups who, in turn, reduce the raw material to suit their own needs, and pass the remnants onward (Renfrew 1977:77–79). Binford (1979) suggests that the production of stone tools would thus be executed in a staged manner whereby the manufacturing process would take place in episodes. He further suggests that staged production may be related to travel junctures where lithic items are partially processed at one site, packed away, then further processed at the next site along the travel route of the particular group (Binford 1979:268). At each subsequent site there would be different lithic debitage produced (Binford 1979; Newman 1994). With greater distance from the lithic source, the smaller the lithic core supply becomes.

Due to the lack of evidence for early-stage reduction we think that the Phillip's Garden data best fit down-the-line exchange. The generally low Ramah chert artefact frequencies, and the low frequency, small size, and low weight of the Ramah chert debitage (Tables 10.2 and 10.3) correspond to Renfrew's (1977) model of down-the-line exchange. Loring and Cox (1986:78) note for Labrador that the spatial constraints of a linear coast encourage down-the-line exchange. Considering the vast distance between Ramah Bay and Phillip's Garden, our conclusions are also consistent in terms of costs and benefits related to raw material acquisition.

The acquisition of exotic raw materials like Ramah chert could be interpreted as a visible manifestation of distant and complex social networking. Loring and Cox (1986:78) argue that the presence of large quantities of Ramah chert in Groswater Palaeoeskimo site assemblages in Labrador reflects the vast social networks of these small, dispersed and mobile populations. Based on data observed with Ju/'hoansi *hxaro* exchange in Namibia and Botswana, Kelly (1995:188) argues that trade and exchange networks established and strengthened social ties among different hunter–gatherer groups (see also Wiessner 1982). These social links and exchange networks were spread over large distances, sometimes in excess of 200 km (Kelly 1995; Nagle 1984; Wiessner 1982), or over 800 km in the case of the Labrador Groswater (Loring and Cox 1986). When resources fluctuated, exchange and reciprocity provided an ostensible motive for visiting others (Halstead and O'Shea 1989; Kelly 1995:188). Down-the-line exchange would almost certainly have been embedded in these broader social relations and communication networks and therefore would have been the most likely means of social communication amongst Arctic and Subarctic hunter–gatherers like the Dorset.

## *Social Dynamics of Ramah Chert Use*

The temporal and technological dynamics of Ramah chert use at Phillip's Garden reflect patterns in broader social dynamics. Given that Ramah chert is an exotic material on the island of Newfoundland and that, as we have argued, it reached Newfoundland through down-the-line exchange, then the presence of Ramah chert at Phillip's Garden indicates the existence of an extended exchange network. As previously noted, this exchange would have been rooted within broader social relations and communication networks between related groups. The maintenance of social relations and communication between Dorset Palaeoeskimo groups in Labrador and in insular Newfoundland may have been affected by a number of cultural and environmental factors.

Recent Indians or, following Stopp (2008), Late Precontact Amerindians, occupied central and southern Labrador at the same time that Dorset Palaeoeskimos occupied Phillip's Garden. These Amerindian populations were members of what archaeologists call the Daniel Rattle complex (Loring 1989, 1992; Stopp 2008). Based on the scarcity of Dorset sites in the Strait of Belle Isle, McGhee and Tuck (1975) and Robbins (1989) argued that the Recent Indian occupation of southern Labrador was a barrier to Dorset occupation of that region. However, as Tuck (1982:210), Loring (1992, 2002) and Stopp (2008:112) have more recently pointed out, given that the majority of Daniel Rattle lithic assemblages are comprised of Ramah chert there must have been some form of interaction or exchange between these Amerindian groups in central and southern Labrador and the Dorset Palaeoeskimo groups who were the near-exclusive occupants of northern Labrador, including Ramah Bay. Renouf et al. (2000) and Renouf (2003b) argue that in Newfoundland the different economic patterns of Dorset Palaeoeskimos and Recent Indians provided the basis for mutually beneficial cooperation. If so, this might also have been the case within Labrador and between Labrador and Newfoundland. In this light, the Daniel Rattle groups would have facilitated rather than inhibited social relations and communication.

Bell and Renouf (2008) argue that Phillip's Garden was a strategically located site that facilitated Dorset and possibly Dorset-Recent Indian social relations across the Strait of Belle Isle. They argue that maintenance of this complex network of social relations and communication would have been affected by environmental conditions. They link the abandonment of Phillip's Garden to increasing temperatures at 1500–1100 cal BP that are seen in chironomid (Rosenberg et al. 2005) and dinoflagellate cyst (Levac 2003) data from Port au Choix and Bay of Islands (Fig. 10.1), respectively (see also Bell and Renouf, Chap. 2). They argue that increased temperatures would have affected the timing and distribution of sea ice and the harp seal herds that relied on that ice. Since the harp seal hunt was the economic basis of Phillip's Garden, a change in seal herd predictability, abundance and/or availability would seriously undermine that foundation. That it did in fact occur is suggested by data from Hodgetts et al. (2003:113) that demonstrate a decline in reliance on harp seal in the latter part of the middle phase and throughout the late phase of site occupation.

This is consistent with changes in Phillip's Garden house construction which show a decrease in size from the middle to late phase of site occupation. The middle phase dwellings examined (House 18, House 2 and House 17) are all substantial rectangular, square or lobate structures (for images see Renouf, Chap. 7). As argued by Renouf (2003a), this reflects an upfront investment of time and effort into building a structure that was designed to be used on a regular basis over an extended period of time. However, the early phase dwelling Feature 1 and the late phase dwelling Feature 55 are both relatively small and less formalized oval structures. McGuire and Schiffer (1983) and Binford (1990) connect shape and size of dwellings to hunter–gatherer mobility, arguing that mobile hunter–gatherers use easily constructed oval or circular dwellings and less mobile hunter–gatherers use more substantially constructed dwellings that are often square or rectangular (see also Renouf 2003a). They argue that it would be expected that more highly mobile groups would construct low cost dwellings (e.g. Feature 1, Feature 55) and less mobile groups would invest in more substantial structures (e.g. House 18, House 2, House 17). McGuire and Schiffer (1983) argue that circular/oval dwellings were quick and easy to build and were more appropriate for highly mobile hunter–gatherers, while rectangular structures required much more time and effort to build and thus were more suited to less mobile groups. If this is correct, the shift at Phillip's Garden from the middle phase houses which are large (*c.* 100 $m^2$), substantially constructed, and rectangular or trilobate to the single example of late phase dwellings which is small (*c.* 30 $m^2$), circular and less substantially constructed, may indicate an increase in mobility during the late phase of site occupation. In other words, we argue that a decrease in occupational intensity at Phillip's Garden during 1500–1100 cal BP goes hand in hand with an increase in residential mobility.

An increase in residential mobility during the late occupation phase of Phillip's Garden would have allowed Dorset Palaeoeskimo groups in Newfoundland to move to resources as necessary (Renouf 1999, 2003b; Renouf et al. 2000). Many Dorset Palaeoeskimo sites on the island of Newfoundland have a proportion of Ramah chert tools and/or debitage (Anstey 2008; Eastaugh 2002:51; Erwin

2004:7–8; Loring 2002:172–173; Renouf 1999:413–414; Robbins 1985; Sawicki 1983). Many Dorset Palaeoeskimo sites in Labrador and the Quebec Lower North Shore have a proportion of Cow Head chert from northwestern Newfoundland (Fig. 10.1) (Fitzhugh 1972; Nagle 1984, 1985, 1986; Pintal 1998). Thus, in order for both of these foreign lithic materials to have reached each respective region, there must have been some form of interaction between groups.

The total lithic assemblage for Feature 55 is the smallest of all dwellings sampled. The size of the lithic assemblages for the early phase dwellings is also small relative to the middle phase lithic assemblages. The relatively small size of early and late phase lithic assemblages and the large size of middle phase lithic assemblages may be linked to periods of higher mobility and lower mobility, respectively. It is suggested, then, that the higher proportion of Ramah chert in the late occupation phase is linked to increased mobility and communication between Dorset Palaeoeskimo populations in Newfoundland and Labrador; communications may have included Labrador Recent Indian groups as well. This is consistent with Renouf's (1999) argument that during periods of resource scarcity or instability hunter–gatherer groups on the island of Newfoundland would have buffered this risk by maintaining links to kin and non-kin not only throughout Newfoundland but in Labrador as well. She makes the point that during periods of resource scarcity or unpredictability in Newfoundland, social links to Labrador would become indispensable. If this is the case it would be expected that access to Ramah chert would increase, thus explaining the higher proportion of Ramah chert in the late phase occupation of Phillip's Garden.

Our data refine Bell and Renouf's (2008) argument that the collapse of Phillip's Garden severed an important communication link between Newfoundland and Labrador thereby contributing to the Dorset abandonment of the island of Newfoundland. Although in that case one might expect to see a decrease in communication and residential mobility towards the end of Phillip's Garden use, our data suggest that mobility and communication with Labrador intensified during the last centuries of occupation. However, this makes perfect sense given that ties with Labrador populations must have been crucial in the context of an increasingly uncertain resource base.

## Conclusions

This chapter examines Ramah chert use at the Dorset Palaeoeskimo site of Phillip's Garden and discusses the temporal, technological and social dynamics of use. It is posited that the relative frequencies of Ramah chert in our sample of six dwellings which together span the early, middle and late phase of site occupation reflect changing intensity of communication and exchange. Ramah chert artefact and debitage frequencies and type indicate that Dorset Palaeoeskimo groups received finished Ramah chert tools or preforms at Phillip's Garden. We argue that the increase in frequency of Ramah chert tools and debitage in the single assemblage from the late phase of site occupation indicates an increase in communication with Labrador. Based on attributes of house size, shape and construction, we argue that this was coupled with an increase in residential mobility at the same time. This is consistent with faunal and other data that suggest a decline in exploitation of harp seal at this time, possibly linked to changing ice conditions and warming temperatures. Our data support Bell and Renouf's (2008) hypothesis that Phillip's Garden was a social and exchange hub for Dorset groups on the island of Newfoundland and that increasing temperatures in northwestern Newfoundland during 1500–1100 cal BP undermined harp seal exploitation towards the late phase of site occupation. While they argue that the eventual abandonment of Phillip's Garden weakened or broke the all-important communication network that extended across the Strait of Belle Isle, our data suggest that just prior to this break communication with Labrador was at its highest.

**Acknowledgements** Multi-year funding for the Northern Peninsula Archaeology and Landscape History Project, of which the Port au Choix Archaeology Project is a part, has been provided by the Social Sciences and Humanities Research Council of Canada, the National Science and Engineering Research Council, NSERC, Parks Canada, the Canada Research Chairs Program, Memorial University and the Provincial Archaeology Office. Charles Conway, Memorial University Geography Department, drafted the map. We would also like to thank John Erwin, Dominique Lavers, Stephen Loring and Patricia Wells for their helpful and insightful comments on earlier drafts of this paper.

## References

Ahler, S.A. 1989. Mass analysis of flaking debris: Studying the forest rather than the tree. *Archeological Papers of the American Anthropological Association* 1(1): 85–118.

Andrefsky, W. 2001. Emerging directions in debitage analysis. In *Lithic debitage: Context, form, meaning*, ed. W. Andrefsky, 2–14. Salt Lake City: University of Utah Press.

Andrefsky, W. 2005. *Lithics: Macroscopic approaches to analysis*. Cambridge: Cambridge University Press.

Anstey, R.J. 2008. Ramah Chert Use at Phillip's Garden (EeBi-1), Port au Choix, Newfoundland. Honour's dissertation, Department of Anthropology and Archaeology, Memorial University of Newfoundland, St. John's.

Bell, T., and M.A.P. Renouf. 2008. The domino effect: Culture change and environmental change in Newfoundland, 1500-1100 cal BP. *The Northern Review* 28: 72–94.

Binford, L.R. 1979. Organization and formation processes: Looking at curated technologies. *Journal of Anthropological Research* 35(3): 255–273.

Binford, L.R. 1990. Mobility, housing and environment: A comparative study. *Journal of Anthropological Research* 46(2): 119–152.

Cogswell, A. 2006. House 18 and the Middle Phase Occupation at Phillip's Garden (EeBi-1). Master's thesis, Department of Anthropology, Memorial University of Newfoundland, St. John's.

Coniglio, M. 1987. Biogenic cherts in the Cow Head group (Cambro-Ordovician), western Newfoundland. *Sedimentology* 34: 813–823.

Cox, S.L. 1978. Palaeo-Eskimo occupations of the north Labrador coast. *Arctic Anthropology* 15(2): 96–118.

Cox, S.L. 2003. Palaeoeskimo structures in the Okak region of Labrador. *Études/Inuit/Studies* 27(1–2): 417–433.

DNE. 1990. *Dictionary of Newfoundland English*, ed. G.M. Story, W.J. Kirwin and J.D.A. Widdowson. Toronto: University of Toronto Press.

Eastaugh, E.J.H. 2002. The Dorset Palaeoeskimo Site at Point Riche, Newfoundland: An Intra-site Analysis. Master's thesis, Department of Anthropology, Memorial University of Newfoundland, St. John's.

Erwin, J.C. 1995. An Intrasite Analysis of Phillip's Garden: A Middle Dorset Palaeo-Eskimo Site at Port au Choix, Newfoundland. Master's thesis, Memorial University of Newfoundland, St. John's.

Erwin, J.C. 2004. Interim Report for the 2003 Field Season: The Cow Cove Excavations. On file, Provincial Archaeology Office, Government of Newfoundland and Labrador, St. John's.

Fitzhugh, W.W. 1972. *Environmental archeology and cultural systems in Hamilton Inlet, Labrador: Survey of the central Labrador coast from 3000 B.C. to the present.* Washington, D.C.: Smithsonian Institution Press.

Gramly, R.M. 1978. Lithic source areas in northern Labrador. *Arctic Anthropology* 15(2): 36–47.

Halstead, P., and J. O'Shea (eds.). 1989. *Bad year economics: Cultural responses to risk and uncertainty*. Cambridge: Cambridge University Press.

Harp, E. 1976. Dorset settlement patterns in Newfoundland and southeastern Hudson Bay. In *Eastern Arctic prehistory: Paleoeskimo problems*, ed. M.S. Maxwell, 119–138. Washington, D.C.: Society for American Archaeology.

Hodgetts, L.M., M.A.P. Renouf, M.S. Murray, D. McCuaig-Balkwill, and L. Howse. 2003. Changing subsistence practices at the Dorset Paleoeskimo site of Phillip's Garden, Newfoundland. *Arctic Anthropology* 40(1): 106–120.

Kelly, R.L. 1995. *The Foraging spectrum: Diversity in hunter-gatherer lifeways*. Washington, D.C.: Smithsonian Institution Press.

Kennett, B. 1985. A Comparative Study of Two Lithic Assemblages from Port au Choix. Term paper submitted for Anthropology 6182, Memorial University of Newfoundland.

Knapp, R. 2008. An Analysis of Tabular Slate Tools from Phillip's Garden (EeBi-1), a Dorset Palaeoeskimo Site in Northwestern Newfoundland. Master's thesis, Department of Anthropology, Memorial University of Newfoundland, St. John's.

Kooyman, B.P. 2000. *Understanding stone tools and archaeological sites*. Calgary: University of Calgary Press.

Lazenby, M.E.C. 1980. Prehistoric sources of chert in Northern Labrador: Field work and preliminary analyses. *Arctic* 33(3): 628–645.

LeBlanc, D., I. Duval, and J.-F. Moreau. 2010. Geochemical signature of Mistassini quartzite and Ramah chert artefacts and quarries, Québec/Labrador, Canada. In *Ancient mines and quarries: A trans-Atlantic perspective*, ed. M. Brewer-LaPorta, A. Burke, and D. Field, 67–84. Oxbow: Oxbow Books.

LeBlanc, S. 2000. Middle Dorset (1900-1100 BP) regional variability on the island of Newfoundland and in Saint-Pierre et Miquelon. In *Identities and cultural contacts in the Arctic*, ed. M. Appelt, J. Berglund, and H.C. Gulløv, 97–105. Copenhagen: The Danish National Museum and Danish Polar Center.

LeBlanc, S. 2008. Middle Dorset Variability and Regional Cultural Traditions: a Case Study from Newfoundland and Saint-Pierre and Miquelon. PhD dissertation, Department of Anthropology, University of Alberta, Edmonton.

Levac, E. 2003. Palynological records from Bay of Islands, Newfoundland: Direct correlation of Holocene palaeo-oceanographic and climate changes. *Palynology* 27: 135–154.

Loring, S. 1989. Réserve d'outils de la période Intermédiare sur la côte du Labrador. *Recherches Amérindiennes au Québec* 19(2–3): 45–57.

Loring, S. 1992. Princes and Princesses of Ragged Fame: Innu Archaeology and Ethnohistory in Labrador. PhD dissertation, Department of Anthropology, University of Massachusetts, Amherst.

Loring, S. 2002. And they took away the stones from Ramah: Lithic raw material sourcing and eastern Arctic archaeology. In *Honoring our elders: A history of eastern Arctic archaeology*, ed. W.W. Fitzhugh, S. Loring, and D. Odess, 163–185. Washington, D.C.: Arctic Studies Center.

Loring, S., and S.L. Cox. 1986. The Postville Pentecostal Groswater site, Kaipokok Bay, Labrador. In *Palaeo-Eskimo cultures in Newfoundland, Labrador and Ungava*, 65–93. Reports in Archaeology No. 1, Memorial University of Newfoundland, St. John's.

McGhee, R., and J.A. Tuck. 1975. *An Archaic sequence from the Strait of Belle Isle, Labrador*. Mercury Series Archaeology Paper 34, National Museums of Canada, Ottawa.

McGuire, R.H., and M.B. Schiffer. 1983. A theory of architectural design. *Journal of Anthropological Archaeology* 2: 227–303.

Morgan, W.C. 1975. Geology of the Precambrian Ramah Group and Basement Rocks in the Nachvak Fiord-Saglek Fiord Area, North Labrador. Department of Energy, Mines and Resources, Ottawa.

Murray, M.S. 1992. Beyond the Laundry List: The Analysis of Faunal Remains from a Dorset Dwelling at Phillip's Garden (EeBi 1), Port au Choix, Newfoundland. Master's thesis, Department of Anthropology, Memorial University of Newfoundland, St. John's.

Nagle. C.L. 1984. Lithic Raw Materials Procurement and Exchange in Dorset Culture along the Labrador Coast. PhD dissertation, Department of Anthropology, Brandeis University, Waltham.

Nagle, C.L. 1985. Lithic raw materials resource studies in Newfoundland and Labrador: A progress report. In *Archaeology in Newfoundland and Labrador 1984*, ed. J.S. Thomson and C. Thomson, 86–121. Historic Resources Division, Government of Newfoundland and Labrador, St. John's.

Nagle. C.L. 1986. Flaked stone procurement and distribution in Dorset culture sites along the Labrador coast. In *Palaeo-Eskimo cultures in Newfoundland, Labrador and Ungava*, 95–110. Reports in Archaeology No. 1, Memorial University of Newfoundland, St. John's.

Newman, J.R. 1994. The effects of distance on lithic material reduction technology. *Journal of Field Archaeology* 21(4): 491–501.

O'Brien, S.R., P.A. Mayewski, L.D. Meeker, D.A. Meese, M.S. Twickler, and S.I. Whitlow. 1995. Complexity of Holocene climate as reconstructed from a Greenland ice core. *Science* 270: 1962–1964.

Odell, G.H. 2003. *Lithic analysis*. Manuals in archaeological method, theory and technique. New York: Springer.

Pecora, A.M. 2001. Chipped stone tool production strategies and lithic debitage. In *Lithic debitage: Context, form, meaning*, ed. W. Andrefsky, 173–190. Salt Lake City: University of Utah Press.

Pintal, J.-Y. 1998. *Aux frontières de la mer: La préhistoire de Blanc-Sablon*. Les Publications du Québec, Dossiers 102. Collections Patrimoines et Municipalité de Blanc-Sablon, Québec.

Renfrew, C. 1977. Alternative models for exchange and spatial distribution. In *Exchange systems in prehistory*, ed. T.K. Earle and J.E. Ericson, 71–90. New York: Academic Press.

Renouf, M.A.P. 1986. 1985 Excavations at Phillip's Garden and Point Riche, Port au Choix Historic Park. In *Archaeology in Newfoundland and Labrador 1985*, ed. J.S. Thomson and C. Thomson, 159–195. Historic Resources Division, Government of Newfoundland and Labrador, St. John's.

Renouf, M.A.P. 1987. Archaeological Excavations at the Port au Choix National Historic Park: Report of 1986 Field Activities. On file, Parks Canada, Archaeology, Atlantic Region, Halifax.

Renouf, M.A.P. 1993. Palaeoeskimo seal hunters at Port au Choix, northwestern Newfoundland. *Newfoundland Studies* 9(2): 188–210.

Renouf, M.A.P. 1999. Prehistory of Newfoundland hunter-gatherers: Extinctions or adaptations? *World Archaeology* 30(3): 403–420.

Renouf, M.A.P. 2002. *Archaeology at Port au Choix, northwestern Newfoundland 1990–1992*. Occasional Papers in Northeastern Archaeology 12. St. John's: Copetown Press.

Renouf, M.A.P. 2003a. A review of Palaeoeskimo dwelling structures in Newfoundland and Labrador. *Études/Inuit/Studies* 27(1–2): 375–416.

Renouf, M.A.P. 2003b. Hunter-gatherer interactions: Mutualism and resource partitioning on the island of Newfoundland. *Before Farming* 1(4): 1–16.

Renouf, M.A.P. 2006. Re-investigating a middle phase Dorset dwelling at Phillip's Garden, Port au Choix, Newfoundland. In *Dynamics of northern societies*, Proceedings of the SILA/NABO Conference on Arctic and North Atlantic Archaeology, Copenhagen, May 10th–14th, 2004, ed. J. Arneborg and B. Grønnow, 119–128. Publications from the National Museum Studies in Archaeology and History Vol. 10, Copenhagen.

Renouf, M.A.P. 2007. Re-excavating House 17 at Phillip's Garden, Port au Choix. *Provincial Archaeology Office Archaeology Review* 5: 57–61.

Renouf, M.A.P. 2009. Dorset Palaeoeskimo whalebone use at Phillip's Garden, Port au Choix. In *On the track of the Thule culture, from Bering Strait to East Greenland*, Proceedings of the SILA Conference "The Thule Culture – New Perspectives in Inuit Prehistory", ed. B. Grønnow, 91–104. Publications from the National Museum Studies in Archaeology and History Vol. 15, Copenhagen.

Renouf, M.A.P., and M. Murray. 1999. Two winter dwellings at Phillip's Garden, a Dorset site in northwestern Newfoundland. *Arctic Anthropology* 36(1–2): 118–132.

Renouf, M.A.P., and T. Bell. 2008. Dorset Palaeoeskimo skin processing at Phillip's Garden, Port au Choix, northwestern Newfoundland. *Arctic* 61(1): 35–47.

Renouf, M.A.P., T. Bell, and M. Teal. 2000. Making contact: Recent Indians and Palaeoeskimos on the island of Newfoundland. In *Identities and cultural contacts in the Arctic*, ed. M. Appelt, J. Berglund, and H.C. Gulløv, 106–119. Copenhagen: The Danish National Museum and Danish Polar Center.

Robbins, D.T. 1985. Stock Cove, Trinity Bay: The Dorset Eskimo Occupation of Newfoundland from a Southeastern Perspective. Master's thesis, Department of Anthropology, Memorial University of Newfoundland, St. John's.

Robbins, D.T. 1986. "Newfoundland Dorset" culture? In *Palaeo-Eskimo cultures in Newfoundland, Labrador and Ungava*, 119–124. Reports in Archaeology No. 1, Memorial University of Newfoundland, St. John's.

Robbins, D.T. 1989. Regards archéologiques sur les Béothuks de Terre-Neuve. *Recherches Amérindiennes au Québec* 19(2–3): 21–32.

Rosenberg, S.M., I.R. Walker, and J.B. Macpherson. 2005. Environmental changes at Port au Choix as reconstructed by faunal midges. *Newfoundland and Labrador Studies* 20(1): 57–73.

Rutherford, D.E., and R.K. Stephens 1991. Geological Approaches to Prehistoric Trade: Physical and Chemical Characterization of Metachert from the Ramah Group, Labrador (Ramah Chert). Final Research Grant Report prepared for the Institute of Social and Economic Research, Memorial University of Newfoundland, St. John's.

Sawicki, A.I. 1983. Palaeo-Eskimo Occupations in Bonavista Bay, Newfoundland. Master's thesis, Department of Anthropology, Memorial University of Newfoundland, St. John's.

Shott, M.J. 1994. Size and form in the analysis of flake debris: Review and recent approaches. *Journal of Archaeological Method and Theory* 1(1): 69–110.

Stopp, M.P. 2008. FbAx–01: A Daniel Rattle hearth in southern Labrador. *Canadian Journal of Archaeology* 32: 96–127.

Stuiver, M., and P. Reimer. 1993. Extended 14C database and revised CALIB radiocarbon calibration program. *Radiocarbon* 35: 215–230.

Tuck, J.A. 1976. *Newfoundland and Labrador prehistory*. National Museums of Canada, Ottawa.

Tuck, J.A. 1982. Prehistoric archaeology in Atlantic Canada since 1975. *Canadian Journal of Archaeology* 6: 201–218.

Wiessner, P. 1982. Risk, reciprocity and social influences on !Kung San economics. In *Politics and history in band societies*, ed. E. Leacock and R.B. Lee, 61–84. Cambridge: Cambridge University Press.

# Chapter 11
# Whitecoats, Beaters and Turners: Dorset Palaeoeskimo Harp Seal Hunting from Phillip's Garden, Port au Choix

**Maribeth S. Murray**

## Introduction

Exploitation of the harp seal (*Phoca groenlandica*) has been a significant component of socio-ecological systems in Newfoundland and Labrador for thousands of years. Whether one talks of the prehistoric Palaeoeskimo and Amerindian peoples, or the later European settler groups, the biannual migration of the harp seal herd has long been important in demarcating local and regional subsistence and settlement activities as evidenced by the abundance of harp seal remains in archaeological deposits along the outer coasts of Newfoundland and Labrador (Auger 1986; Cox and Speiss 1980; Jordan 1986; Pastore 1986; Pintal 1989; Renouf 2000; Renouf and Murray 1999; Speiss 1978), and by ancient technologies that reflect decision-making with respect to harp seal exploitation (LeBlanc 2000). In recent times, the importance of the harp seal in the cultural and economic life of Newfoundlanders is well documented in various sources, and colourfully expressed in a complex and detailed folk taxonomy (Dictionary of Newfoundland English – hereafter referenced as DNE 1990; Ryan 1994), and in songs and poetry (Ryan and Small 1978).

> Ho! We be the Sealers of Newfoundland!
> We clear from a snowy shore,
> Out into the gale with our steam and sail,
> Where tempest and tumult roar.
> We battle the floe as we northward go,
> North, from a frozen strand!
> Through lead, through bay, we fight our way,
> We Sealers of Newfoundland (from The Sealers of Newfoundland, G. A. England 1969 as reprinted in Ryan and Small 1978:108).

M.S. Murray (✉)
Department of Anthropology, University of Alaska Fairbanks, P.O. Box 757720, Fairbanks, Alaska 99708, USA
e-mail: msmurray@alaska.edu

M.A.P. Renouf (ed.), *The Cultural Landscapes of Port au Choix*, Interdisciplinary Contributions to Archaeology, DOI 10.1007/978-1-4419-8324-4_11,

In the prehistoric setting the quantity of harp seal remains in some contexts has led to the suggestion that seal hunting success played a critical role in human and cultural survival, especially in concert with fluctuations in caribou populations (Tuck and Pastore 1985; but see Renouf 1999; Schwarz 1994 for a different perspective). Among these ancient harp sealers were the Newfoundland Dorset Palaeoeskimos and, while Dorset methods of harp sealing are still poorly understood, the practice is perhaps best known from archaeological sites along the west coast of the island, and especially from the large Dorset Palaeoeskimo site of Phillip's Garden (EeBi-1) at Port au Choix, reviewed in Renouf (Chap. 7).

Many years of research at Phillip's Garden have produced a wealth of archaeofauna from multiple contexts that make it clear that Phillip's Garden was an important centre for Dorset harp seal hunting along the west coast of Newfoundland (e.g. Harp 1964; Renouf 2000, 2006). In recent years questions about harp seal use there have become increasingly sophisticated, as have methods used to address them (see for example Hodgetts 2002, 2005; Hodgetts et al. 2003; Murray 1992; Renouf and Murray 1999). Of particular interest is the extent to which harp seals served as a resource that may have enabled year-round occupation of Phillip's Garden, or at least a fall through spring occupation as reflected in the season(s) of the hunt (Harp 1964; Hodgetts 2002, 2005; Hodgetts et al. 2003; Murray 1992; Renouf 2000; Renouf and Murray 1999; Schwarz 1994). Determination of the seasonality or the timing of the hunt and any changes in this over the history of site occupation will provide information on changes in seal biogeography and about Dorset methods of sealing, including the use of open water technology or focused hunting at haul-out locations on the sea ice. Under any circumstances sealing is a precarious and dangerous activity that requires a good measure of skill. Hunting harp seals amid shifting ice floes or in open water requires a clear understanding of conditions conducive to success and also involves a solid appreciation of harp seal behaviour and perhaps even biology.

## Harp Seal Biology and Behaviour

> Their swimming movements are enrhythmic and beautiful, and the curves of their body are very pleasing, and when we watch them and study their habits, we are unconsciously influenced by the fact that they are big bundles of adaptation (Munn 1923:1).

### *Seasonal Distribution*

Harp seals are the most abundant marine mammal in the northwest Atlantic with a 1994 population estimate of between 4.5 and 4.8 million (Shelton et al. 1996). Estimates of the modern distribution and descriptions of their migratory behaviour come from surveys of whelping and moulting concentrations, catch data, tag returns and anecdotal observations (Stenson and Sjare 1997:1). Of particular interest here is the northwest Atlantic stock. In this population seasonal movements vary among individuals and years, but generally individuals spend the summer in Arctic waters, some as far north as the Thule district of Greenland, and in Lancaster Sound, Canada. Others summer as far west as Hudson Bay, Canada. In these places the seals spend their time feeding and accumulating blubber. During November/December (fall/early winter) when the herd is composed of adult and sexually immature animals swimming in loose groups ahead of the advancing pack ice, the seals migrate south along the Labrador coast. There they divide into two herds, one which goes to the front, a breeding area off the coast of northeast Newfoundland, and a second which moves into the Strait of Belle Isle (Sergeant 1991) (Fig. 11.1). Once in the Strait the seals further divide with one group remaining off the coasts of Newfoundland and Labrador in what is called the Mecatina patch, and another moving into the Gulf of St. Lawrence to a breeding patch off the Magdalen Islands (Sergeant 1991).

Satellite transmitter studies indicate that there is considerable seasonal and geographic variation in the activities of individual seals, and that some seals spend much of their time in offshore areas (Stenson and Sjare 1997:7). In late February/early March (mid/late winter) the seals form large whelping concentrations off southern Labrador and northeast Newfoundland (Stenson and Sjare 1997:2). The herds are stratified at this time with females and pups hauled out separately from immature seals and male seals (Sergeant 1991). During whelping and breeding the seals reduce their feeding activities and then, following breeding, they briefly

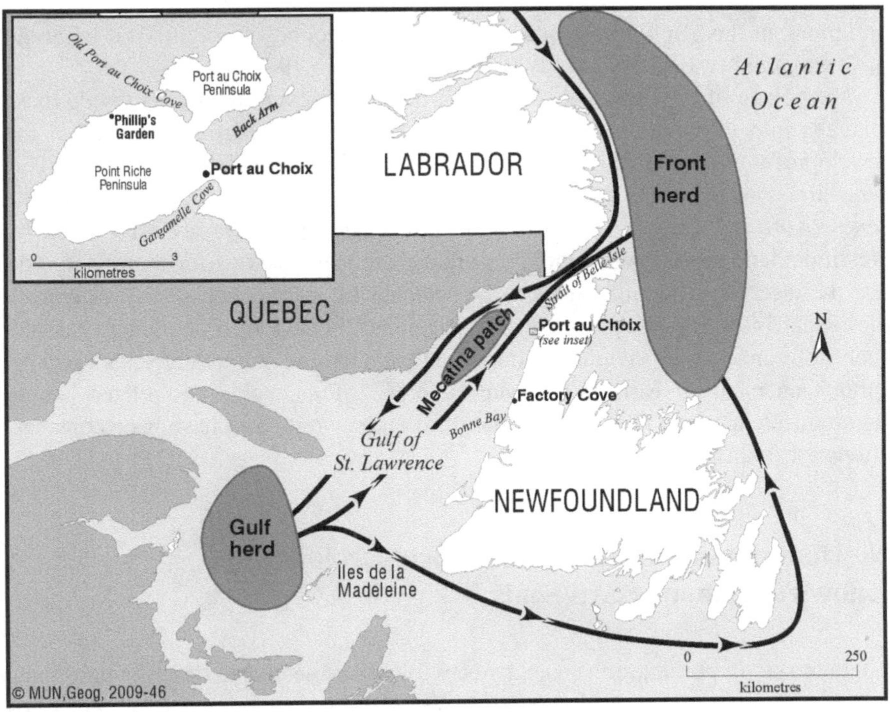


**Fig. 11.1** Map of Newfoundland and Labrador showing harp seal distribution, and Newfoundland place and site names mentioned in the text

disperse. During this time the female seals are probably feeding to replenish energy stores lost during the whelp (Chabot et al. 1996:15).

From mid-April to mid-May (late winter/early spring) the seals re-assemble into large moulting concentrations on the sea ice, and feeding is again reduced (Chabot et al. 1996:15). When the moult is completed the seals resume feeding. Their most important prey species, capelin and northern shrimp, are concentrated in areas of marine upwelling along the edges of the deep-water Esquimau channel that runs up the Strait of Belle Isle (LeBlanc 2000; Sergeant 1991). The Point Riche Peninsula is the closest point of land along the length of this channel and provides human hunters with comparatively easy access to the herds (Hodgetts et al. 2003; LeBlanc 2000:26). The herds remain in Newfoundland waters into June after which they migrate north to their Arctic feeding grounds (Stenson and Sjare 1997:2).

## *Harp Seal Growth, Development and Body Condition*

Harp seals are minimally sexually dimorphic; males grow longer than females, averaging between 170.4 and 168.8 cm while females average between 164.1 and 165.3 cm (Chabot et al. 1996; Hammill et al. 1995). Growth in length stops at about age eight with adult body length probably a reflection of conditions encountered during gestation, lactation and the first few years of life (Chabot et al. 1996:25). Harp seals have a thick blubber layer in the winter, although there are monthly variations in this. Body mass peaks in February, declines through April by about 30% and another 10–15% in May (Chabot et al. 1996:19–20).

The pelage of the harp seal changes throughout life and biologists divide these changes into three categories: spotted, spotted-harp and harp. These reflect the development of the saddle or harp pattern across the back of the seal as it ages. This transition does not occur at a particular age, and the pattern is different between the sexes (Roff and Bowen 1986:557). These changes in pelage are well-noted by Newfoundland sealers who employ a greater number of terms for them, and who use these terms to identify seals to be selected for harvest as well as to convey biological information about seal life stages and behaviour. For modern hunters such differences in behaviour are important as they make immature seals easier to shoot than matures (Roff and Bowen 1986:563). Presumably age-related behavioural differences and coat and blubber differences would also have been considerations for prehistoric hunters.

## Newfoundland Taxonomy and Local Ecological Knowledge about Harp Seals

There are six important terms used by Newfoundland sealers to describe harp seals in various stages of life: whitecoats, ragged-jackets or raggy jacks, beaters, bedlamers, turners and harps. These are defined below and also presented in use to provide a context for understanding how these terms convey traditional ecological knowledge of seal behaviour, body and fur condition, and, ultimately, utility to human hunters.

## *Whitecoat*

> Young harp seal with white fur, soon shed (DNE 1990:610).

The term whitecoat is used to designate seal age, coat/hair condition and aesthetics, and information about quality of seal oil.

For example:

> On the floating fields of Arctic ice the seals bring forth their young about the end of February. In 4–5 weeks these 'white coats,' as the young are called, are in the best condition for being taken and their fat then yields the finest oil (Harvey 1897 cited in DNE 1990:610),

and

> The young one is the white coat; when 'tis pupped 'tis called a pup. Then after she gets nursed a little bit she begins to perk up an' she gets a beautiful coat on it. ['Tis a] beautiful thing to look at (P. Saunders, Botwood T43/4-64 cited in DNE 1990:610),

and

> The white coat of the seal pups is actually foetal hair that is fast for only about 10 days. Commercial sealers preferred pups that were 2 or 3 days old as newborns had little value (Fisher 1955:511).

## *Ragged-Jacket/Ragajack/Raggy Jack*

> A young harp seal, undergoing colour change from 'white-coat' to 'bedlamer' stage (DNE 1990:402).

## *Beater*

> A harp seal just past the 'whitecoat' stage and migrating north from the breeding grounds on the ice floes of Newfoundland (DNE 1990:35).

These terms are closely linked as a seal may be both a ragged-jacket and a beater, at least for a period of a few weeks following weaning. The terms convey information about seal behaviour, quality of coat, aesthetics, seal stage of life and time of year. For example, with respect to ragged-jackets:

> As soon as they starts taking to the water this white coat starts to come off, and they're called a ragged jacket, because there's the new fur coming on. 'Tis old dark fur, and their beautiful white coat is falling off (P. Saunders, Botwood T43/4-64 cited in DNE 1990:403),

and

> A young harp seal undergoing its first moult from a whitecoat to a beater, beginning at about 12 days to 2 weeks (*Decks Awash* 1978 cited in DNE 1990:403),

and

> Skin buyers call it a *shedder* (Firestone 1967 cited in DNE 1990:403),

and

> The young seals…now begin naturally and fearlessly to take to swimming and diving…and learning without instruction all other sea-necessities of a full seal-life…Now they become 'beaters' of the Seal Hunt, and of certain surety every little head will then be pointed for the North (Greene 1933:78–9 cited in DNE 1990:36),

and

> The beater coat is short and spotted, similar to the coats of older animals, but thicker and softer. In the late 1940s the hair fast white coat was worth double the hair fast beater coat but by 1955 the beater pelt was worth slightly more than the white coat (Fisher 1955:511).

## *Bedlamer*

Munn (1923:15) claims that the term bedlamer is derived from the early Jersey settlers in the Strait of Belle Isle who apparently used nets or traps to capture seals. The DNE (1990) provides several definitions of the term including: (1) lunatic (Oxford English Dictionary 1933); (2) a troublesome person or animal (English Dialect Dictionary 1898); (3) an immature seal, esp. a harp seal approaching breeding age (DNE 1990:37).

The application of the term bedlamer to seals of a particular age is strongly linked to changes in pelage, stage of life and observations of seal behaviour as evidenced by the following statements:

> The Bedlamer Quite dusky without any mark they themselves tell you that the Bedlamer is the young harp (Banks [1766] 1971:145 cited in DNE 1990:37),

and

> When 12 months old the males…are still scarcely to be distinguished from the females, and during that season they are called 'bedlamers' (Jukes 1842:i, 310 cited in DNE 1990:37),

and

> A juvenile harp seal from about 1–5 years of age which has a spotted coat (*Decks Awash* 1978 cited in DNE 1990:38),

and

> The old Harp and Hood seals are undoubtedly marshalled and conform to rules in migration South and North. The Harps keep comparatively near the Shore and the Hoods a few miles off. The giddy bedlamers alone break the rules of the road (Munn 1923:9),

and

> These young Bedlamer seals appear to be free of the strict herd-control that comes with later days. They seem to be allowed to swim, and fish, and herd by themselves, and indeed to live as they choose; whereas some kind of an almost military discipline seems to exist among the adult members of these seal communities (Greene 1933:74 cited in DNE 1990:38).

## *Turners and Harps*

> A young harp seal undergoing a change to the darker markings of the adult stage (DNE 1990:589).

The terms turner and harp reflect not only changes in pelage but also perceptions of seal chronological age. For example:

> This year they're a bellamer, small bellamer, and the next year they're a big bellamer. What we calls a turner seal is turnin' from a bellamer to a harp (F. Hynes, Change Islands T80/4-64 cited in DNE 1990:38),

and

> The next year, the third year, they're a turner. An' the next year when they'll be 4 years old in the spring, that winter they're a harp; there's a plain harp on their back (D. Knight, Jackson's Cove T210/11-65 cited in DNE 1990:589),

and

> A turner seal – that's a three yeared seal – [is good] for legs [of a skin boot] (T. Bath, Horse Islands T391/2-67 cited in DNE 1990:589).

# Harp Seals at Feature 1, Phillip's Garden

Harp seal archaeofauna from Phillip's Garden, house Feature 1 (Fig. 11.2), was originally described in Murray (1992) in an effort to understand seasonality of the hunt as reflected in one dwelling at one point during the occupation of the site. To accomplish this, seal skeletal elements were grouped into five age categories based on degree of epiphyseal fusion, the presence or absence of juvenile cortex and bone character development. These age categories were determined as follows: (1) Uncertain – no ageable features; (2) Juvenile – epiphyses unfused, under-developed morphology, porous juvenile cortex; (3) Immature – unfused (one or more epiphyses), juvenile cortex only on margins of diaphyses; (4) Immature+ – fused with fusion line visible; (5) Adult – well fused, fusion lines blurred or absent (Murray 1992).

The total sample of pinniped remains analysed was 3,039 specimens of which the vast majority were identifiable only to as small Phocidae (Table 11.1). Species of identified Phocidae included harp seal, harbour seal (*Phoca vitulina*), bearded seal (*Erignathus barbatus*), grey seal (*Halichoerus grypus*) and hooded seal

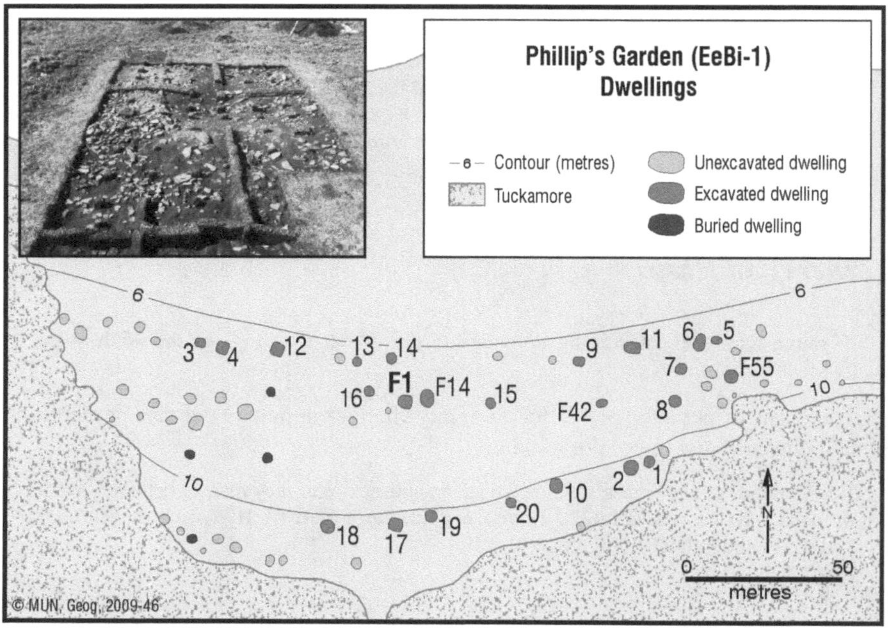


**Fig. 11.2** Map of Phillip's Garden with inset photograph of dwelling Feature 1

**Table 11.1** Identified Pinniped remains in Feature 1, expressed as NISP

| | NISP | | |
|---|---|---|---|
| Feature 1 | Murray (1992) | Renouf and Murray (1999) | This chapter |
| Harp seal | 178 | 178 | 184 |
| Harbour seal | 2 | 2 | 2 |
| Bearded seal | 8 | 8 | 8 |
| Grey seal | 4 | 4 | 4 |
| Hooded seal | 2 | 2 | 3 |
| Small Phocidae | 2,845 | 2,845 | 2,847 |
| Large seal | – | – | 22 |
| Total | 3,039 | 3,039 | 3,070 |

(*Cystophora cristata*). In my original interpretation of this assemblage I argued that the faunal remains from Feature 1 reflected a summer (June) through early/mid-winter (January) occupation with 95% of the seal bone coming from immature, immature+, or adult individuals, as defined above. Eighty-three percent of seal bone identified to species was harp seal, and the only time of year when juvenile harp seals (as defined above) are not present, but immature and adults are, is December and early January (early/mid-winter). Evidence from other deposits at the site suggested the presence of seals of all ages and thus spring hunting in other contexts (Murray 1992:73).

Subsequently Renouf and I re-analysed and published the Feature 1 fauna (Renouf and Murray 1999) and argued, based primarily on ethnographic accounts and some 32 fragments of apparently neonatal bone, that "although both December and March hunts are possible, the latter is more likely as this is when the seals more often would have been most accessible" (Renouf and Murray 1999:127). Residents of Port au Choix note that in December (early winter), the seal herds tend to be closer to the Quebec side of the Strait of Belle Isle. During this time the Strait is largely open water. Depending upon the year, as conditions do vary, beginning as early as March (late winter) or into April and May (late winter/early spring), the seals are generally located on the ice, close offshore of Phillip's Garden (Renouf and Murray 1999:127; see Renouf Chap. 7). Although the nature of Palaeoeskimo boating technology and ability to hunt and travel on open water is far from understood, this description of seasonal variability in harp seal location within the Strait during their migrations (Labrador side vs. Newfoundland side) has led some to suggest that the seals would have been inaccessible in late fall/early winter (December) – at least to those hunters travelling on foot (LeBlanc 2000:24).

Both previous studies of the Feature 1 seal assemblage suffer because, as is common when epiphyseal fusion sequences are unknown, skeletal element age and biological age are conflated in order to make inferences about both animal and human behaviour. However in 2000, Storå published epiphyseal fusion sequences for four species of pinniped including the harp seal. The fusion schedule for harp seals, based on the examination of 40 specimens (Storå 2002:215), makes possible the recovery of data on seal stage of life at time of death and prompted another examination of the archaeofauna from Feature 1. Harp seal skeletal elements can be placed into eight skeletal age groups based upon the fusion of specific epiphyses (unfused, fusing and fused), and then correlated with four stages of sexual maturity (age groups) or life stages and thus directly linked with seal behaviours that concern human hunters. Following Storå (2000:201–208, 2002) these are: (1) Age 1 – Yearlings: seals that are less than 1 year old, skeletal age 1–3; (2) Age 2 – Juveniles: sexually immature seals who are older than 1 year but younger than 4–5 years, skeletal age 4 and 5; (3) Age 3 – Young adults: sexually mature, behaviour related to the breeding cycle, adult body size but growth not complete, skeletal age 6; and (4) Age 4 – Old adults: physically and sexually mature, skeletal ages 7 and 8.

These life stages fit well with traditional Newfoundland terms for describing harp seal life history. Yearlings at various months correspond to whitecoats, ragged-jackets and beaters, while juveniles are bedlamers, and young and old adults are turners and harps.

The skeletal development of seals is comparable to stages of physical and sexual maturity and by skeletal age six through eight all have probably reached the lower limits of adult body size (Storå 2000:206). Because sex differences are not pronounced in harp seals these stages apply to both sexes and, of course, the application of the fusion schedule to archaeological assemblages assumes similar rates of skeletal development in modern and ancient specimens (Storå 2000:208). In this case, following Storå (2000), I used earliest age of complete fusion, latest observed age for unfused epiphysis, and earliest and latest ages of observed fusing. For example, if an anterior phalange one was fused on both the proximal and distal ends I assumed that it was 4.75 years of age or older as 4.75 years is the latest observed age for an unfused proximal epiphysis for this particular element (Storå 2000). Thus these assigned ages for these skeletal elements are more likely to err slightly on the older side than on the younger.

Table 11.1 summarizes the archaeofaunal sample used in this study. The few additional elements come from units along the rear platform of Feature 1 (Renouf and Murray 1999) that were not included in the original faunal analysis (Murray 1992) for a total NISP of 3,070.

As harp seals are minimally sexually dimorphic, no attempt is made to sex the specimens and no bacula were recovered that would positively identify male seals. However, both sexes generally reach sexual maturity before physical maturity – somewhere between ages four and eight (King 1983; Sergeant 1991). The majority of specimens identifiable as to species belong to harp seals. Therefore the remaining specimens identified as "small seal" (small Phocidae) are aged using the harp seal schedule. This does introduce the possibility of some error as a few elements may belong to harbour seal or grey seal (species which overlap in size with harp seals); however this is probably minimal. Where possible, identified elements from other seal species are also aged.

## Results

Table 11.2 summarizes the age and skeletal element data on phocid species other than harp seal recovered from Feature 1. With the exception of the harbour seal humerus which was aged following Storå (2000), all specimens were aged based on size, presence, or absence of juvenile cortex, and degree of epiphyseal fusion. There are no fusion schedules for bearded seals and hooded seals, and the grey seal remains are all teeth. Therefore it is not possible to link the elements from these larger species to specific life history stages. Here I have reclassified elements originally defined as Immature and Immature+ as Juvenile (J), and Young Adult+ (YA+), respectively.

The remains of seal species other than harp provide little information on seasonality or on hunting techniques. A few individuals were probably taken fortuitously during the harp seal hunt. This is a likely scenario for bearded and hooded seals, both of which are ice-loving species (King 1983). Bearded seals have been observed off the Point Riche Peninsula during the spring (Northcott and Phillips 1976:25) and hooded seals whelp in the same locations as the harps seals although they tend to maintain separate patches (Maxwell et al. 1967:93). The period of either gearing up for or winding down from a major harp seal hunt was probably a more opportune time for catching the occasional grey and harbour seal. The former are most common in the Port au Choix area during spring and summer, although they are sometimes present from as early as February until as late as November (Beck 1983), while the latter are available year round although prior to 1925 they were most common from spring through fall (Northcott and Phillips 1976:23).

**Table 11.2** Identified species in Feature 1, excluding harp seal

| Phocid | Skeletal element | Age | Criteria for ageing |
|---|---|---|---|
| Bearded seal | Auditory Bulla | YA+ | Size |
| | Carpal 2 | YA+ | Size |
| | Phalange 1 | A | Both epiphyses fused |
| | Mandible | YA+ | Size |
| | Phalange 2 | A | Both epiphyses fused |
| | Phalange 2 | J | Proximal epiphysis unfused |
| | Phalange 2 | J | Proximal epiphysis unfused |
| | Phalange 3 | A | Both epiphyses fused |
| Grey seal | Incisor | A | Root closed |
| | Incisor | A | Root closed |
| | Incisor | U | Incomplete |
| | Post canine | U | Incomplete |
| Hooded seal | Radius | I | Proximal epiphysis unfused |
| | Femur | I | Epiphyses unfused |
| | Phalange 2 | I | Proximal epiphyses unfused |
| Harbour seal | Auditory Bulla | YA+ | Size |
| | Humerus | J | Skeletal age 2 (Storå 2000) |

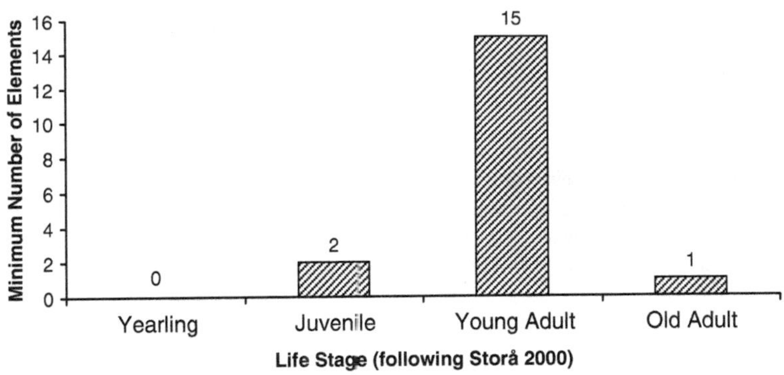


**Fig. 11.3** Harp seals, Feature 1, Phillip's Garden. Yearling = whitecoats, ragged-jackets, beaters; juvenile = bedlamers; young adult = turners, harps; old adults = harps

## *Harp Seal*

There are 184 specimens of harp seal bone of which 24 could be aged; six ulnae and 18 humeri (Fig. 11.3). Data are presented as Minimum Number of Elements. There are no elements from age group 1 (yearlings), only two elements from age group 2 (juveniles) and only one element from age group 4 (old adults). Most ageable elements ($n = 15$) are age group 3, the bones of young adults. These are from seals that were probably sexually mature, and had developed adult behaviour patterns. To Newfoundland sealers they would be recognized as turners or harps, with pelage beginning to show or showing the harp pattern.

## *Small Phocidae: Probably Harp Seal*

A total of 1,391 elements from "small seal" was aged (Fig. 11.4). Because most of the small seal that could be identified to species is harp seal, it is assumed here that the majority of the unidentifiable small seal remains are also harp.

There are 26 elements from age group 1, yearlings. These elements were classed as age group 1 based on fusion data as well as the presence of juvenile bone cortex. However, no elements in this category were sufficiently amorphous or small enough to belong to foetal or neonatal individuals. Newfoundland sealers would probably have recognized these individuals as ragged-jackets and beaters. Age group 2 is the largest category with 641 elements presumably from juvenile seals, animals that were probably sexually immature, and would likely be called bedlamers by sealers today. Age group 3 contains 329 elements and age group 4 contains 340. These specimens are from young adult and old adult individuals, with both groups being sexually mature and probably having pelage identifiable as either turners or harps. Figure 11.5 illustrates the distribution of age classes among the aged skeletal elements for the small seal category. All body parts are well represented. For the original description of small seal body part representation and density dependent taphonomic issues refer to Renouf and Murray (1999).

## Discussion

And then when springtime is coming
And white coats to the water will take
You hearken at dawn to the splashing
Of old seals down in the wake. (from A Sealers Love Letter by Frank Neville, February 1940, reprinted from The Family Fireside in Ryan and Small 1978:91).

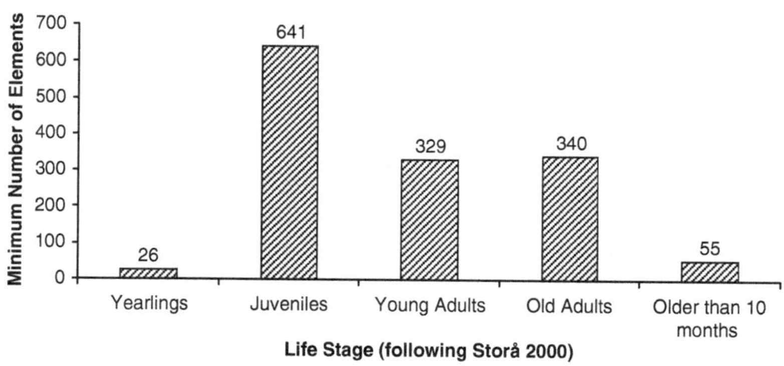


**Fig. 11.4** Small seals, Feature 1, Phillip's Garden. Yearling = whitecoats, ragged-jackets, beaters; juvenile = bedlamers; young adult = turners, harps; old adults = harps; older than 10 months = fragmentary/incomplete specimens where age was unable to be further refined

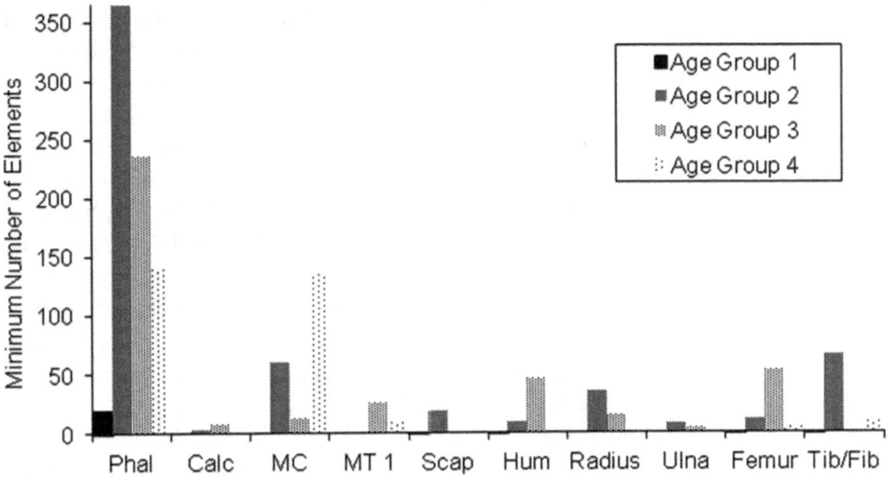


**Fig. 11.5** Distribution among age groups of aged skeletal elements of small seal, Feature 1, Phillip's Garden, excluding fragmentary specimens. Age group 1 = yearling (whitecoats, ragged-jackets, beaters); age group 2 = juvenile (bedlamers); age group 3 = young adult (turners, harps); age group 4 = old adult (harps). *Phal* phalanges; *Calc* calcaneus; *MC* metacarpals; *MT* metatarsals; *Scap* scapula; *Hum* humerus; *Tib/Fib* tibia/fibula

It is probably safe to assume that most of the unidentifiable seal remains are harp seal but it remains true that we cannot be certain. Probably there are some elements from other species mixed in, but generally in such low numbers that the overall results are not affected. There are few skeletal elements from seal pups and probably none from whitecoats, that is, seals under two weeks of age. The assemblage contains a fairly even distribution of bedlamers and breeding-age adults, or turners and harps. This mortality profile suggests several possibilities for seasonal hunting: (1) a hunt during the December migration (early winter); (2) a hunt in February and March (mid/late winter) before the April moult; (3) an April/May (late winter/early spring) hunt during the moult when all age classes are available on the ice; (4) or some combination of these.

## A December Hunt

With respect to yearling seals the inability to identify species is a special problem because even with a seal ageing schedule, the identification of these individuals is important for determining season of death. The presence of foetal harp seal remains would be proof positive of early winter hunting; if sealers took female seals as opportunity allowed in December then we would expect at least some of those females to be pregnant and at least some of those foetal remains to be preserved in the Port au Choix assemblage, especially given the generally acknowledged excellent preservation at the site (see Hodgetts et al. 2003).

In the case of Feature 1 there are no foetal/neonatal remains; however this does not mean that an early winter hunt during the December migration can be ruled out. While some ethnographic and contemporary biogeographic data indicate that the seals tend towards the Labrador side of the Strait at this time of the year, there is good evidence that both Palaeoeskimo and other land-based sealers (sealers that hunt from the shore in small open boats) were hunting from Newfoundland during both migration phases. At Phillip's Garden itself, Hodgetts (2005) argues, based on measurements of femora recovered from midden Features 2, 49 and 73, that at least some sealing was done during December as evidenced by the presence of 9–10 month-old individuals in these deposits. Although there is a possibility that the femora may come from harbour or hooded seals this seems unlikely given that harp seals comprise the vast majority of identifiable specimens in each of the three samples examined (Hodgetts 2005:98–101).

Away from Port au Choix there is precedent for December land-based hunting in both Newfoundland and in Labrador. For example, in northeastern Newfoundland during the nineteenth century a land-based hunt was conducted during the December migration (Sanger 1977) and as Schwarz (1994:66) points out, at least the early (Groswater) Palaeoeskimo migrants from Labrador to Newfoundland would have been accustomed to land-based harp seal hunting and thus open water hunting and hunting amid shifting pack ice. During both phases of the migration along this stretch of the Labrador coast the seals are in the water and amid the pack ice rather than hauled out on the ice whelping – yet site deposits, both Groswater and Dorset, contain substantial quantities of harp seal (cf. Cox and Speiss 1980; Speiss 1978) suggesting the skill and technology necessary for this type of hunting.

## *A Late Winter/Early Spring Hunt*

If the few elements from age group 1, yearlings, are harp seals then they are likely ragged-jackets or beaters rather than whitecoats, suggesting at least some hunting several weeks after the February/March whelp which might have occurred further south and further out into the Strait of Belle Isle of southern Labrador as it does today. Hunting at this time is consistent with the mix of bedlamer, turner and harp seals in the assemblage as the bedlamers do not haul out with the adults until whelping is finished. This occurs about two weeks after the white coats are weaned – the point at which the whitecoats begin to turn to ragged-jackets. A hunt before the moult might also have required the use of both open water techniques (for bedlamers) and sea-ice hunting techniques (for turners and harps). It is impossible to say if hunting occurred before or after the moult, although before the moult the seals would have been more attractive retaining good quality coats, and yet to experience the up to 20% of body weight loss that occurs during moulting. In any case, if hunting occurred during the spring it appears that pups were not targeted even though they would have been easy prey. At this time their fine white coats would have been shedding, the quality of their fat would have declined, and as is common after weaning, there would have been some weight loss. Compared to the bedlamers, turners and harps, the return on ragged-jackets especially, and possibly also beaters, may not have been worth the effort expended. The deliberate selection of adults' harp seals headed north in the spring was also suggested as the preferred subsistence strategy among the earlier Groswater Palaeoeskimos at the site of Factory Cove (DlBk-3), about 150 km to the south of Port au Choix (Auger 1986:113), although in light of new seal bone ageing possibilities this collection should be probably be re-examined.

## Conclusions

The archaeofuanal data from Feature 1 and other features at Phillip's Garden (e.g. Hodgetts 2005) suggest that there was little to no harp sealing during the whelp. This is perhaps not surprising as today the seals whelp further south and further offshore than the Point Riche Peninsula. However, the present-day pattern of seal availability may not always have been the same in the past and assuming such temporal uniformity in distribution may mask variability that is important to our improved understanding of the flexibility and resilience that enabled Dorset peoples to persist in Newfoundland and elsewhere for millennia. Changes in environmental conditions, including changes in water temperature, changes in ice cover, movement of prey, weather and climate may result in annual variation in the timing of migration and variation in the distribution and movements of the harp seals (Sergeant 1991; Stenson and Sjare 1997:7). For example, in the early 1900s harp seals concentrated around the southwest portion of Greenland during the summer; however warming Arctic waters led to an extension of open water and by the 1950s the seals were summering as far north as the Thule district (Fisher 1955:507). This range extension changed the timing of the fall migration to the south, with seals arriving in the Strait of Belle Island in January rather than November although it did not appear to affect the February/March timing of the whelp (Fisher 1955:508). Since the 1970s the northwest Atlantic has experienced three anomalously cold periods, the most recent from 1989 to 1993. These periods of lower temperature are associated with more persistent and stronger winds from the northwest, increased ice growth and transport, increased ice cover in southern areas and delayed ice melt in the spring and summer (Colbourne et al. 1994:311), presumably with related impacts on the timing and location of the herds at any point during the migration. Any and all of these factors could have influenced herd location during migration and whelping in the past and changed accessibility to the herds from Port au Choix on a seasonal, annual and/or a decadal basis, as is well illustrated by the following:

> On the 20th of March we were jammed for a day about ten miles west-north-west of Point Riche, on good sealing ground, where many a load in previous years was killed, but there wasn't a seal to be seen (Smith 1936:101 as cited in DNE 1990:456).

Certainly the historic seal fishery was not always predictable, and there are documented failures in 1800 and 1850, with the former event lasting for 10 years and the latter lasting for 15 years. During these times the seals "disappeared" from the north shore of the Gulf of St. Lawrence, sealing posts closed down and sealing companies abandoned the region (Munn 1923:19). Similarly at times in the past harp seals have whelped, at least in small numbers quite close to Port au Choix. For example, on March 20 and 21 in 1962 when the Strait of Belle Isle was choked with ice, approximately 120 whitecoats were taken in the area (Sergeant 1991:41). This variability in seal availability and location of whelping patches makes it difficult to base interpretation of Dorset sealing activities on recent observations of seal movements and availability (cf. LeBlanc 2000) or to definitively assign Dorset seal hunting to a particular month. Rather it illustrates that seasonality should be conceptualized with shifting temporal parameters, and it reinforces the notion that Dorset hunters would have to be finely attuned to these shifts resulting from changing environmental conditions in order to be successful in any given year (cf. Jordan 1986), and over the long-run, flexible in their adaptive strategies, including those related to mobility, procurement and social relations (cf. Renouf 1999, 2006).

## Future Research

Some of the uncertainties in this work stem from an inability to identify seal elements as to species and especially those elements from foetal and neonatal individuals. DNA analysis and isotopic studies could help to resolve this problem by providing irrefutable evidence for the presence of harp or seal pups of other species. For example, DNA studies on salmon bones have demonstrated potential for distinguishing different species on the northwest coast (Yang et al. 2004) and via isotopic proxy data, patterns of variability in human consumption of salmon (Cannon et al. 1999). In concert DNA and stable isotope analysis of some of the yearling seal remains from deposits at Phillip's Garden may be warranted to help identify the presence of whitecoats. At Phillip's Garden and other Dorset sealing sites in Newfoundland, such methods may provide more definitive answers to questions about seasonality, duration of occupations and methods of hunting as well as shedding light on the palaeobiogeography of the harp seal.

**Acknowledgements** Thanks to Charles Conway for drafting Figs. 11.1 and 11.2. Thanks also to two anonymous reviewers for their helpful comments on an earlier draft of this manuscript.

## References

Auger, R. 1986. Factory Cove: An early Palaeo-Eskimo component from the west coast of Newfoundland. In *Palaeo-Eskimo cultures in Newfoundland, Labrador and Ungava*, 111–117. Reports in Archaeology No. 1, Memorial University of Newfoundland, St. John's.

Beck, B. 1983. *The harbour seal in Canada*. Department of Fisheries and Oceans Canada, Ottawa.

Cannon, A., H.P. Schwarcz, and M. Knyf. 1999. Marine-based subsistence trends and the stable isotope analysis of dog bones from Namu, British Columbia. *Journal of Archaeological Science* 26: 399–407.

Chabot, D., G.B. Stenson, and N.G. Cadigan. 1996. Short- and long-term fluctuations in the size and condition of harp seal *Phoca groenlandica* in the northwest Atlantic. *NAFO Science Council Studies* 26: 15–32.

Colbourne, E., S. Narayanan, and S. Prinsenberg. 1994. Climate changes and environmental conditions in the northwest Atlantic, 1970–1993. *ICES Marine Science Symposium* 198: 311–322.

Cox, S.L., and A. Speiss. 1980. Dorset settlement and subsistence in northern Labrador. *Arctic* 33(3): 659–669.

DNE 1990. *Dictionary of Newfoundland English*, ed. G.M. Story, W.J. Kirwin, and J.D.A. Widdowson. Toronto: University of Toronto Press.

Fisher, H.D. 1955. Utilization of Atlantic harp seal populations. *Transactions of the North American Wildlife Conference* 20: 507–518.

Hammill, M.O., M.C.S. Kingsley, G.G. Beck, and T.G. Smith. 1995. Growth and conditions in the northwest Atlantic harp seal. *Canadian Journal of Fisheries and Aquatic Sciences* 52: 478–488.

Harp, E. 1964. *The cultural affinities of the Newfoundland Dorset Eskimo*. National Museums of Canada Bulletin 200, Ottawa.

Hodgetts, L.M. 2002. Dorset Palaeoeskimo harp seal exploitation at Phillip's Garden (EeBi 1), northwestern Newfoundland. In *The exploitation and cultural importance of sea mammals*, Proceedings of the 9th Conference of the International Council of Archaeozoology, Durham August 2002, ed. G. Monks, 62–76. Oxford: Oxbow Books.

Hodgetts, L.M. 2005. Using bone measurements to determine the season of harp seal hunting at the Dorset Palaeoeskimo site of Phillip's Garden. *Newfoundland and Labrador Studies* 20(1): 91–106.

Hodgetts, L.M., M.A.P. Renouf, M.S. Murray, D. McCuaig-Balwill, and L. Howse. 2003. Changing subsistence practices at the Dorset Paleoeskimo site of Phillip's Garden, Newfoundland. *Arctic Anthropology* 40(1): 106–120.

Jordan, R.H. 1986. Palaeo-Eskimos in Atlantic Canada: A regional comparison of Newfoundland and Labrador Middle Dorset. In *Palaeo-Eskimo cultures in Newfoundland, Labrador and Ungava*, 135–150. Reports in Archaeology No. 1, Memorial University of Newfoundland, St. John's.

King, J.E. 1983. *Seals of the world*. Oxford: Oxford University Press.

LeBlanc, S. 2000. Groswater technological organization: A decision-making approach. *Arctic Anthropology* 37(2): 23–37.

Maxwell, G., J. Stidworthy, and D. Williams. 1967. *Seals of the world*. London: Constable and Company Ltd.

Munn, W.A. 1923. Chafe's sealing book: A history of the Newfoundland seal fishery from the earliest available records down to and including the voyage of 1923. In *Introductory review*, ed. H.M. Mosdell, 1–30. The Trade Printers and Publishers, Ltd, St. John's..

Murray, J.A.H., Henry Bradley, W.A. Craigie, and C.T. Onions (eds.) 1933. *Oxford English Dictionary*. Oxford: Clarendon Press.

Murray, M.S. 1992. Beyond the Laundry List: The Analysis of Faunal Remains from a Dorset Dwelling at Phillip's Garden (EeBi 1), Port au Choix, Newfoundland. Master's thesis, Department of Anthropology, Memorial University of Newfoundland, St. John's.

Northcott, T.H., and F.R. Phillips. 1976. The Land and Sea Mammals of Port au Choix National Historic Park, Newfoundland. On file, Parks Canada and the Dept. of Indian and Northern Affairs.

Pastore, R.T. 1986. The spatial distribution of late Palaeo-Eskimo sites on the island of Newfoundland. In *Palaeo-Eskimo cultures in Newfoundland, Labrador, and Ungava*, 125–133. Reports in Archaeology No. 1, Memorial University of Newfoundland, St. John's.

Pintal, J.-Y. 1989. Contributions à la préhistoire récente de Blanc-Sablon. *Recherche Amérindiennes au Québec* 19(2): 33–44.

Renouf, M.A.P. 1999. Prehistory of Newfoundland hunter-gatherers: Extinctions or adaptations? *World Archaeology* 30(3): 403–420.

Renouf, M.A.P. 2000. Symbolism and subsistence: Seals and caribou at Port au Choix, northwestern Newfoundland. In *Animal bones, human societies*, ed. P. Rowley-Conwy, 65–73. Oxford: Oxbow Books.

Renouf, M.A.P. 2006. Re-investigating a middle phase Dorset dwelling at Phillip's Garden, Port au Choix, Newfoundland. In *Dynamics of northern societies*, Proceedings of the SILA/NABO Conference on Arctic and North Atlantic Archaeology, ed. J. Arneborg and B. Grønnow, 121–130. Publications from the National Museum Studies in Archaeology and History Vol. 10, Copenhagen.

Renouf, M.A.P., and M.S. Murray. 1999. Two winter dwellings at Phillip's Garden, a Dorset site in northwestern Newfoundland. *Arctic Anthropology* 36(1–2): 118–132.

Roff, D., and W.D. Bowen. 1986. Further analysis of population trends in the northwest Atlantic harp seal (*Phoca groenlandica*) from 1967 to 1985. *Canadian Journal of Fisheries and Aquatic Sciences* 43: 553–564.

Ryan, S. 1994. *The ice hunters: A history of Newfoundland sealing to 1914*. Newfoundland History Series 8. Breakwater Books, St. John's.

Ryan, S., and L. Small. 1978. *Haulin' rope and gaff: Songs and poetry in the history of the Newfoundland seal fishery*. Breakwater Books, St. John's.

Sanger, C.W. 1977. The evolution of sealing and the spread of settlement in northeastern Newfoundland. In *The peopling of Newfoundland: Essays in historical geography*, ed. J.J. Mannion, 136–151. Institute of Social and Economic Research, St. John's.

Schwarz, F. 1994. Paleo-Eskimo and Recent Indian subsistence and settlement patterns on the island of Newfoundland. *Northeast Anthropology* 47: 55–70.

Sergeant, D.E. 1991. *Harp seals, man, and ice*. Canadian Special Publications of Fisheries and Aquatic Sciences No. 114. Department of Fisheries and Oceans, Ottawa.

Shelton, P.A., G.B. Stenson, B. Sjare, and W.G. Warren. 1996. Model estimates of harp seal numbers-at-age for the northwest Atlantic. *NAFO Science Council Studies* 26: 1–14.

Speiss, A. 1978. Zooarchaeological evidence bearing on the Nain area Middle Dorset subsistence-settlement cycle. *Arctic Anthropology* 15(2): 48–60.

Stenson, G.B., and B. Sjare. 1997. Seasonal distribution of harp seals, *Phoca groenlandica,* in the northwest Atlantic. *ICES C.M. 1997/CC:10 (Biology and Behaviour II)*:1–23.

Storå, J. 2000. Skeletal development in the grey seal *Halichoerus grypus*, the ringed seal *Phoca hispida botnica*, the harbour seal *Phoca vitulina vitulina*, and the harp seal *Phoca groenlandica*: Epiphyseal fusion and life history. *Archaeozoologia* 11: 199–222.

Storå J. 2002. Neolithic seal exploitation on the Åland Islands in the Baltic Sea on the basis of epiphyseal fusion data and metric studies. *International Journal of Osteoarchaeology* 12(1): 49–64.

Tuck, J.A., and R.T. Pastore. 1985. A nice place to visit, but…prehistoric extinctions on the island of Newfoundland. *Canadian Journal of Archaeology* 9(1): 69–79.

Wright, J. (ed.). 1898. *The English Dialect Dictionary*. London: Oxford University Press.

Yang, D.Y., A. Cannon, and S. Saunders. 2004. DNA species identification of archaeological salmon bone from the Pacific northwest coast of North America. *Journal of Archaeological Science* 31(5): 619–631.

# Chapter 12
# Aspects of Dorset Palaeoeskimo Mortuary Behaviour on the Northern Peninsula of Newfoundland

**Stuart C. Brown**

## Introduction

This chapter briefly summarizes salvage excavations at the Dorset Palaeoeskimo burial sites of Crow Head Cave (EeBi-4) and the Gargamelle Rockshelter (EeBi-21) on the Port au Choix Peninsula (Fig. 12.1) yielding evidence which modestly supplements the meagre database for Palaeoeskimo mortuary behaviour (Brown 1988). These data are analysed in conjunction with other Palaeoeskimo burials from the eastern Arctic to suggest a number of characteristics of Dorset Palaeoeskimo mortuary behaviour in general and Newfoundland Dorset burial practices in particular. These data indicate that the regional nature of the Newfoundland Dorset culture, first suggested by Harp (1964), may be further expressed in terms of mortuary behaviour.

Elmer Harp, Jr. began the first of several seasons of archaeological investigations in the Port au Choix area in 1949 (Harp 1951). Further seasons followed in 1950 and 1961–1963 during which Harp excavated extensively within the Phillip's Garden site (Fig. 12.1) and also located other Dorset occurrences within the present Port au Choix National Historic Site (Harp 1964). In addition, Harp was presented with skeletal material and associated artefacts from a multiple Dorset Palaeoeskimo burial in a rockshelter in Gargamelle Cove, which was published as Harp and Hughes (1968) (Fig. 12.1). This was an important publication inasmuch as it represented the largest collection of Dorset skeletal remains then known.

As the township of Port au Choix expanded, further accidental discoveries of prehistoric Amerindian burials were made. This drew the attention of James Tuck who conducted three seasons of excavation (1967–1969) at an extensive Maritime Archaic Indian burial ground within the township area (Tuck 1976). During the 1968 season, two local collectors, Eugene and Gordon Billard, discovered Crow Head Cave (Fig. 12.1) and, unfortunately, dug through the deposits, locating a number of Palaeoeskimo burials and associated grave goods. The skeletal material

S.C. Brown (✉)
Department of Archaeology, Memorial University, St. John's, NL, Canada, A1C 5S7
e-mail: sbrown@mun.ca

M.A.P. Renouf (ed.), *The Cultural Landscapes of Port au Choix*, Interdisciplinary Contributions to Archaeology, DOI 10.1007/978-1-4419-8324-4_12,

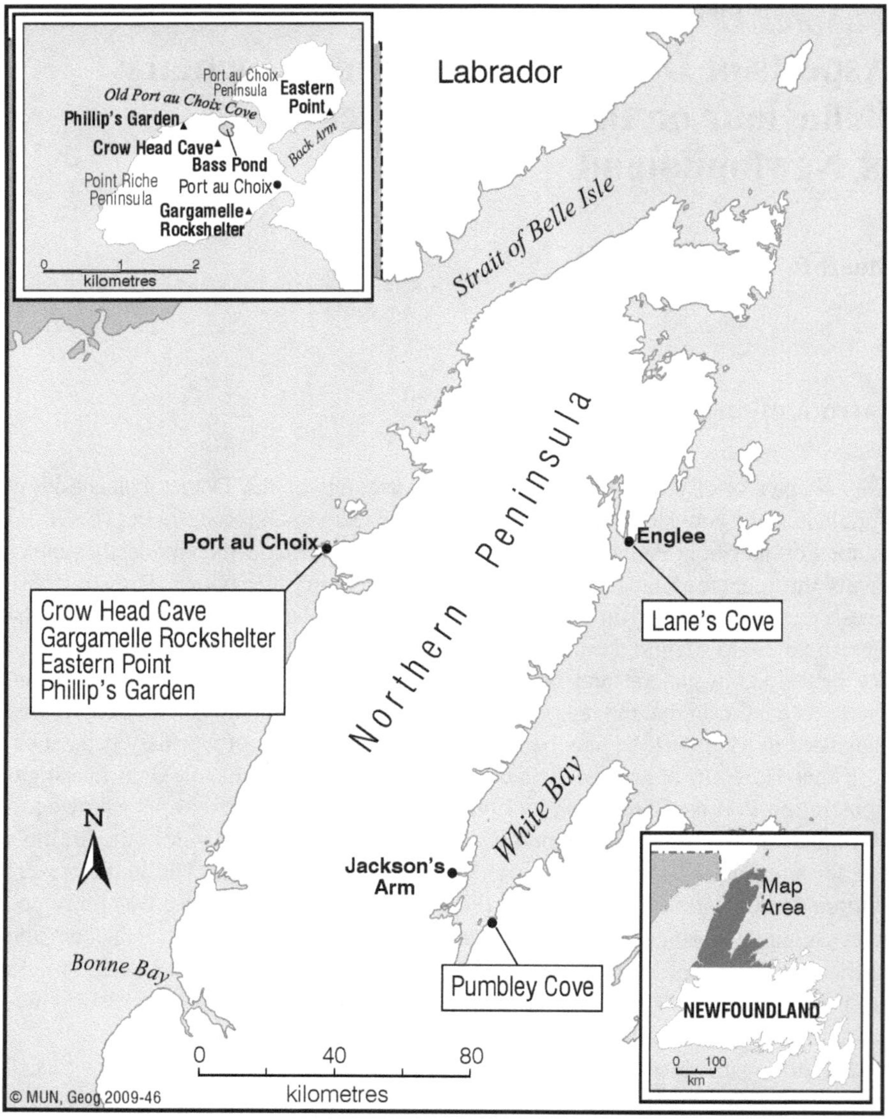


**Fig. 12.1** Map of sites and place names from the Northern Peninsula of Newfoundland, mentioned in the text

was donated to Memorial University and Tuck was permitted to inspect the artefacts which the Billards retained in their possession. It was not until 1971, however, that the Billards revealed quite how extensive their collection from Crow Head Cave was. At that time, Eugene Billard showed Tuck "a collection of similar material but much more impressive by virtue of its size, variety and excellent state of preservation" (Anderson and Tuck 1974:92). Fortunately, Tuck was at least permitted to photograph the Billards' collection (Fig. 12.2); its present whereabouts are unknown. In the interim, other individuals apparently excavated in the cave and further cultural material was evidently removed (Anderson and Tuck 1974:92).

**Fig. 12.2** Artefacts from the 1968 discovery of Crow Head Cave (Photo: J.A. Tuck)

In 1986 I joined the Port au Choix Archaeology Project to direct salvage excavations at Crow Head Cave and the Gargamelle Rockshelter (Fig. 12.1). As our present knowledge of Palaeoeskimo physical anthropology and mortuary behaviour is minimal (but see Koch et al. 1996; Lynnerup et al. 2003) a salvage excavation of even these drastically disturbed sites was thought to be of considerable potential significance. Despite the severity of the disturbance at both sites, a large collection of artefacts, ecofacts and fragmentary human remains was obtained.

## Crow Head Cave

Crow Head Cave is located 10 m below the highest point of the escarpment that bisects the Point Riche Peninsula at the juncture of the sheer rock face and a steep talus slope which is covered with dense boreal vegetation. Though the cave is only 35 m above sea level (asl), it provides an excellent vantage point looking out over Bass Pond, Old Port au Choix Cove and the sea beyond. The cave mouth is approximately 5 m wide with an original elevation of 1.8 m (Fig. 12.3a). The interior measures 6.3 by 3.9 m for a total area of <25 m$^2$. The stratigraphy comprises a deep basal sterile deposit of angular cryoclastic rubble overlain by a thin layer of water- and wind-borne dark brown soil. The cave was looted on more than one occasion and the entrance has been blasted with explosives.

Point provenience recording was conducted throughout the excavation of the cave but, in such a highly disturbed situation, it can be assumed that virtually nothing

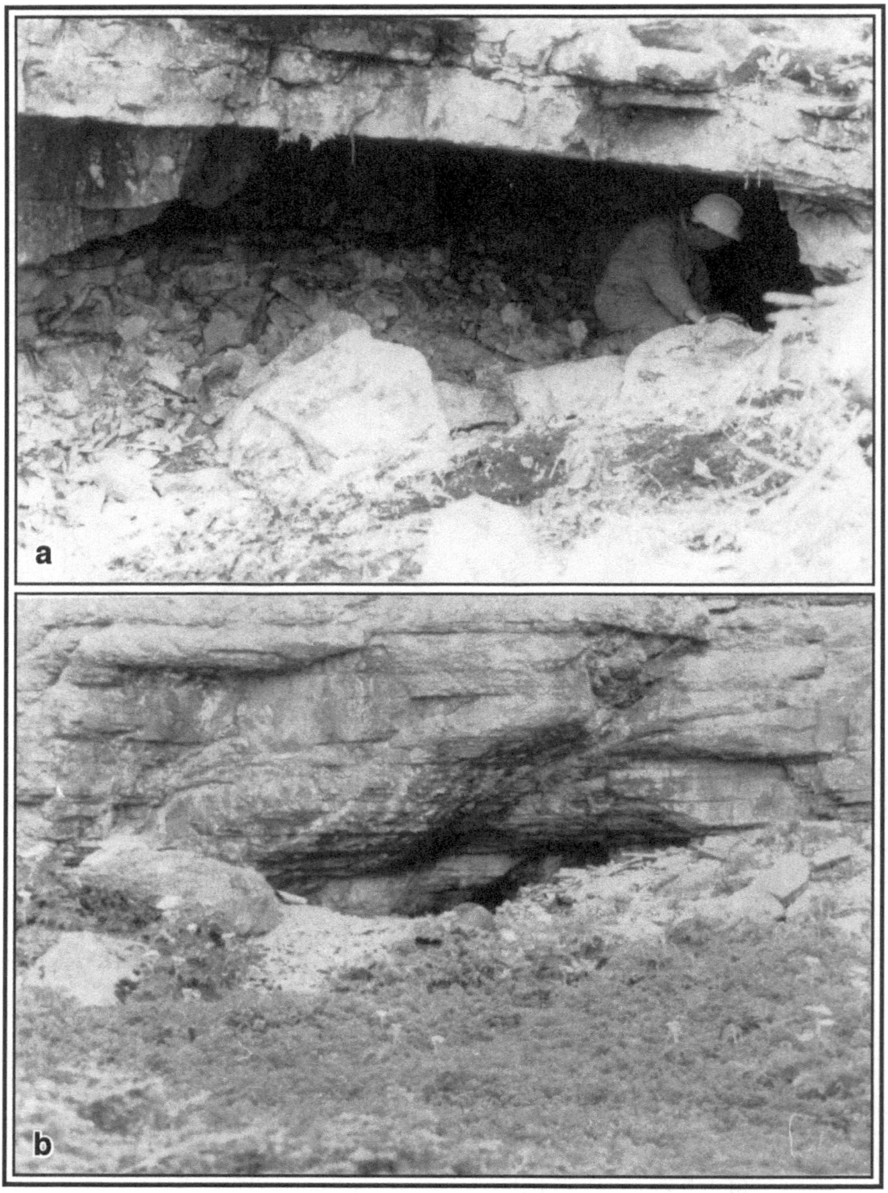


**Fig. 12.3** (**a**) Entrance of Crow Head Cave before excavation. (**b**) Gargamelle Rockshelter (Photos: S. Brown)

was in situ. Nevertheless, analysis of the gross spatial distribution of artefacts and skeletal fragments indicates that six disarticulated individuals, comprising one adult female, four juveniles and one infant, were buried in the cave in a number of temporally and spatially separate episodes (Brown 1988:59). With a few exceptions, the artefacts from the cave are typical of Middle Dorset. Over 90 artefacts were found (Brown 1988:183), including bifaces, tip-fluted endblades, chert and quartz microblades and nephrite burin-like tools. Bone artefacts included a barbed fish spear, harpoon heads, harpoon foreshafts, various points, burin support pieces, eyed needles and decorated non-functional items. A few pieces of carved ivory, including points and pendants, were also found. Figure 12.4a shows a range of artefacts found.

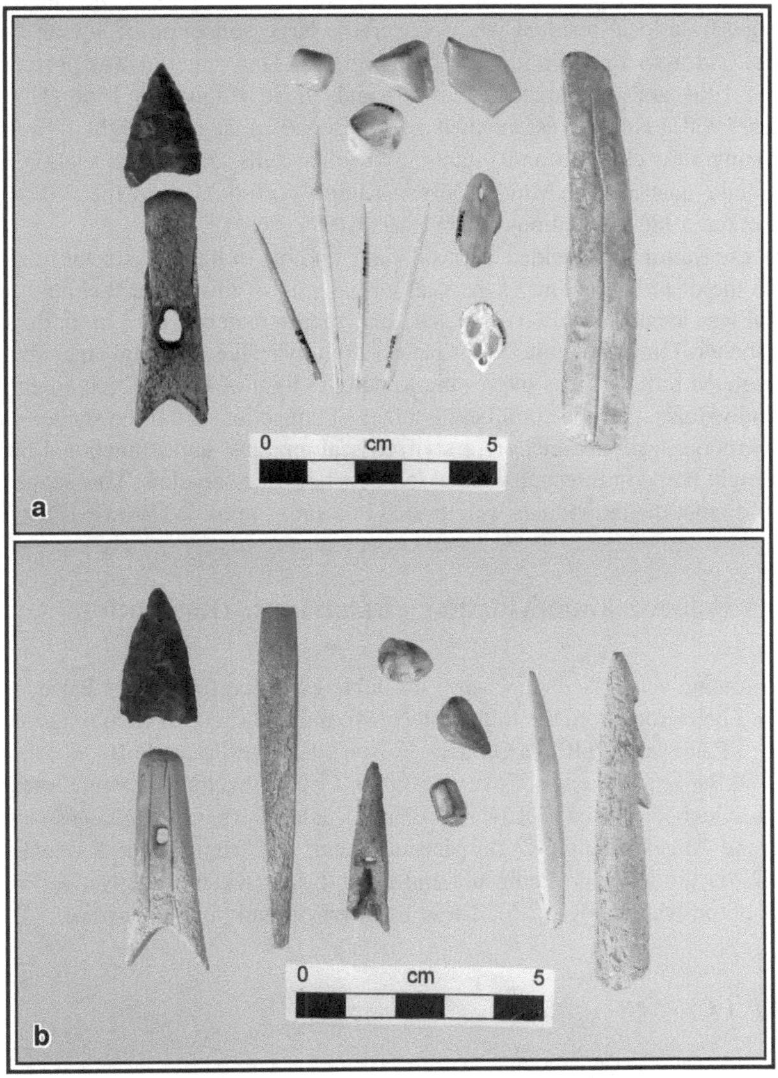


**Fig. 12.4** (**a**) Artefacts from Crow Head Cave: a harpoon head and endblade, four bone needles, four cut ivory pieces, an ivory pendant, a grooved bone piece and a ground periwinkle shell similar to those from the Maritime Archaic burial ground at Port au Choix. (**b**) Artefacts from Gargamelle Rockshelter: a harpoon head and endblade, a harpoon foreshaft, a Groswater-type harpoon head, three cut ivory pieces, an awl and a barbed bone point

## Gargamelle Rockshelter

Gargamelle Rockshelter is located on the northeast shore of Gargamelle Cove at the base of a 6 m limestone sea-cliff, at approximately 9 m asl. The width of the rock-shelter is a little over 7 m; maximum depth is 4 m and maximum height of the overhang a little less than 3 m (Fig. 12.3b). The first test trenches quickly established that the thin deposit was completely disturbed. In 1953 the shelter was excavated by a local resident who recovered a large collection of human skeletal material and associated lithic and bone artefacts. This material was presented to Harp in 1961 and published by himself and David Hughes in 1968 (Harp and Hughes 1968). Hughes established the presence of at least eight individuals, comprising four children and infants, and four adults. Harp's correlation of the Gargamelle material with Middle Dorset is amply corroborated by the >90 artefacts found in our 1986 excavations (Brown 1988:199) (Fig. 12.4b).

The excavation also yielded a considerable number of human skeletal fragments. Despite the disturbed nature of the site, virtually all of this artefactual and skeletal material was located within a restricted area, approximately 1 by 2 m, in the centre of the shelter. This very limited distribution strongly suggests a single mass burial of disarticulated remains. It is interesting to note, in light of Hughes' identification of eight individuals, that the combined artefactual collections from the shelter yielded eight harpoon head amulets, eight stylized seal amulets, eight functional harpoon heads, eight harpoon foreshafts and at least six bear head amulets. This suggests the possibility that the individuals were buried in a single episode (Brown 1988:79).

## Dorset Palaeoeskimo Mortuary Behaviour: Description

The following sections review other available evidence for Dorset Palaeoeskimo mortuary behaviour from the following Newfoundland sites, all from or close to the Northern Peninsula: Phillip's Garden (EeBi-1), Eastern Point (EeBi-10), Pumbley Cove (DkBe-4) and Lane's Cove (EeBa-2) (Fig. 12.1); the following sites from Ungava, Quebec: Imaha (JaEj-1) on Pamiok Island, Tyara (KbFk-7) on Sugluk Island and Angekok (JlGu-2) on Mansel Island; and from Alarnerk (NhHd-1) at Igloolik, on the Melville Peninsula and the T-1 site (KkHh-1) at Native Point on Southampton Island (Fig. 12.5). These finds vary greatly in their nature.

### *Phillip's Garden*

Harp discovered an infant skeleton buried in an upright foetal position in a stone-capped pit, approximately 0.6 m diameter and 0.6 m deep, in the centre of House 12 at Phillip's Garden. The infant was accompanied by a wide range of grave goods and was in close association with an adult human mandible (Harp and Hughes 1968:17–18).

**Fig. 12.5** Map of sites and place names from the eastern Arctic mentioned in the text

## *Eastern Point*

The burials at Eastern Point, referred to as Back Arm by Howley (1915:328–330), were discovered by a local resident in a rock crevice in 1904 and comprised two mandibles and one maxilla along with a fairly large collection of artefacts. The skeletal remains and associated artefacts were seen, described and drawn by Howley (1915:Plate XXIV), later purchased by Alfred Kidder in 1910 and donated to the Peabody Museum, Harvard University, where they were studied by de Laguna and also by Wintemberg (1939:86, note 112).

## *Pumbley Cove*

In 1959 two residents of Jackson's Arm discovered skeletal fragments in a small crevice in a cliff in Pumbley Cove near the head of White Bay on the north coast of Newfoundland. These finds, later donated to Memorial University, included an adult male cranium with a mandible, several long bone fragments and a few artefacts (Anderson and Tuck 1974:93, Plate 4m–p).

## *Lane's Cove*

In the late 1960s an Englee resident discovered some human remains and a number of associated items in a nearly vertical rock crevice in limestone cliffs in Lane's Cove on the eastern side of the Northern Peninsula. The skeleton proved to be that of a child aged about 8 years and was substantially complete. The only associated artefact was a polished slate projectile point (Anderson and Tuck 1974:91).

## *Imaha Site*

In 1957, while investigating three house rings at the Imaha site, W.E. Taylor, Jr., discovered a stone cist burial on the low flank of a rock outcrop just to the southeast of one of the house rings. The cist is described as a large, roughly oval, feature constructed from massive schistose slabs and granite boulders. The interior, a narrow rectangle, was approximately 1.9 m in length, 0.5–0.7 m in width, with a maximum height of 0.7 m. The orientation of the long axis was northeast-to-southwest. Inside were the skull, mandible and various post-cranial elements of an adult male in a very disarticulated state, possibly the result of small mammal activity. The find is interpreted as a single extended burial on the ground surface with a stone vault built around the body. The artefacts from the cist and from the house rings were relatively undiagnostic and the individual pieces were uncharacteristically large for a Palaeoeskimo tool industry. Taylor described them as "suggestive" of Dorset culture, a cultural attribution which is only provisionally accepted here (Laughlin and Taylor 1960:5, Plate 3a–d).[1]

## *Tyara Site*

In 1958 Taylor discovered a human mandible and three rib fragments in a culture layer of black humus in the Early Dorset Tyara site (Taylor 1968:46). The mandible

[1] Editor's note: since the writing of this paper, others have argued that the Imaha skeletal material is Thule rather than Dorset, based on radiocarbon dates, blood group and stable isotope chemistry (Hayes et al. 2005).

was initially identified as a young adult female (Oschinsky 1960:214) but later reclassified as a young adult male (Oschinsky 1964:Plates 4–5). There was no evidence of any burial feature such as a pit, cist, mound or cairn and Taylor interpreted the find as a corpse "exposed on that old surface and ravaged by animals and near-British weather" (Taylor 1968:79). Within the same level and in close proximity, Taylor also located three small carved ivory "swimming bear" figurines and a fragmented ivory sheath containing two of the carvings (Taylor 1968:79, 120, Plate 24f–i). Given the proximity of these objects to the skeletal fragments, the rarity of "art pieces" in Dorset sites in this region and the fact that there was only one culture layer at the site, it is plausible to interpret these objects as grave goods.

## *Angekok*

Also in 1958, Taylor recovered the mandible and various post-cranial elements of a young adult male from midden deposits in the Late Dorset Angekok site. The presence of ten maxillary teeth signals that the cranium was initially in association with the mandible but subsequently separated by some human or natural factor. In the absence of a detailed excavation report it is impossible to assess the significance of the absence of the cranium. If this was a burial, then the absence must presumably have been a deliberate human act. However, if this was an exposed corpse, then the absence could be explained by a number of taphonomic factors. Evidently no artefacts could be associated with the skeletal remains (Maxwell 1985:158; Oschinsky 1960:212).

## *Alarnerk*

Jorgen Meldgaard excavated 205 Dorset houses and 18 graves at the Alarnerk site at Igoolik (Meldgaard 1955, 1960a, b). The few skeletal remains included one complete and two fragmentary mandibles, a cranial fragment and a long bone; all but the child's mandible and long bone were found in midden deposits (Lynnerup et al. 2003:353; Meldgaard 1960a:598). Summary description of these graves divides them into three types (Lynnerup et al. 2003:350; Maxwell 1985:158, 241; Meldgaard 1960a:589). The most recent burials were found in the settlement site at the highest elevation, 21–22 m asl, and are described as chamber-like stone structures. These were rectangular cists of stone slabs, in some cases 1–1.4 m long, paved with flat stones and covered with large boulders. Most had been disturbed and had been re-used for meat storage. No human bones remained so it is uncertain whether the burials had been articulated or not. An earlier and more common type, at 15–19 m asl, was a simple rectangular or circular pit approximately 1 m in diameter and dug about 0.5 m into gravel with a few stones around the perimeter. There were no human bones in this case as well; however, since the pits are too small for a human body, the burials must have been disarticulated. Artefacts, including both broken and unfinished specimens, were strewn in these graves without apparent order. Associated non-functional items included miniature harpoon heads and finely carved pieces. The third and possibly earliest type of inhumation was a single undisturbed burial in a 0.8 m deep pit containing a child's mandible, a single adult long bone and a scatter of grave goods. The pit was covered with a 0.5 m high mound of gravel.

### *T-1 Site*

Excavations at the Early Dorset T-1 site on Southampton Island in 1954 by Collins failed to locate any graves that could be assigned to this phase (Collins 1956a, b, 1957). Those graves that were investigated in or near the occupation areas all proved to be of comparatively recent Sadlermiut origin. However, five human crania were recovered from the surface of the site and should probably be assigned to the Early Dorset phase. Their advanced state of weathering differed markedly from exposed Sadlermiut remains and was comparable to weathering on Dorset bone artefacts and faunal bone on the surface of the site. It could be argued therefore that in the Early Dorset phase burial did not involve the construction of cairns, cists or any other form of conspicuous grave marker. The absence of burials or of human bone, apart from the five isolated crania, in the Early Dorset deposits is not necessarily significant since Collins excavated less than 1% of the 12 ha over which this site is discontinuously spread.

## Dorset Palaeoeskimo Mortuary Behaviour: Analysis

The previous section demonstrates that there was a good deal of variation in Dorset mortuary behaviour. However, the limitations of the archaeological database cannot be overemphasized – it is exceedingly small and thinly spread over a vast expanse of time and space and the difficulties of analysis are compounded by serendipitous finds by amateurs, limited reporting by professionals and uncertain phase attributions. Any inferences from this database must therefore be made with great caution.

### *Primary and Secondary Burial*

One apparent regularity in this catalogue of Dorset Palaeoeskimo burials is the disarticulated nature of most of the skeletal material. The only apparent exceptions to this pattern are the infant burial from Phillip's Garden and the child burial at Lane's Cove. The adult burial at the Imaha site may be another exception if small mammal activity is invoked as an explanation for the scattered and incomplete

**Table 12.1** Summary of primary and secondary Dorset Palaeoeskimo burials

| Site | Skeletal description | Articulated/disarticulated |
|---|---|---|
| Phillip's Garden | Complete infant skeleton | Articulated |
| | Adult mandible | Disarticulated |
| Crow Head Cave | 6+ individuals; many missing skeletal elements, esp. mandibles | Disarticulated |
| Gargamelle Rockshelter | 8–9 individuals; many missing skeletal elements, esp. mandibles | All disarticulated? |
| Eastern Point | 2 adult mandibles | Disarticulated |
| Pumbley Cove | Adult cranium | Disarticulated |
| | Long bones | Disarticulated |
| Lane's Cove | Complete child skeleton | Articulated |
| Imaha | Scattered skeleton; elements missing | Disarticulated? |
| Tyara | Adult mandible | Disarticulated |
| | Rib fragments | Disarticulated |
| Angekok | Adult mandible | Disarticulated |
| | Some post-cranial elements | Disarticulated |
| T-1 | 5 adult crania | Disarticulated |
| Alarnerk | 2 mandibles and 1 cranial element in separate midden contexts | Disarticulated |
| | Infant mandible and adult longbone in gravel mound pit | Disarticulated |

? = indicates that there is some doubt about whether the disarticulation was primary

skeleton. In all other cases, there is either clear or strongly presumptive evidence for the prior exposure of the corpse and the secondary burial of selected or remaining disarticulated skeletal elements (Table 12.1).

## *Differential Occurrence of Skeletal Elements*

In the absence of detailed bone catalogues from almost all of the sites, it is impossible to explore the differential occurrence of skeletal elements. However, it is evident that in many cases there seems to be an inverse relationship between the occurrence of crania and mandibles. Mandibles occur without crania at Phillip's Garden, Back Arm, Tyara, Angekok, and in both the gravel mound burial and the stone-circle pit burials at Alarnerk. Crania occur without mandibles at Pumbley Cove, Lane's Cove (maxillary fragment) and the T-1 site. At Alarnerk, the three mandibles and cranial fragment occurred separately (Lynnerup et al. 2003:353). The majority of mandibles are missing at Crow Head Cave and Gargamelle Rockshelter, given the number of individuals identified at both sites. In short, it appears that there are factors arising from either Dorset mortuary behaviour or taphonomic processes, or a combination of both, that frequently resulted in mandibles and crania occurring in burial contexts in isolation from each other. The potential implications of this phenomenon are discussed below.

## *Diachronic Variation*

The diachronic distribution of the database is too uneven to permit any firm observations about shifts in Dorset mortuary behaviour through time. The Early Dorset phase is represented at only three sites, Tyara, T-1 and Alarnerk and the Late Dorset phase is represented at only two sites, Angekok and Alarnerk. The remaining occurrences can be firmly or tentatively attributed to the Middle Dorset phase. It could be provisionally argued that a progression in complexity is evident from this distribution (Table 12.2), from simple exposure with no further treatment to exposure followed by burial in more secure loci such as caves, crevices, cairns and cists.

## *Regional Variation*

When the database is viewed from a regional perspective, disposal of the dead in the Dorset culture seems to fall into two broad behavioural categories. With the single exception of the house pit burial at Port au Choix, all known examples of intra-settlement burials come from the Arctic, while all known examples of burials isolated from settlements come from Newfoundland. Moreover, of the six known burial loci in Newfoundland, five are in caves, crevices or rockshelters.

There are a number of aspects that must be considered in seeking an explanation of this apparent regularity. First, there is the problem of research objectives and the manner in which these data have been acquired. All five Newfoundland examples

**Table 12.2** Diachronic variation in burial methods

| Phase | Disposal methods | Site |
|---|---|---|
| Early Dorset | Simple exposure; no definite evidence of further treatment | |
| | Midden deposit | Tyara |
| | Midden deposits | Alarnerk |
| | Surface finds | T-1 |
| Middle Dorset | Exposure; later treatment | |
| | Crevice burial | Back Arm |
| | Crevice burial | Pumbley Cove |
| | Crevice burial | Lane's Cove |
| | Cave burial | Crow Head |
| | Rockshelter | Gargamelle |
| | Slab-covered pit | Phillip's Garden |
| Late Dorset | Exposure; no definite evidence of further treatment | |
| | Midden deposit | Angekok |
| | Midden deposit | Alarnerk |
| | Gravel mound pit | Alarnerk |
| | Exposure; later treatment | |
| | Gravel mound pit | Alarnerk |
| | Chamber burials | Alarnerk |

of cave or rock crevice burial were located initially by inquisitive local residents and not through professional investigation. In the Arctic, where modern population is very much smaller and more dispersed, it can be expected that fewer discoveries would be made in this manner. Moreover, Palaeoeskimo archaeology has largely concentrated on the investigation of occupation sites detected from surface scatters of cultural material. Burial loci have been investigated either through accidental discovery in the course of such excavations or because the burial has had some obvious above-ground marker. The systematic investigation of caves, crevices and rockshelters is almost never mentioned in excavation reports and it is therefore possible that such a focus in future research might produce a more balanced picture. However, Robert McGhee investigated several caves on the Grinnell Peninsula, Devon Island, located close to Dorset occupation sites and found no traces of burials (Anderson and Tuck 1974:94). It is possible, therefore, that the selection of caves, crevices and rockshelters for burial purposes is a phenomenon restricted to Newfoundland, another aspect of the "regional nature" of the Dorset culture in its most southerly extension (Harp 1964). However, given the widespread Middle Dorset occupation throughout Newfoundland, it is striking that all of these burials come from the Northern Peninsula. The curiosity of the modern population for investigating similar loci in other parts of Newfoundland has been rewarded so far only with Beothuk finds. Robbins (1986) has already drawn attention to sub-regional phenomena in the Dorset occupation of Newfoundland (see also LeBlanc 2000). It is possible, therefore, that this mode of burial may eventually prove to be specific to the northwest of Newfoundland and not characteristic of the remainder of Newfoundland, or of more northerly areas of the Subarctic and Arctic.

## *Burial Orientation*

Orientation of the skull of the deceased to face west or north is ethnographically documented among a number of Inuit groups (Petersen 1966/67:261). However, no consistent pattern of burial orientation emerges from the Dorset burials described above. At the Imaha site, the long axis of the cist was oriented in a northeast-to-southwest direction. At Gargamelle, the linear arrangement of the bones suggested to the discoverers a northwest-to-southeast orientation (Harp and Hughes 1968). At all other sites, burial orientation is either not reported or is likely to have been determined primarily by the nature and specific circumstances of the burial.

## *Grave Goods*

In all cases, with the possible exceptions of the Imaha and Tyara burials where the association between the skeleton and nearby artefacts is open to doubt, Dorset burials are provided with grave goods. In incidence, grave goods may range from few to many. In nature, they range from the purely functional (harpoon heads, endblades, foreshafts, bone needles, an oval steatite lamp, snow goggles, etc.) to the symbolic-functional (models of harpoon heads, etc.) to the purely symbolic (swimming-bear, seal, walrus amulets, etc.). Functional categories represented include both subsistence activities (marine and terrestrial hunting equipment) and maintenance activities (needles, skin scrapers, steatite cooking-pots, etc.), some of which were undoubtedly gender specific in Dorset culture.

Special attention should perhaps be drawn to the small oval steatite lamp in the Crow Head assemblage (Fig. 12.2). The lamp is most obviously understood to be in some way related to mortuary activities. However, it cannot be simply assumed that its function was strictly as a grave good; among the Alaskan Eskimo of Nunivak Island, south Alaska, it was the custom to keep a lamp burning constantly from before the time of death until after the burial and to place the lamp on top of the grave (Curtis, cited in Lantis 1947:17). This raises the possibility that the Crow Head lamp functioned in the context of mourning rather than that of grave goods.

Functional artefacts in Dorset burials also include both whole and broken specimens. Examples of broken artefacts are too numerous to list here individually but they occur in the assemblages from Gargamelle Rockshelter, Crow Head Cave, Eastern Point, Pumbley Cove and the Alarnerk pit burials. Broken artefacts include a wide range of lithic tools (endblades, bifaces, knives, scrapers, burin-like tools and whetstones) and bone and ivory pieces. There are three possible hypotheses to explain this: (1) these artefacts were broken in normal use and subsequently included in the burial; (2) they were purposefully broken as part of a mortuary ritual; or (3) they were broken by taphonomic processes after having been included in the burial. In some cases, the explanation is clear. For example, the needles from Crow Head Cave and Gargamelle Rockshelter are mostly broken at the eye-hole (Brown 1988:129, 155), a breakage pattern that clearly results from use rather than deliberate destruction or taphonomic processes. In other cases, like the infant burial at Phillip's Garden, the grave goods were not broken (Harp and Hughes 1968:18).

Miniature copies of artefacts in Dorset burials in Newfoundland are so far definitely known only from Gargamelle Rockshelter where a number of miniature harpoon heads occur. Two specimens are completely detailed Newfoundland Closed types and may have been fully functional implements (Harp and Hughes 1968:Plate 4k) while the others are described by Harp as "facsimiles" of harpoon heads that may have been used as amulets or buttons since none of the specimens has a blade slot or a proper socket (Harp and Hughes 1968:Plate 7a–h). It cannot be determined at present whether these particular specimens were amulets or buttons worn during the lifetime of the deceased individual as Harp (1970) suggests, or whether they were copies made exclusively as grave goods.

Miniature harpoon heads of similar type were also possibly present in the Eastern Point burial (cf. Howley 1915:Plate XXIV:18). In addition, Meldgaard reports miniature non-functional harpoon heads from the stone-ringed pit burials at Alarnerk (Maxwell 1985:158). Finally, the 1968 collection from Crow Head Cave included a miniature bone sled runner (Fig. 12.2) with a reconstructed width of about 9 mm. The practice of placing miniature copies of functional items in graves is ethnohistorically documented among many Inuit groups (Petersen 1966/67:265). In such cases, the miniatures are models of items that had belonged to the deceased but had been distributed to other members of the group before the death of the owner. Such a practice among the Dorset might account for the presence of the miniature sled runner among the Crow Head collection and it is possible that part of an item could sometimes stand for the whole.

Skilfully carved and incised symbolic items present in Dorset graves include artefacts, variously described as amulets or pendants, which depict a variety of animals, such as walrus, seal, bear and caribou, in a realistic or highly stylized manner. The most parsimonious interpretation of the presence of these artefacts is that they are the normal accoutrements of living individuals and have no necessary relationship to mortuary practices. Their presence or absence seems neither related to age or sex patterning, nor to regional or diachronic factors. Such carved symbolic items appear in the Igloolik sequence in the Middle Dorset phase (Meldgaard 1955:175), comparable items are known from the Early Dorset phase at Tyara and they are well known from the Middle Dorset phase in Newfoundland.

It impossible at this point to determine whether there are any patterned relationships between the nature of functional grave goods in any particular burial and the age, sex and status of the deceased. The adult male mandible in Phillip's Garden was accompanied by a steatite cooking-pot (Harp and Hughes 1968:18), an item that might be expected to correlate more with female burials. Furthermore, although a number of infant and child burials are known, none apparently included items clearly related to non-adult status. There is no obvious relationship between the age of the deceased and the grave goods associated with the child burial at Alarnerk (Lynnerup et al. 2003), the infant burial at Phillip's Garden (Harp and Hughes 1968) and the child burial at Lane's Cove. Similarly, infants and children were buried at Crow Head and Gargamelle and there are no artefacts in either assemblage that can be unequivocally related to a non-adult status. This suggests that if grave goods were envisioned as symbolic of items the deceased might need in a future existence, then the deceased in that future existence was understood to be an adult, no matter what their age at death.

Food offerings do not seem to be part of Dorset mortuary behaviour. Some faunal remains were discovered with the Lane's Cove burial and substantial faunal remains were found at Crow Head Cave; however, since the rock crevice and cave may have been used by carnivores, no relationship can be posited between the faunal material and the burials. Some animal bones were scattered amongst the hearth stones found in the Alarnerk mound burial (Lynnerup et al. 2003:352). It remains possible, however, that food offerings associated with Dorset burials were of an entirely perishable nature, or were deposited at the graveside rather than in the grave; the latter alternative is a pattern which was widespread among recent Inuit (Petersen 1966/67:263).

## Palaeoeskimo Strategies for Disposing of the Dead

In the foregoing section, various aspects of the archaeological evidence for Dorset mortuary behaviour were inductively analysed for evidence of the presence or absence of patterning. This section takes the deductive approach to interpret observed variation and regularities in Dorset burial behaviour. In particular, this section deductively analyses the disposal of the dead in Arctic and Subarctic environments as an objective problem which could be solved by a limited number of strategies. Explanatory models derived from the ethnographic literature assist to plausibly reconstruct behaviour and cultural significance and, where possible, to suggest methods by which these hypotheses might be tested in future research. However, it should be emphasized that the limited and frequently ambiguous database does not permit rigorous testing.

### *Basic Strategies for the Disposal of the Dead*

Six basic strategies can be employed for the disposal of the dead in Arctic and Subarctic environments. These are: (1) cremation, (2) burial in water or abandonment on the ice without subsequent retrieval, (3) abandonment on land, (4) primary subsurface burial, (5) primary surface burial in cairns, cists, or caves, and (6) exposure on the ice or land and subsequent secondary burial. It is important to emphasize that these options could have been differentially employed according to a variety of factors ranging from season of the year to the status, age or sex of the deceased.

#### Cremation

In most areas within the Inuit range, the lack of sufficient quantities of combustible materials has precluded cremation as a viable option. In fact, cremation as a regular practice is documented only among the Siberian Eskimo and the Chukchi who had access to forest resources or considerable quantities of driftwood; it occurred rarely among the Alaskan Eskimo and only under special circumstances (Petersen 1966/67:261). In the southern part of their range in southern Labrador and Newfoundland, the Dorset also had access to forest resources more than adequate for cremation and the lack of any evidence for this mode of disposal of the dead probably signifies that it was not an option ideologically sanctioned by the traditions of a people whose origins lay in treeless Arctic regions.

#### Burial in Water or Abandonment on the Ice

Exposure of corpses on the ice is a potential strategy that is ethnohistorically known to have been a pan-Arctic trait (Petersen 1966/67:259). However, the corpse was usually later retrieved for burial. Abandonment of a corpse on the ice with no subsequent treatment was practiced by only a few groups such as the Copper and Netsilik Inuit (Jenness 1928:174; Rasmussen 1931:263), and a survey of the ethnohistoric and archaeological evidence for Inuit burial practices cites no examples of burial in water (Petersen 1966/67). Nevertheless, however rare or unknown these practices may have been among recent Arctic groups, they remain potential strategies for prehistoric groups. However, they are strategies that leave no archaeological trace. Employment of such strategies of disposal of the dead in prehistoric circumstances may be strongly inferred through the direct historical approach or loosely inferred by the absence of prehistoric human remains in all other contexts. Thus the use of such strategies in the past *may* account for relative scarcity of Dorset human remains, but there is apparently no method for directly testing such a hypothesis.

## Abandonment on Land

Another objective strategy for disposal of the dead is abandonment on land. A number of alternatives present themselves in this overall strategy: (1) the corpse could be abandoned in a room of a house which was then blocked off with occupation continuing in the rest of the structure, (2) the corpse could be left in the dwelling and the site abandoned, or (3) the corpse could be abandoned at a distance from a settlement with no special treatment and no marker of the place where it was deposited.

The first strategy is known only among the Eskimos on Kodiak Island and the Aleutians (Lantis 1947:11). It would hardly have been an option for the Dorset Palaeoeskimos since all known Dorset house forms were simple rectangular or circular structures lacking side rooms.

The second strategy is documented among the Copper Inuit who, during summer at least, would leave dead individuals in their summer tent and abandon the site (Jenness 1928:174). Such a method of disposal of the dead would theoretically be detectable by the following criteria: (1) finding human remains on a former occupation site, (2) the remains would be located on the living floor within a house pit or, where detectable, a tent ring, and (3) there is a high probability that as a result of carnivore activity the remains would be disarticulated and would display characteristic patterns of carnivore destruction and modification (Binford 1981:35). The only examples of Dorset burials that might correspond to this pattern are the finds of isolated human mandibles and post-cranial fragments in midden deposits at Tyara, Angekok and Alarnerk.

The third strategy is ethnographically documented among the Igloolik Inuit who in winter would take the shrouded corpse some distance from the settlement and simply heap snow over it. In summer, the "burial" was sometimes marked by piling stones over the body in an irregular fashion. In both cases, however, carnivores usually dug out the corpse and it was dismembered and eaten (Mathiassen 1928:229). This practice, particularly the winter variant, would obviously be extraordinarily difficult to detect archaeologically and, not surprisingly, none of the known Dorset burials fit this pattern.

## Primary Sub-Surface Burial

Immediate burial of a corpse in the earth in Subarctic and Arctic areas poses difficulties depending on the area and season. Even in the most southerly part of the Dorset culture range, such an option would be possible only after ground thaw in the early summer, while in permafrost regions it would be impossible in winter and a very slow and laborious alternative in summer. The only circumstances in which these observations would not hold would be if the burial was located in a recently occupied house pit or tent ring where the warmth of human occupation had been sufficient to cause at least shallow ground thaw. Such burials, if they were sufficiently deep to avoid being excavated by carnivores, would be fully articulated. However, unless they were located in occupation features or had some surface marker, it is unlikely that they would be located by archaeological research. The articulated infant burial in House 12 at Phillip's Garden falls into the category of a sub-surface primary burial (Harp and Hughes 1968:17). Its location within a house pit is apparently anomalous since, apart from the adult male mandible in the same house pit, no further burials were found by Harp who excavated seven house pits in full and nine in part, or by Renouf who has excavated an additional three house pits in full (Harp 1964; Renouf 2003). The anomalous location of this burial is perhaps to be explained as a winter burial although extensive testing of the site in areas outside house pits as well as within house pits would be required to establish whether or not this burial is anomalous in its specific location. None of the other known Dorset burials correspond to the pattern of primary articulated sub-surface burials.

## Primary Surface Burial in Cairns, Cists, or Caves

Another option is that of primary surface burial of the corpse in a cairn or cist. There is, however, no convincing evidence that the Dorset employed such a burial strategy. The chamber burials at Alarnerk were likely for disarticulated and therefore secondary burial. There is a longer stone cist at the Imaha site, but the cultural attribution of the burial to the Dorset period is not compelling and it is not certain that the skeleton was articulated.

A variant form of this general strategy could involve the simple exposure of a corpse in a cave. In such a case, the archaeological detection of the burial would have a much higher degree of probability and burials of this type, sometimes consisting of the stacked bodies of several individuals, are known from the Thule period (Hansen et al. 1991; Meldgaard 1955). Where there had been subsequent carnivore activity, it might be extremely difficult to distinguish this form of burial from that of a secondary disarticulated burial in a cave. It is possible, therefore, that some or all of the burials in Crow Head Cave and the Gargamelle Rockshelter could be explained as primary inhumations which were subsequently disturbed by carnivore activity. However, this possibility is unlikely, as the archaeological context of the human bone fragments found in our excavation of Crow Head Cave suggests that already disarticulated human remains were deposited, possibly in shallow pits, and covered with stone slabs and rubble (Brown 1988:59).

## Exposure on the Ice or Land with Subsequent Burial

Exposure of the corpse on land or on the ice with the subsequent collection and burial of residual skeletal material could manifest itself in a wide variety of burial forms. Since it is likely that a corpse exposed in the Arctic and Subarctic would quickly be dismembered by carnivores, the main diagnostic indicators in most, if not all, cases would be: (1) disarticulation of the remains, (2) evidence of carnivore destruction and modification on residual bone elements, (3) absence of bone elements most susceptible to carnivore activity, and (4) evidence of post-exposure treatment of the human remains. These few and rather inconclusive indicators leave room for much variation in the location and specific form of the final burial. The latter might be located in, near or distant from the settlement and the specific mode of interment could involve anything from a simple pit to a cairn, cist, cave, crevice or rockshelter. Moreover, some bone elements might receive special treatment and be separated from the rest of the residual bone. Exposure of the dead followed by subsequent treatment of the residual bone material is copiously documented in the ethnographic literature on recent Inuit and Eskimo groups and from archaeological data on prehistoric Eskimo groups in Alaska (Petersen 1966/67).

Most of the known Dorset burials apparently fit this general pattern. One observed regularity is that, apart from the infant burial at Phillip's Garden and the child burial at Lane's Cove, and possibly the Imaha burial, all Dorset burials were apparently disarticulated and had therefore undergone an initial period of exposure during which it is likely that the corpses were subjected to a degree of carnivore activity.

Another observed regularity in Dorset burials to which attention has been drawn above is that crania and mandibles seldom occur together. This dichotomy can be explained either as the result of taphonomic processes, or as a result of some form of mortuary behaviour. The presence of mandibles in isolation is perhaps the easiest aspect to explain by taphonomic processes; mandibles are dense, resistant elements and therefore might be the only major residual bone element to survive after a long period of exposure. It could also be suggested that mandibles frequently occur in isolation from crania because the latter are the focus of some form of mortuary behaviour that would result in their absence from burials. A practice of this nature has been documented among the Alaskan Eskimo of Nunivak Island where, a few years after the burial of a body, the skull (probably the cranium which presumably would be detached from the mandible at this stage) was removed from the grave and placed on a nearby high point facing east (Curtis, cited in Lantis 1947:18; parenthesis mine). Some such analogous behaviour might account for the presence of isolated mandibles in Dorset sites, in particular for the specimen at Angekok where the crania was in close association with the mandible long enough for the maxillary teeth to fall out. The removal of crania from burials to be placed on the surface might also account for the presence of the five crania on the surface of the T-1 site on Southampton Island. In this context of differential treatment of crania, it is interesting to recall the phenomenon of the *tupilak* or "harmful being," which among the Greenland Inuit assumes the form of a repulsively ugly manmade effigy which was created in order to stalk and kill an enemy and eat their entrails

(Petersen 1964:73). These effigies could be composed of a wide variety of materials including a mixture of bones from various animal species whose separate identities the *tupilak* could assume at will. Instances are known of a child's cranium packed in hide and brought to life through sorcery as well as an adult human skull matched with a seal mandible (Hansen et al. 1991:62, Figure 44). The manmade *tupilak* effigy is largely a Greenland phenomenon but it is apparently related to a much wider *tupilak* mythology of a vengeful spirit among the Igloolik, Caribou and Central Inuit and possibly more remotely to the "windigo/witiko" cannibalistic spirit known among the Cree, Ojibwa and Naskapi (Petersen 1964:78).

Somewhat harder to explain is the absence of mandibles in burials in which crania or cranial fragments are present. This can hardly be explained by some diagenetic attritional process in the post-burial environment which selected against the densest element in the human skeleton but allowed other elements to survive. Once again, two possible explanations present themselves. Either the mandible was removed by carnivores during the period of exposure and was not available for burial, or mandibles were culturally treated in a special manner which may sometimes have involved separate burial. There is, however, only one instance, the House 12 burial at Phillip's Garden, in which one can be certain that a human mandible was buried without any other associated skeletal element. Although only two mandibles were reported from the crevice burial at Eastern Point, the site was not professionally excavated and less recognizable skeletal elements could easily have been missed. In all other cases where mandibles occur separately from crania, they are accompanied by at least a few post-cranial elements and this militates against the hypothesis that mandibles were the focus of some special cultural treatment and supports the hypothesis of carnivore scavenging.

However, a fascinating pattern emerges from the analysis of phocid elements in three midden deposits associated with the Groswater Palaeoeskimo occupations from Phillip's Garden West and Phillip's Garden East (Wells 2002, Chap. 4). When these phocid assemblages are statistically investigated using percentage minimal animal unit (%MAU) analysis and by scatterplots of MAU frequencies plotted against bone mineral density values, it becomes clear that seal crania and mandibles were differentially treated, sometimes to quite a dramatic extent. It is suggested that this differential treatment is a direct result of ritual activities (Renouf 2000; Wells 2002:153). There are, in fact, many ethnographically and archaeologically documented examples in which Arctic peoples accorded special treatment to seal crania (see Murray 2000:58; Renouf 2000:65; and Wells 2002:151 for citations).

## Propitiation of the Dead

A common characteristic of ethnohistorically documented mortuary behaviour amongst Arctic groups is the proliferation of rules and taboos designed to placate the deceased and to protect the survivors from contamination or ghosts (Petersen 1966/67). Most of this behaviour has little, if any, effect on the patterning of material culture and would be extremely difficult to document in prehistoric situations.

However, it has been suggested that some of this behaviour does apparently have material correlates. For example, Petersen (1966/67:260) suggests that, while stone cairns and cists were used to protect a corpse against carnivores, stone rings which offered no actual protection for the corpse were a symbolic protection for the living against the dead. However, among the Netsilik, once a person's soul has left their body, their corpse was considered to be "a frame in the actual sense of the word, a case that is no longer of any use to anybody, and therefore nothing is done to protect it against beasts of prey or roving dogs" (Rasmussen 1931:263). While the Netsilik considered the human remains to be of no particular significance, the stones placed about a corpse were called *ilo'fra*, literally "that in which something has been" and they were considered to mark a "kind of holy spot" to which the soul of the dead person occasionally returned (Rasmussen 1931:263).

Stone rings, cairns and cists were all variant forms of Dorset burials along with slab-covered burials and interments in cave and crevices. If further investigation shows that most Dorset burials were of residual disarticulated skeletal elements, then it follows that Petersen's inference that stone cairns and cists were for the protection of the corpse against carnivores can hardly be the correct interpretation of those features in this prehistoric situation. At a minimum level of inference, such constructed features may be understood as simple grave markers. It is also possible that interment in stone-ringed pits, slab-covered pits, cairns, cists, caves and crevices had more elaborate significance as a means of protection for the living and of emphasizing the separation between the living and the dead. Irrespective of the actual explanation, the presence of grave constructions which had no protective function for the corpse would therefore seem to imply a continuing relationship between the survivors and the deceased.

## Conclusions

The limitations of the archaeological database require that conclusions about Dorset Palaeoeskimo mortuary behaviour be framed only in the most tentative manner. Nevertheless, after carefully reviewing the available evidence, some plausible reconstructions can be suggested in summary form.

1. Disposal of the dead normally involved exposure leading to disarticulation of the remains, probably as a result of carnivore activity.
2. Ethnographic comparisons suggest that the practice of exposure was possibly sanctioned by the belief that the devouring of the corpse would lead to the quick liberation of the soul.
3. Residual disarticulated remains were subsequently collected and buried. It is possible that the process of disarticulation may have been completed using a cutting tool.
4. The form of subsequent burial varied widely but there is some evidence of regional patterning in that most known Dorset burials in northwestern Newfoundland were in caves, crevices or rockshelters. Other regions lack consistent patterning but this may merely reflect the limited database.

5. There is no strong evidence for consistent orientation of burials.
6. The inclusion of grave goods was a normal part of mortuary behaviour.
7. Grave goods normally included functional items related to both subsistence and maintenance activities as well as items which probably had symbolic religious value.
8. Miniature models of functional types also occur but their significance is obscure.
9. Both whole and broken artefacts were included in burials but it cannot be determined at present whether mortuary behaviour included the intentional destruction of artefacts.
10. In sub-adult burials, the grave goods are adult in nature.
11. There is no strong evidence for the inclusion of food items in burials.
12. The differential occurrence of crania and mandibles in burials suggests the possibility that crania may sometimes have been accorded special treatment.
13. Apart from the addition of chamber burials in the Late Dorset phase at Alarnerk, there is no strong evidence for diachronic variation in Dorset mortuary behaviour.
14. Primary sub-surface burial is evidenced by only one example, the infant burial at Phillip's Garden and may therefore have been an anomalous strategy used only for infants.
15. With the possible exception of the Imaha burial whose phase attribution is uncertain, there is no evidence for primary surface burial in cists.
16. There is no evidence for the practice of cremation.
17. Methods of disposal of the dead such as abandonment on the ice or burial in water may have been practiced but this cannot be tested since they leave no archaeological evidence.
18. If most Dorset burials prove to be secondary and disarticulated, burial constructions such as stone rings, cairns and cists cannot be interpreted as protection for the corpse. Even at the most minimal level of inference, their construction implies a continuing relationship between the survivors and the deceased which may have taken the form of visitations to the grave and further mortuary behaviour to memorialize the deceased.
19. It is possible that the Dorset understood the dead to be a potential source of harm and that this apprehension is manifested in the use of stone rings, cairns, cists and rock crevices as a symbolic prophylaxis against harm.

In the words of Stephen J. Gould (1988:16) " …we must have frameworks to discipline our thoughts." If nothing else, it is hoped that these reconstructions will prove to be of some heuristic value in stimulating further discussion of the topic and in promoting future research on Palaeoeskimo mortuary behaviour.

My conclusions can be summed up by a brief anecdote. In the summer of 1987, Dr. Renouf and I had the great pleasure of visiting the Harps at Dartmouth College in Hanover, New Hampshire. During our stay, we were frequently entertained by Elmer's stories about Arctic fieldwork in general and Port au Choix in particular. On occasions, Elaine would register an amused objection to hearing any given story "yet again," accompanied by a threat to retaliate with some sensational tale of her own. One evening at a dinner in the Hanover Inn with the Harps at either end of a table filled with anthropological faculty from the College, Elmer launched into another story. At one point in his discourse, Elaine interrupted with some evident surprise, "Elmer, I haven't heard that story before!" And from the depths of a marriage that had then flourished over half a century, he replied graciously, "I don't give it all out at once, babe!" Neither, it seems, do the Palaeoeskimo.

**Acknowledgements** I would like to thank Charles Conway for drafting Figs. 12.1 and 12.5, James Tuck for kind permission to use his photograph in Fig. 12.2, and Chris Hammond, Memorial University Photographic Services, for photographing the artefacts in Fig. 12.4. I would also like to thank the Port au Choix Archaeology Project for inviting me to join the 1986 field season.

## References

Anderson, J.E., and J.A. Tuck. 1974. Osteology of the Dorset people. *Man in the Northeast* 8: 89–97.

Binford, L.R. 1981. *Bones: Ancient men and modern myths*. New York: Academic Press.

Brown, S.C. 1988. Archaeological Investigations at Crow Head Cave and the Gargamelle Rockshelter in the Port au Choix National Historic Park, Newfoundland: Report of 1986 Field Activities. On file, Parks Canada, Archaeology, Atlantic Region, Halifax.

Collins, H.B. 1956a. Archaeological investigations on Southampton and Coats Islands, N.W.T. In *Annual report of the National Museum of Canada for 1954-55*, 82–113. National Museums of Canada Bulletin 142, Ottawa.

Collins, H.B. 1956b. The T-1 site at Native Point, Southampton Island, N.W.T. *Anthropological Papers of the University of Alaska* 4: 63–89.

Collins, H.B. 1957. Archaeological investigations on Southampton and Walrus Islands, N.W.T. In *Annual report of the National Museum of Canada for 1955-56*, 22–61. National Museums of Canada Bulletin 147, Ottawa.

Gould, S.J. 1988. A novel notion of Neanderthal. *Natural History* 6: 16–21.

Hansen, J.P.H., J. Ledgaard, and J. Nordqvist (eds.). 1991. *The Greenland mummies*. Washington, DC.: Smithsonian Institution Press.

Harp, E. 1951. An archaeological survey in the Strait of Belle Isle area. *American Antiquity* 16(3): 203–220.

Harp, E. 1964. *The cultural affinities of the Newfoundland Dorset Eskimo*. National Museums of Canada Bulletin 200, Ottawa.

Harp, E. 1970. Late Dorset Eskimo art from Newfoundland. *Folk* 12: 109–124.

Harp, E., and D.R. Hughes. 1968. Five prehistoric burials from Port au Choix. *Polar Notes* 8: 1–47.

Hayes, M.G., J.B. Coltrain, and D.H. O'Rourke. 2005. Molecular archaeology of the Dorset, Thule, and Sadlermiut. In *Contributions to the study of the Dorset Palaeo-Eskimos*, ed. P.D. Sutherland, 10–32. Mercury Series Archaeology Paper 167, Canadian Museum of Civilization, Gatineau.

Howley, J.P. 1915. *The Beothucks or Red Indians: The aboriginal inhabitants of Newfoundland*. Cambridge: Cambridge University Press.

Jenness, D. 1928. Ethnological problems of Arctic America. In *Problems of polar research: A series of papers by thirty-one authors*, 167–175. American Geographical Society Special Publication, No. 7, American Geographical Society, New York.

Koch, A., B. Frøhlich, N. Lynnerup, and J.P. Hart Hansen. 1996. The bones from Qeqertasussuk: The earliest human remains from Greenland. In *The Paleo-Eskimo cultures of Greenland: New perspectives in Greenlandic archaeology*, ed. B. Grønnow and J. Pind, 35–37. Danish Polar Center Publications No. 1, The Danish Polar Center, Copenhagen.

Lantis, M. 1947. *Alaskan Eskimo ceremonialism*. Seattle: University of Washington.

Laughlin, W.S., and W.E. Taylor, Jr. 1960. A Cape Dorset culture site on the west coast of Ungava Bay. In *Contributions to Anthropology, 1958*, 1–28. National Museums of Canada Bulletin 167, Ottawa.

LeBlanc, S. 2000. Middle Dorset (1900-1100 BP) regional variability on the island of Newfoundland and in Saint-Pierre et Miquelon. In *Identities and cultural contacts in the Arctic*, ed. M. Appelt, J. Berglund and H.C. Gulløv, 97–105. Copenhagen: The Danish National Museum and Danish Polar Center.

Lynnerup, N., J. Meldgaard, J. Jakobsen, M. Appelt, A. Koch, and B. Frølich. 2003. Human Dorset remains from Igloolik, Canada. *Arctic* 56(4): 349–358.

Mathiassen, T. 1928. *Material culture of the Iglulik Eskimos*. Report of the Fifth Thule Expedition 1921-24, Vol. 6(1). Gyldeddalske Boghandel, Nordisk Forlag, Copenhagen.

Maxwell, M.S. 1985. *Prehistory of the eastern Arctic*. New York: Academic Press.

Meldgaard, J. 1955. Dorset kulturen: Den Dansk-Amerikanske ekspedition til Arktisk Canada 1954. *KUML* 1955: 158–177.

Meldgaard, J. 1960a. Prehistoric sequences in the eastern Arctic as elucidated by stratified sites at Igloolik. In *Men and cultures: Selected papers of the fifth international congress of anthropological and ethnological sciences*, ed. A.F.C. Wallace, 588–595. Philadelphia: University of Pennsylvania Press.

Meldgaard, J. 1960b. Origin and evolution of Eskimo culture in the eastern Arctic. *Canadian Geographical Journal* 60: 64–75.

Murray, M.S. 2000. A zooarchaeological approach to Arctic prehistory. In *Animal bones, human societies*, ed. P. Rowley-Conwy, 58–64. Oxford: Oxbow Books.

Oschinsky, L. 1960. Two recently discovered human mandibles from Cape Dorset Sites on Sugluk and Mansel Islands. *Anthropologica* 2: 212–227.

Oschinsky, L. 1964. *The most ancient Eskimos*. Canadian Research Centre for Anthropology, Ottawa.

Petersen, R. 1964. The Greenland tupilak. *Folk* 6(2): 73–101.

Petersen, R. 1966/67. Burial-forms and death cult among the Eskimos. *Folk* 8–9: 259–280.

Rasmussen, K. 1931. *The Netsilik Eskimos: Social life and spiritual culture*. Report of the fifth Thule Expedition 1921-1924, Vol. 8(1–2). Glydendalske Boghandel, Nordisk Forlag, Copenhagen.

Renouf, M.A.P. 2000. Symbolism and subsistence: Seals and caribou at Port au Choix, northwestern Newfoundland. In *Animal bones, human societies*, ed. P. Rowley-Conwy, 65–72. Oxford: Oxbow Books.

Renouf, M.A.P. 2003. A review of Palaeoeskimo dwelling structures in Newfoundland and Labrador. *Études/Inuit/Studies* 27(1–2): 375–416.

Robbins, D.T. 1986. "Newfoundland Dorset" culture? In *Palaeo-Eskimo cultures in Newfoundland, Labrador and Ungava*, 119–124. Reports in Archaeology No. 1, Memorial University of Newfoundland, St. John's.

Taylor, W.E., Jr. 1968. *The Arnapik and Tyara sites: An archaeological study of Dorset culture origins*. Memoirs of the Society for American Archaeology 22, University of Utah Printing Service, Salt Lake City.

Tuck, J.A. 1976. *Ancient people of Port au Choix*. Newfoundland Social and Economic Studies 17, Institute of Social and Economic Research, Memorial University of Newfoundland, St. John's.

Wells, P.J. 2002. An Analysis of Faunal Remains from Two Groswater Palaeoeskimo Sites at Port au Choix, Northwestern Newfoundland: Phillip's Garden West (EeBi-11) and Phillip's Garden East (EeBi-1). Master's thesis, Department of Anthropology, Memorial University of Newfoundland, St. John's.

Wintemberg, W.J. 1939. Eskimo sites of the Dorset culture in Newfoundland, Part I. *American Antiquity* 5(1): 83–102.

# Chapter 13
# In the Woods: The Cow Head Complex Occupation of the Gould Site, Port au Choix

**M.A.P. Renouf, Michael A. Teal, and Trevor Bell**

## Introduction

This chapter presents Recent Indian Cow Head data from the Gould site (EeBi-42), Port au Choix. The Cow Head complex (hereafter abbreviated to Cow Head) appears to be centred on the Northern Peninsula of Newfoundland (Fig. 13.1), and its current characterization is based on published material from two Northern Peninsula sites, Spearbank (DlBk-1) (Hartery 2007; Tuck 1978) and Peat Garden (EgBf-6) (Hartery 2007). Material from the recently excavated St. Pauls Bay-2 site (DlBk-6), also on the Northern Peninsula, supports this characterization (Lavers 2009). In addition, Cow Head bifaces have been identified at L'Anse aux Meadows (EjAv-1) (Ingstad 1977) and in two surface collections at Portland Creek Pond (EbBj-4, EbBj-5) (Biggin 1985; Thomson 1987). On the Quebec Lower North Shore, sites comparable to Cow Head (EiBg-85, EiBg-86) are found at the Blanc Sablon River (Pintal 1998). Elsewhere in Newfoundland, individual Cow Head bifaces are identified in Recent Indian contexts at a small number of Bonavista Bay sites (Austin 1980, 1984; Carignan 1975, 1977), and Cow Head projectile points are identified in private collections from three sites near Burgeo (Rast 1998, 1999) (Fig. 13.1).

Spearbank, Peat Garden and St. Pauls Bay-2 are all lithic workshop sites. They are situated at the shore where large amounts of lithic debitage, cores and preforms were associated with multiple cobble hearths. By contrast, the Gould site (EeBi-42) is set back 350 m from the current shore in what we call the coastal margin, by which we mean the wooded area near but not at the shore. There are two Cow Head components at the Gould site. The older component comprises two cooking pits associated with few artefacts, and the younger component centres on a dwelling with abundant and diverse cultural material including >300 ceramic sherds, the only such assemblage found in Newfoundland (Teal 2001). In this chapter, we argue that the Gould site was a residential base where 1–2 families camped together

M.A.P. Renouf (✉)
Department of Archaeology, Memorial University, St. John's, NL, Canada, A1C 5S7
e-mail: mapr@mun.ca

M.A.P. Renouf (ed.), *The Cultural Landscapes of Port au Choix*, Interdisciplinary Contributions to Archaeology, DOI 10.1007/978-1-4419-8324-4_13,

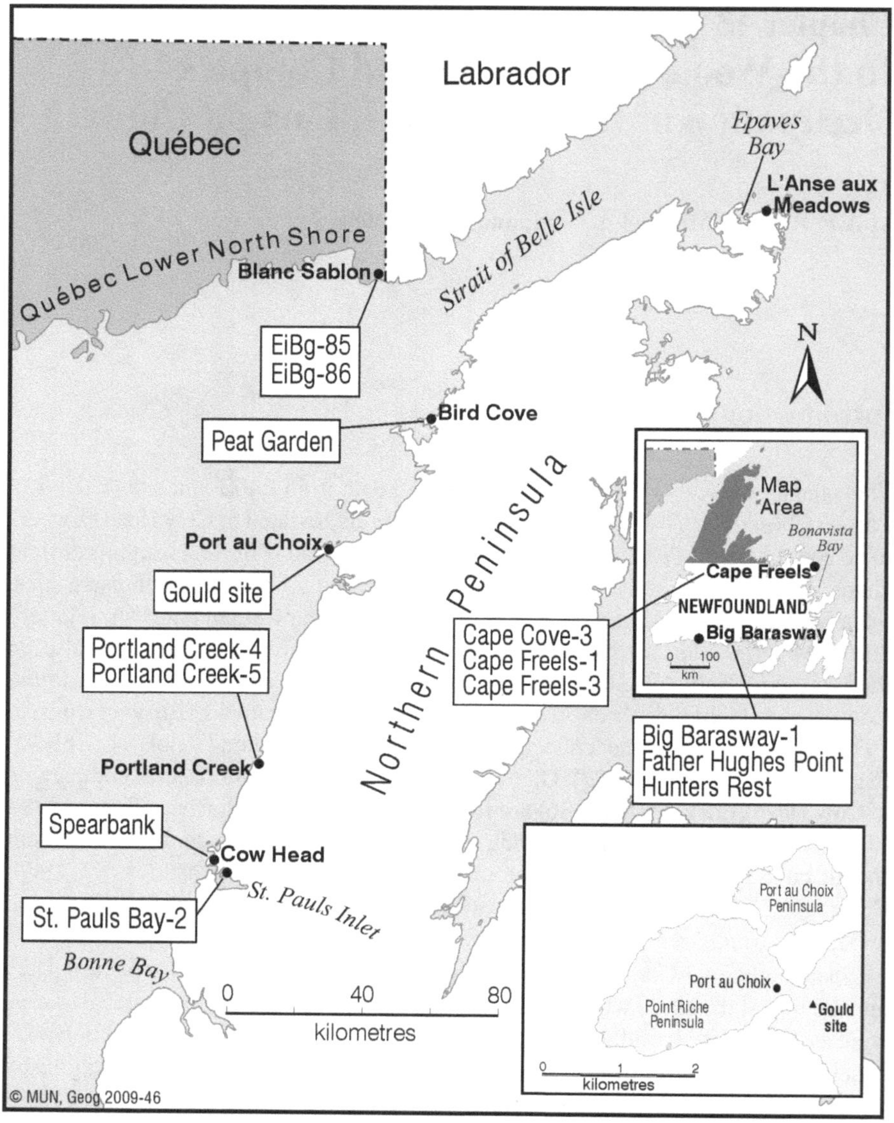


**Fig. 13.1** Sites and places mentioned in the text

to undertake a wide range of subsistence and domestic activities. The archaeological signature from this site broadens the characterization of Cow Head, which up until now has been based almost entirely on material from lithic workshops.

We identify the wooded coastal margin as a preferred setting for Cow Head residential bases, in contrast to lithic workshops which are situated at the coast where outcrops and cobbles are easily accessible. If we wish to find other Cow Head residential bases, we need to re-focus our site surveys away from the shoreline to the more heavily vegetated coastal margin. This focus on the coastal margin adds a new geographic dimension to existing models of Recent Indian site location in Newfoundland.

## Context

### *The Recent Indian Period in Newfoundland*

The Recent Indian period in Newfoundland is divided into three chronological complexes: Cow Head (2110-930 cal BP[1]), Beaches (1900-800 cal BP) and Little Passage (1170-300 cal BP). These chronological ranges are based on currently available radiocarbon dates summarized in Holly (2002), Hull (2002) and Hartery (2001, 2007). Archaeological data demonstrate that Little Passage people were the ancestors of the Beothuks, the indigenous people of Newfoundland at the time of European contact (Pastore 1992). It is generally accepted that this ancestry can be pushed back as far as the Beaches complex (Pastore 1992; Schwarz 1984). The relationship between Cow Head and Beaches is unclear at this time; there is no evidence to suggest a link. Holly (2002), Hull (2002) and Hartery (2007) suggest that overlapping dates between the two complexes indicate contemporaneity rather than sequential occupation as once thought (Tuck 1988).

### *Cow Head Sites on the Northern Peninsula*

The most recent and comprehensive characterization of Cow Head complex material culture is found in Hartery (2001, 2007) based on re-examination of material from the Spearbank site (Tuck 1978) and her own excavations at Peat Garden (Hartery and Rast 2001). Hartery (2007:12–14) describes three forms of Cow Head projectile points which we condense into two (Fig. 13.2): a short, broad-bladed variety with a straight or expanding-stem and a narrow-bladed contracting-stemmed form with a straight, rounded or pointed base. Other characteristic material includes small, bipointed bifaces; broad ovate or lanceolate unstemmed bifaces with a rounded or straight base; narrow lanceolate unstemmed bifaces with a straight base; large flake side scrapers; and *pièces-esquillées*, small core-like pieces that are battered at both ends.

The Spearbank site is on the sheltered inner shore of the Cow Head peninsula in the community of that name (Fig. 13.1). The Cow Head peninsula is the source of

[1]Cal BP=calendar years before present based on calibrations using Calib 6.0html (Stuiver and Reimer 1993) and represented by the one sigma probability range.

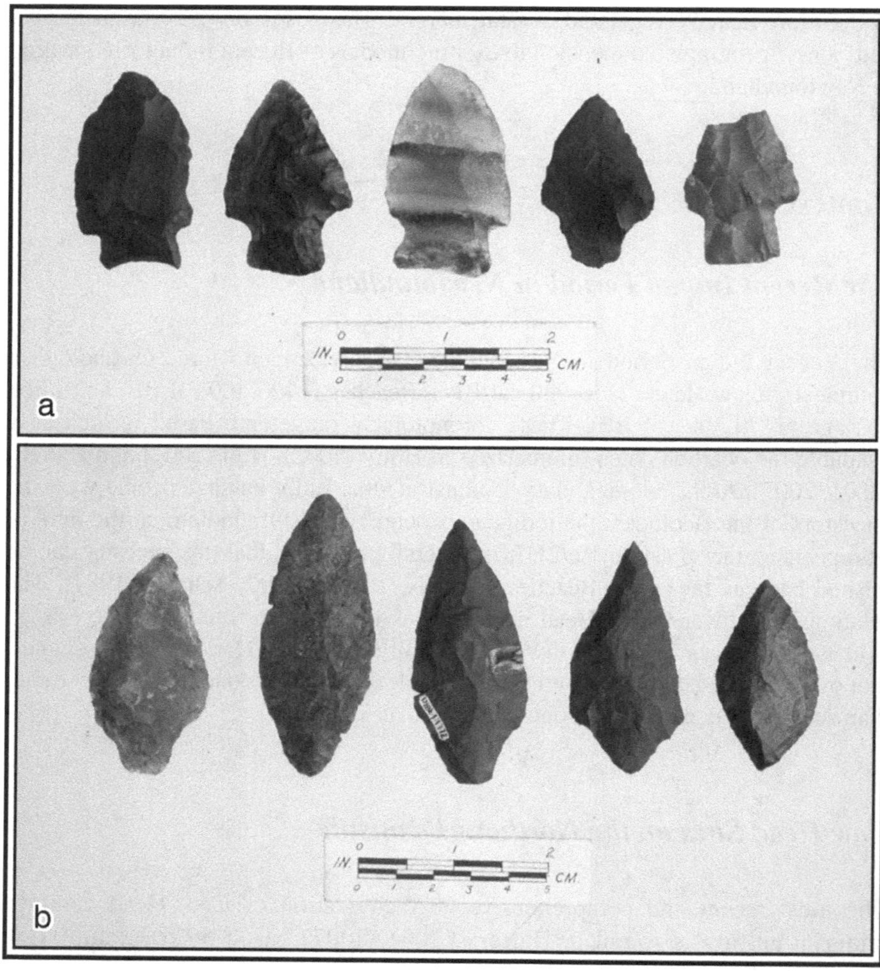


**Fig. 13.2** Projectile points from the Spearbank site. (**a**) Broad-bladed stemmed points. (**b**) Narrow-bladed stemmed points

fine-grained Cow Head chert, which occurs as outcrops and beach cobbles and which was a favoured raw material for Amerindian and Palaeoeskimo groups. Spearbank is a stratified multi-component site with well-separated cultural strata and is only steps away from the cobble-strewn beach. Five Cow Head strata in two site areas date from 1995 ± 90 BP[2] (DAL-275) to 1600 ± 95 BP (S-1953); there is also a single younger date of 995 ± 85 BP (DAL-324) (Hartery 2007:65–66), which might pertain to a corner-notched Little Passage point from the same site area.

[2] BP = radiocarbon years before present.

Tuck (1988) and Hartery (2007) conclude that Spearbank is a lithic quarry/workshop site based on the large amount of raw material, cores and debitage, together with few finished artefacts, associated with 22 unformalized cobble hearths. Hartery (2007:11) estimates that of the 1,482 artefacts that she studied from the site, which excludes the large and as yet unstudied debitage collection, 42% are cores, core fragments, chert cobbles and hammerstones. Raw material was predominantly fine-grained chert visually similar to Cow Head cobbles available on the nearby beach. There is no direct evidence of site seasonality although Hartery (2007:32) notes that access to raw material would be easier during the snow-free period of the year.

Peat Garden is on the sheltered inner shore of the Dog Peninsula near the community of Bird Cove (Fig. 13.1). Cow Head material overlies a Groswater Palaeoeskimo component. The Cow Head component is identified on the basis of characteristic ovate and lanceolate bifaces and flake scrapers in addition to one complete and three fragments of Cow Head projectile points. An unusual long, narrow, straight-based biface was found, which expanded the Cow Head lithic inventory (Hartery 2007:26). A series of 11 dates range almost continuously from 1795±45 BP (BGS 2170) to 1153±40 BP (BGS 2174) (Hartery 2007:65–66).

Hartery interprets Peat Garden as a lithic workshop where some domestic activities also took place. There were ten hearths, most of which were shallow bowl-like depressions associated with fire-cracked rock, charcoal and abundant debitage. Excavation recovered 37,806 flakes from the site (Hartery 2007:25), and a >70% sample of the total flake assemblage showed that all stages of tool manufacture were represented (Hartery and Rast 2001:8). Of the total of 551 artefacts there were 15 preforms and 81 cores and core fragments (Hartery 2007:28). The raw material was a coarse-grained, pitted, white-to-grey chert of unknown source. The quantity of workshop debitage suggests that it cannot be far away.

Domestic activities are reflected in two clay-lined hearths interpreted as small cooking pits based on size, shape, some highly fragmented bone, a few artefacts and the absence of lithic workshop debris (Hartery 2007:29). Bird, seal and mammal were identified among the few hundred burned, fragmented bones (Hartery 2007:29). Based on site location and resource availability Hartery (2007:32) suggests that Peat Garden was likely occupied during spring-summer.

St. Pauls Bay-2 is on the sheltered inner shore of St. Pauls Inlet. It is very close to outcrops and beach cobbles of Cow Head chert (Coniglio 1987). The site was first reported by Penney (1989), and more recent excavations (Lavers 2009) exposed four Recent Indian cobble hearths associated with large amounts of debitage, cores and preforms; beneath this was a Groswater Palaeoeskimo component. There was a low, 3 m by 2 m, gravel platform which Lavers (2010) interprets as a possible dwelling floor, based on its construction and that it was entirely free of debris. Four Recent Indian dates range from 1390±70 BP (Beta 211320) to 1100±70 BP (Beta 252627). A Cow Head complex component is identified on the basis of broad, ovate bifaces and two short, broad-bladed, stemmed projectile points. Lavers (2009) collected over 50,000 flakes from 20 m$^2$ along with at least 211 cores or core fragments, 170 preforms and 11 hammerstones. Lavers (2008, personal communication) notes that most flakes were primary and secondary. Raw material is fine-grained chert visually similar to Cow Head outcrops at the nearby beach. Lavers (2009) interprets St. Pauls Bay-2 as a quarry-workshop with some domestic activities. There is no basis to establish season of occupation beyond observing that chert outcrops and cobbles were likely more accessible when there was no cover of snow or ice.

L'Anse aux Meadows is in a small exposed cove that faces west into Epaves Bay at the tip of the Northern Peninsula (Fig. 13.1). Although the site is best known for its Norse occupation (Ingstad 1977; Wallace 1991), there are Palaeoeskimo and Amerindian components as well (Eldjárn 1977; Kristensen et al. 2009; Kristsensen and Renouf 2009; Petré 1977; Wallace 1989, 2003, 2006). These occupations are on two terraces through which Black Duck Brook winds its way to the sea. Wallace groups the Recent Indian material into two components, late Prehistoric A and late Prehistoric B. Tuck (1988:159) and Hartery (2007:40, 49) identify the former as Cow Head on the basis of a dozen or so large, ovate bifaces; lithic material is visually similar to Cow Head chert. Two cooking pits are attributed to the Cow Head complex. Both were large, deep and lined with a thick layer of charcoal and fire-cracked rock; some fire-cracked rocks were spread out on either side where they had been tossed after the pit was opened (Eldjárn 1977:95; Petré 1977:64). The pits are dated to 1170±90 BP (T-368) and 1140±90 BP (T-365). A date of 1165±65 BP (S-1352) comes from a third activity area with a small number of flakes and bifaces. The absence of lithic workshop material distinguishes the Cow Head component of L'Anse aux Meadows from Spearbank, Peat Garden and St. Pauls Bay-2. The narrow range of artefacts and presence of cooking pits suggest that it might have been a seasonal camp focused on as yet undetermined resources (see Kristensen 2010).

Portland Creek-4 and Portland Creek-5 are situated near the mouth of two different salmon streams that run out of Portland Creek Pond to the ocean a short distance away. In the mid 1980s, these sites were eroding out of a sand bank (Biggin 1985; Thomson 1987). Both sites are undated and were identified as Cow Head based on a small number of broad, ovate bifaces. Cultural material included fire-cracked rock, charcoal and some flakes; lithic material is visually similar to Cow Head chert. Site locations at the source of salmon fishing streams suggest that these sites were seasonal fishing camps.

## *Other Sites with Cow Head and Similar Material*

Cultural material comparable to Cow Head is found at Blanc Sablon on the Quebec Lower North Shore (Fig. 13.1), attributable to the Flèche Littorale (2500–1500 BP) and Petit Havre (1500–1200 BP) complexes as defined by Pintal (1998). In particular, broad-bladed, straight- and contracting-stemmed points were found at two sites, EiBg-85 and EiBg-86, both at the mouth of the Blanc Sablon River. Two ceramic sherds were found at EiBg-85; this site dates to 1320±80 BP (Beta 28703) (Pintal 1998:207). At both sites flakes and a small number of artefacts occurred around cobble hearths on sand platforms. Some burned bones were found, most of which were seal but which also includes small amounts of bird, beaver, otter and other small game (Pintal 1998, 2003). Pintal (2003) interprets EiBg-85 and EiBg-86 as small residential camps of 1–2 families where the focus was mainly seal hunting.

Some Cow Head material is found at three sites on the prominent headland of northern Bonavista Bay (Fig. 13.1) known as Cape Freels. These sites are comprised of an extensive palimpsest of cobble hearths. Austin (1980, 1984) identifies two Cow Head bifaces from a lithic workshop activity area at Cape Cove-3 (DhAi-7) in what is otherwise a Beaches and Little Passage site. Hartery (2007:35) identifies similar Cow Head bifaces in the artefact collections from nearby Cape Freels-1 (DhAi-1) and Cape Freels-3 (DhAi-3) (Carignan 1977). The location of these sites on a headland well known as an important area for netting harp seals as they swim close to shore on their southward migration (Sanger 1977) suggests these were seal hunting camps.

Rast (1998, 1999) identifies a small number of possible Cow Head projectile points in private artefact collections from three sites near Burgeo (Fig. 13.1): Big Barasway-1 (CjBk-1), Father Hughes Point (CjBk-8) and Hunters Rest (CjBk-10). All sites are within a large, sheltered barasway, on low elevations and two of the three are near streams. Based on details of site location Rast et al. (2004) suggest that these were summer sites.

## The Gould Site

The Gould site is situated in what was until recently fir and spruce woods on a broad, level terrace at 8–10 m asl. The site is adjacent to a small brook that runs to the shoreline approximately 350 m away. At the time of Recent Indian occupation, sea level was similar to present (Bell et al. 2005).

The site has four temporal and two cultural components. The earlier two components are Maritime Archaic Indian (Renouf and Bell, Chap. 3). The two younger components are Recent Indian, the earlier dating from 2080±40 BP to 1870±60 BP and the later from 1520±60 BP to 1480±70 BP (Table 13.1). Based on projectile point form, the younger of the two components has been identified as Cow Head (Teal 2001). There are no diagnostic artefacts from the older component, but its chronology suggests that it too is Cow Head. Site stratigraphy consists of up to 2 m of peat over basal limestone with a black clay-like contact zone between them. The Recent Indian components are entirely within the peat.

The Recent Indian area of the site was approximately 500 $m^2$. There were three activity areas, two older and one younger (Fig. 13.3). A single Recent Indian stratum was characterized by the presence of charcoal and/or fire-cracked rocks. Although in places the charcoal layer was as much as 2 cm thick, more often it formed a layer that was paper-thin. Impressions of tree roots in the peat at the same level or just below the Recent Indian stratum indicated that the location had been wooded at the time of occupation (Renouf and Bell 1999).

**Table 13.1** Charcoal-based radiocarbon dates from the Recent Indian components of the Gould site

| Context | C14 years BP | Lab No. | Calibrated median BP | Cal BP range 1σ | Cal BP range 2σ |
|---|---|---|---|---|---|
| Younger component, refuse pit | 1480±70 | Beta 134150 | 1380 | 1480–1300 | 1520–1290 |
| Younger component, dwelling | 1500±40 | Beta 134156 | 1390 | 1490–1340 | 1520–1310 |
| Younger component, refuse pit | 1520±60 | Beta 108552 | 1410 | 1510–1350 | 1530–1310 |
| Older component, smaller cooking pit | 1870±60 | Beta 134149 | 1810 | 1870–1730 | 1950–1630 |
| Older component, larger cooking pit | 1950±60 | Beta 120796 | 1900 | 1990–1830 | 2040–1730 |
| Older component, larger cooking pit | 2080±40 | Beta 134147 | 2050 | 2110–2000 | 2150–1950 |

Calibrations based on Calib 6.0html (Stuiver and Reimer 1993)

**Fig. 13.3** Aerial view of the Recent Indian area of the Gould site, showing the older and younger components

## *The Older Component*

Each of the two older activity areas was centred on a burned pit feature, which we identified as a cooking pit based on size, shape and pattern of charcoal and fire-cracked rock (Teal 2001). These were similar to the cooking pits at L'Anse aux Meadows and to those described elsewhere in the literature (e.g. Hayden and Cousins 2004; Thomas 2008; Wandsnider 1997). The majority of the lithic material from both pits was medium-grained brown-to-beige chert (Teal 2001:53).

The larger of the two pits was 2.4 m by 1.4 m and 60 cm deep (Fig. 13.4a). It was charcoal-lined and contained many small retouch flakes and a single late-stage preform. Large concentrations of fire-cracked rock were scattered outside the pit as if they had been removed after the pit was opened and tossed to either side. There was a small amount of burned bone, the only identifiable fragment of which was a juvenile beaver phalange (Teal 2001:61). One fragment of animal skin similar to pieces found elsewhere at the site was identified as beaver. Burned fir and spruce needles were abundant throughout the excavation area (Deal 2005). This feature dates to 2080±40 BP and 1950±60 BP (Table 13.1).

To one side of the pit were 11 pieces of wood visually identified as spruce or fir. Most were logs 1–1.5 m long and 20–25 cm wide and were parallel to each other. Their position suggests that they might have been part of a toppled structure. Alternatively they could have been a supply of fuel for the pit.

**Fig. 13.4** (**a**) Large cooking pit of the older Recent Indian component, shown in profile with one half removed. (**b**) Two projectile points from the younger Recent Indian component. (**c**) Ceramic sherds from the younger Recent Indian component. (**d**) Gritty black level that defines the interior depression (outlined) of the reconstructed tent structure of the younger Recent Indian component

The second pit was approximately 35 m away from the first and was smaller, 1.5 m by 1.1 m and 46 cm deep. There were a few fire-cracked rocks, some flakes and a single scraper. The charred interior dated to 1870 ± 60 BP (Table 13.1).

## *The Younger Component*

Material from the younger component was remarkably diverse (Table 13.2), including many projectile points (Fig. 13.4b), bifaces, an axe bit, the periostracum (outermost layer) of freshwater clam valves (Roger Pickavance, 2001, personal communication), several hematite and iron pyrites nodules, the latter likely functioning as fire-starters, a problematic item chemically identified as possible iron slag (Teal 2001:65) and >1,800 tiny retouch flakes. The majority of the material was a medium-grained white-to-grey chert with tiny round and square pits (Fig. 13.4b), visually similar to

**Table 13.2** Summary of artefacts from the Recent Indian components of the Gould site

| Artefact | Older component | Younger component |
|---|---|---|
| Projectile points, complete/near complete | | 15 |
| Projectile point fragments | | 6 |
| Bifaces, complete/near complete | | 8 |
| Biface fragments | | 21 |
| Scraper | 1 | |
| Preforms | 1 | 5 |
| Hammerstones | | 2 |
| Cores | | 2 |
| Axe bit | | 1 |
| Ground stone items or fragments | | 4 |
| Ceramic sherds | | 384 |
| Animal skin fragments | 1 | 20 |
| Clam valves | | 23 |
| Red ochre nodules | | 21 |
| Iron pyrites | | 7 |
| Iron slag? | | 1 |
| Misc. other, e.g. raw material, linear flakes | | 6 |
| Palaeoeskimo lithics | | 21 |
| Flakes | >1,006 | >1,824 |

lithics from Peat Garden. There were also smaller amounts of Ramah chert as well as a distinctive black chert (Fig. 13.4b). There were also >1,650 fragments of burned bone, 90 of which were identifiable to class or lower and included seal, including harp and grey, small and medium-sized mammal, including canid and caribou, bird, including waterfowl, and fish (Teal 2001:121–122).[3] Burned fir and spruce needles were abundant throughout the excavation area (Deal 2005) and there were several highly worn raspberry seeds, which Deal argues were likely ancient rather than modern. There were several small pieces of animal skin, which were identified as beaver on the basis of characteristic guard hair identified on several specimens (Michael Deal, 2000, personal communication). Significantly, there were 384 large and small quartz tempered sherds from a number of dentate-stamped vessels (Fig. 13.4c), some of which had charred residue on them.

The focus of the younger occupation was a shallow oval depression 2.5 m by 3 m and 24 cm deep and clearly delineated by a thin gritty black cultural layer (Figs. 13.4d and 13.5). A hearth was centrally located within this depression and was defined by a platform of burned sand 1.29 m by 80 cm and 7 cm thick; it dated to 1500±40 BP (Table 13.1). The depression was densely filled with cultural material, almost all of which showed signs of burning; the greatest concentration of material was within the hearth, in particular the larger items and almost all the ceramics. The greatest concentration of small retouch flakes was immediately south of the hearth.

[3]Faunal identifications were done by Lisa Hodgetts for Mike Teal's Master's thesis (Teal 2001).

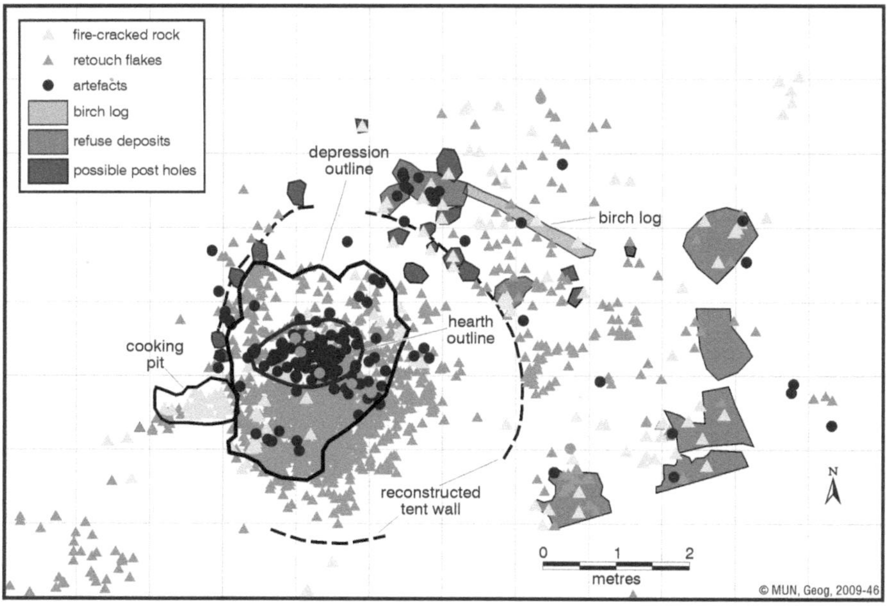


**Fig. 13.5** Map showing the distribution of small objects (e.g. retouch flakes), large objects (e.g. artefacts), refuse pits and possible post-holes that define a tent structure and associated outdoors area

An unexpected inclusion in the depression was a number of Dorset Palaeoeskimo artefacts, most of which occurred in the hearth: seven nephrite burin-like tools, three microblades, one endscraper, five endblades and five stemmed, ground nephrite tools which Renouf et al. (2000:114) identified as scrapers, although they are more usually identified as adzes in Dorset contexts (Maxwell 1985). The microblades and endblades were visually similar to Cow Head chert, which is characteristic of west coast Dorset Palaeoeskimos. A fifth endblade was of the coarse-grained grey chert more characteristic of Recent Indian contexts. The significance of this Dorset material is discussed in Renouf et al. (2000).

Several features occurred outside and to the east of the depression (Fig. 13.5) where they seemed to follow the curve of the depression itself. There were 17 small pits, 15–38 cm in diameter and 5–10 cm deep, which we interpreted as possible post-holes based on size and fill. Although the pits could have been natural, since many followed the perimeter of the depression we think they are not. There were also eight larger charred pits, which we identified as refuse deposits, because they contained fire-cracked rocks and almost all the artefacts found outside the depression; these ranged in size from 29 cm by 45 cm to 1.9 m by 94 cm and were 10–20 cm deep. Dates from two of these are 1520 ± 60 BP and 1480 ± 70 BP (Table 13.1), overlapping with the date from the depression. In addition, two pieces of the same projectile point were mended between the depression and one refuse pit. Small retouch flakes were thinly distributed in an arc outside and to the east of the depression, following its perimeter outline. Whole and fragmented fire-cracked rocks were similarly distributed, and a birch log, 2 m long and 17 cm wide, roughly paralleled the depression perimeter. There was little material found on the west side of the depression aside from a small number of fire-cracked rocks concentrated in a small cooking pit, 1.1 m by 50 cm and 36 cm deep. There was a small number of burned bones in the pit, including seal cranial fragments (Teal 2001).

We argue that the depression was the central area of a dwelling, likely a tent, on the basis of the debris pattern which is consistent with Binford's (1978) model of toss and drop zones around an indoor hearth. According to this model, people sitting around an indoor hearth toss larger items into the fire while smaller items are dropped and left discarded where people sit. By far the greatest concentration of debris occurred within the depression and most of the large items were within the hearth itself (Fig. 13.5). Although smaller debris such as retouch flakes occurred within the hearth, the greatest concentration was immediately outside and to the south. We argue that the sitting area is defined by this distribution of small debris.

Immediately outside the depression, along the east side only, there was a 1 m wide arc devoid of all cultural material except pieces of animal skin; this arc followed the depression's curve (Fig. 13.5). We argue that this was a skin-covered sleeping area against the tent wall. Since this area was 24 cm higher than the central depression, in effect it would have been a raised platform.

We think the tent wall is defined by the distribution of retouch flakes that abuts the outer edge of the proposed sleeping area (Fig. 13.5), where the movement of outdoor debris terminated at the barrier of the tent wall. Outside the wall, this flake distribution formed an arc approximately 4 m wide. Seven of the eight refuse deposits occurred at the arc's outer edge, with the seventh at the outer edge of the tent wall. This is also consistent with Binford's drop and toss model centred on an indoor hearth, where some large items are tossed into the fire while others are carried outdoors where they are tossed aside or, as in this case, into shallow pits.

Following the curves of both the inner depression and the outer tent wall were several proposed post-holes, the outer ring outlining the perimeter of the dwelling.

This debris and post-hole distribution define a roughly circular dwelling 4 m by 4.1 m. As noted, the refuse is distributed only on the east side of the depression while the west side is relatively clear (Fig. 13.5). If the hearth was centrally placed within the dwelling, the interior space would increase to 5.8 m by 4 m. Based on its size, this dwelling would be suitable for 1–2 families.

## Discussion

The diversity of material culture at the younger Recent Indian component of the Gould site suggests that this was a residential base where a wide variety of activities took place. This variety is represented by the remains of a dwelling and the presence of cultural material relating to hunting (projectile points), food processing (bifaces), tool maintenance (small retouch flakes, hammerstones), cooking (hearths, cooking pits, iron pyrite fire-starters, fire-cracked rock, burned bone, burned residue on some ceramic sherds), storage (ceramics), woodcutting (axe fragment) and other activities (red ochre nodules, Dorset artefacts). Similarly, a variety of food resources is represented by the burned bone fragments of waterfowl, harp seal, grey seal and beaver and by clam valves and raspberry seeds. The occurrence of refuse in concentrations outside the dwelling suggests an occupation of sufficient duration to require organized disposal of debris.

In most high latitude regions, residential bases exist in the context of a strategy of logistically organized mobility (Binford 1980; cf. Fitzhugh and Habu 2002). Accordingly, a residential base is a central hub to which various short-term, task-specific sites are connected, for example kill sites, monitoring stations, caches and field camps. A field camp is where a task group stays for a period of days or weeks in order to focus on a particular activity, such as fishing, hunting or lithic quarrying. A field camp is linked to a residential base to which the task group eventually returns. Domestic activities are carried out at the field camp in support of the task group who lives there.

If the Gould site is a residential base, then we suggest that Spearbank, Peat Garden and St. Pauls Bay-2 were field camps focused on lithic quarrying and/or tool manufacture. Evidence of associated domestic activities is found in the multiple hearths at all sites, in the faunal material from Peat Garden and in the possible dwelling floor at St. Pauls Bay-2. This is not to say that these sites were directly linked to the Gould site, but rather it is to contrast their site function to that of the Gould site as a basis for comparing site assemblages and site locations.

There are insufficient data to draw conclusions about Portland Creek-4 and Portland Creek-5 except to note that their location near a salmon-fishing stream suggests the possibility that they are fishing field camps. The two cooking pits from the Cow Head component of L'Anse aux Meadows suggests that it might be a residential base; if so, the restricted artefact assemblage suggests that it was more narrowly focused than the Gould site. A comparison might be EiBg-85 and EiBg-86 which Pintal (2003) interprets as residential bases with a primary but not exclusive focus on seal hunting.

We suggest that the function of the Gould site as a residential base is directly connected to its location in the coastal margin (Fig. 13.6). The coastal margin is the area within approximately 1 km of the coastline but set back from the shore; it is commonly wooded. Sites in the coastal margin are located with respect to marine or littoral resources while at the same time they are adjacent to shelter, timber, firewood, fresh water and freshwater resources. This is in contrast to the Palaeoeskimo preference for exposed outer coastal locations (Pastore 1986; Schwarz 1994) and reflects the boreal adaptation and origins of Cow Head populations. This is consistent with several Flèche Littorale sites in Blanc Sablon, which are about 1 km up the Blanc Sablon River (Pintal 1998:172).

This adds a new dimension to existing models of Recent Indian site location developed by Pastore (1986), Rowley-Conwy (1990), Schwarz (1994) and Holly (2002). The original Pastore (1986) model looked at prehistoric site locations across the inner and outer coastal zones as a basis for comparing Amerindian and

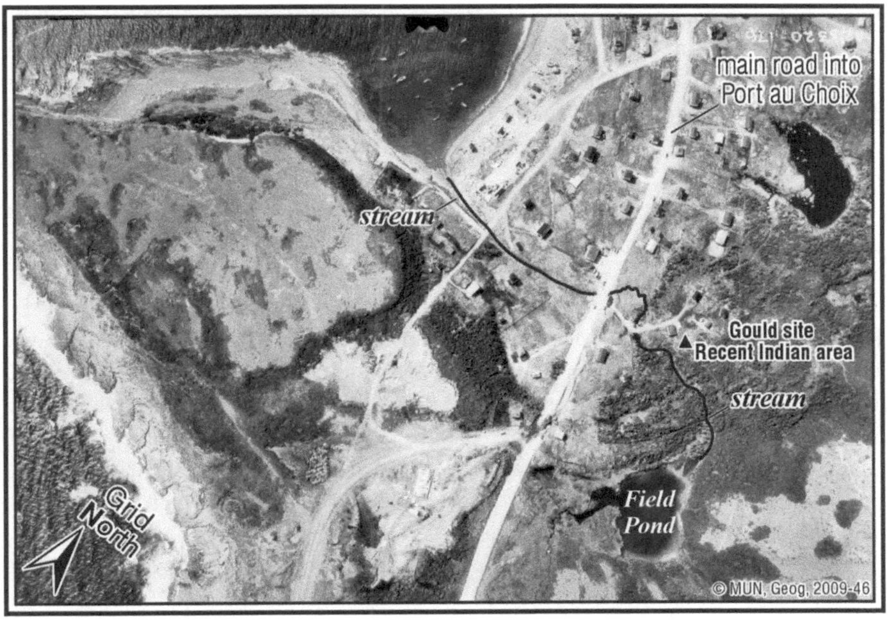


**Fig. 13.6** 1973 air photo of Port au Choix showing the Recent Indian area of the Gould site in the coastal margin. The scale is provided by the main road, houses and boats

Palaeoeskimo site locations and, by inference, economy. Pastore (1986) argued that, in contrast to the outer coastal focus of Palaeoeskimo site distribution, Amerindian sites were more evenly spread between the inner and outer coastal zones, which suggested a generalized marine-terrestrial economy. Schwarz (1994) broadened the comparison to include the interior zone, which he subdivided into deep interior and near coastal; the latter was within striking distance of both the coast and the deep interior by means of an overnight camp (see also Rowley-Conwy 1990). We propose adding the coastal margin to this geographic mix. The coastal margin is a subset of the inner and outer coastal zone and refers to landscape at a larger, higher resolution, spatial scale.

This has practical implications. Most archaeological surveys in Newfoundland operate on the assumption that all the island's inhabitants were adapted to coastal resources to a greater or lesser degree and that the sites from which they exploited these resources were at or near the coast at the time of occupation. As a direct consequence, most archaeological surveys are coastally oriented, except where there has been a concerted effort to redress that imbalance by surveying in the interior (e.g. Erwin and Holly 2006; Holly and Erwin 2009; Schwarz 1987). However, both interior and coastally oriented survey strategies necessarily miss sites in the coastal margin, which is neither inland nor at the shore. Compounding this, where the coastal margin is wooded, sites are very difficult to find. As a case in point, the Recent Indian component of the Gould site was discovered as a result of a survey for Maritime Archaic Indian sites that targeted ancient shorelines on the

Northern Peninsula that are today situated 8–10 m above sea level (Bell et al. 2005) and which were wooded until very recently. Finding a Cow Head component at this elevation and set back 350 m from the shore was entirely serendipitous (Renouf and Bell 2000). Therefore, if we wish to find other Cow Head residential bases like the Gould site, we will have to adjust our survey strategy accordingly.

## Conclusions

The Gould site is currently the only Northern Peninsula Cow Head site securely identified as a residential base. It has a rich and diverse material culture that includes the only ceramic assemblage found so far in Newfoundland as well as the only well-defined Cow Head tent structure; there is no evidence of stone tool manufacture. This site contrasts with the Spearbank, Peat Garden and St. Pauls Bay-2 sites which are dominated by lithic workshop debris and which we argue are field camps rather than residential bases.

These workshop sites are situated at the coast near visible and accessible lithic outcrops. By contrast, the Gould site is located in the wooded coastal margin where sites are neither visible nor easily accessible. Since lithic workshop sites are more visible, Spearbank and Peat Garden were the first to be found and their assemblages formed the basis for the archaeological definition of Cow Head. However, this definition is now expanded to include a wide range of material from the Gould site. This brings us to a new understanding not only of Cow Head material culture but also of the landscape of these populations, in which the coastal margin seems to have played an important role.

**Acknowledgements** Gould site excavations were initially and generously funded by a Memorial University New Initiatives Research Grant from the Office of the Vice-President (Research). Subsequent funding was provided by the Social Sciences and Humanities Research Council of Canada; the Institute of Social and Economic Research, Memorial University; the J.R. Smallwood Foundation for Newfoundland and Labrador Studies, Memorial University; Parks Canada; the Provincial Archaeology Office of the Government of Newfoundland and Labrador; and the Town of Port au Choix. Facilities for analysis were funded by the Canada Foundation for Innovation, the Canada Research Chairs Program and Memorial University. We would like to thank: Lisa Hodgetts for identifying the faunal material from the Gould site, Larry Nolan for developing our GIS software and assisting us in its application, Charles Conway for drafting the figures, Chris Hammond and Dominique Lavers for artefact photographs and the many crew members who spent several summers working on our Gould site excavations. This chapter has benefited greatly from the insightful comments of Don Holly and Jean-Yves Pintal.

## References

Austin, S.J. 1980. Cape Cove Beach (DhAi-5,6,7), Newfoundland: Prehistoric Cultures. Master's thesis, Department of Anthropology, Memorial University of Newfoundland, St. John's.

Austin, S.J. 1984. Maritime Archaic and Recent Indian evidence from Cape Cove Beach, Newfoundland. *Canadian Journal of Archaeology* 8(2): 115–126.

Bell, T., I.R. Smith, and M.A.P. Renouf. 2005. Postglacial sea-level history and coastline change at Port au Choix, Great Northern Peninsula, Newfoundland. *Newfoundland and Labrador Studies* 20(1): 9–31.

Biggin, S. 1985. Prehistoric Indian Sites at Portland Creek, EbBj-4, -5. On file, Provincial Archaeology Office, Government of Newfoundland and Labrador, St. John's.

Binford, L.R. 1978. Dimensional analysis of behaviour and site structure: Learning from an Eskimo hunting stand. *American Antiquity* 43(3): 330–361.

Binford, L.R. 1980. Willow smoke and dogs' tails: Hunter-gatherer settlement systems and archaeological site formation. *American Antiquity* 45(1): 4–20.

Carignan, P. 1975. *The Beaches: A multi-component habitation site in Bonavista Bay*. Mercury Series Archaeology Paper 39, National Museums of Canada, Ottawa.

Carignan, P. 1977. *Beothuck archaeology in Bonavista Bay*. Mercury Series Archaeology Paper 69, National Museums of Canada, Ottawa.

Coniglio, M. 1987. Biogenic cherts in the Cow Head group (Cambro-Ordovician), western Newfoundland. *Sedimentology* 34: 813–823.

Deal, M. 2005. Palaeoethnobotanical research at Port au Choix. *Newfoundland and Labrador Studies* 20(1): 131–156.

Eldjárn, K. 1977. Investigations. In *The discovery of a Norse settlement in America*, ed. A. Stine Ingstad, 87–96. Oslo: Universitetsforlaget.

Erwin, J., and D.H. Holly. 2006. Baie Verte Waterways Project: Birchy Lake Survey and Reconnaissance 2005. On file, Provincial Archaeology Office, Government of Newfoundland and Labrador, St. John's.

Fitzhugh, B., and J. Habu (eds.). 2002. *Beyond foraging and collecting: Evolutionary change in hunter-gatherer settlement systems*. New York: Kluwer Academic/Plenum Publishers.

Hartery, L. 2001. The Cow Head Complex. Master's thesis, Department of Archaeology, University of Calgary, Calgary.

Hartery, L. 2007. *The Cow Head complex and the Recent Indian period in Newfoundland, Labrador and the Quebec Lower North Shore*. Occasional Papers in Northeastern Archaeology 17. St. John's: Copetown Press.

Hartery, L., and T. Rast. 2001. Bird Cove Archaeology Project 2000 Field Season: Final Report. On file, Provincial Archaeology Office, Government of Newfoundland and Labrador, St. John's.

Hayden, B., and S.M. Cousins. 2004. The social dimensions of roasting pits in a winter village site. In *Complex hunter-gatherers: Evolution of prehistoric communities on the Plateau of northwestern North America*, ed. W. Prentice and I. Kuijt, 140–154. Salt Lake City: University of Utah Press.

Holly, D.H. 2002. From Space to Place: An Archaeology and Historical Geography of the Recent Indian Period in Newfoundland. PhD dissertation, Department of Anthropology, Brown University, Providence.

Holly, D.H., and J.C. Erwin. 2009. Terra Incognita, still: Archaeological investigations in the interior of the island of Newfoundland. *Archaeology of Eastern North America* 37: 65–84.

Hull, S.H. 2002. Tanite uet Tshinauetamin? A Trail to Labrador: Recent Indians and the North Cove Site. Master's thesis, Department of Anthropology, Memorial University of Newfoundland, St. John's.

Ingstad, A.S. 1977. *The discovery of a Norse settlement in America*. Oslo: Universitetsforlaget.

Kristensen, T.J. 2010. An Archaeological and Ethnohistorical Investigation of Bird Use during the Recent Indian Period in Newfoundland and Labrador. Master's thesis, Department of Archaeology, Memorial University of Newfoundland, St. John's.

Kristensen, T.J., and M.A.P. Renouf. 2009. Final Report on the 2008 Excavations at L'Anse aux Meadows National Historic Site, Newfoundland (EjAv-01). On file, Parks Canada, Western Newfoundland and Labrador, Rocky Harbour.

Kristensen, T.J., J. Curtis, and M.A.P. Renouf. 2009. Archaeological survey and excavations at L'anse aux Meadows, National Historic Site of Canada. *Provincial Archaeology Office Archaeology Review* 7: 71–77.

Lavers, D. 2009. St. Pauls Archaeology Project: St. Pauls Bay-2 (DlBk-06) summary of 2008 fieldwork. *Provincial Archaeology Office Archaeology Review* 7: 77–80.

Lavers, D. 2010. The Recent Indian Cow Head Complex Occupation of the Northern Peninsula, Newfoundland: A Geochemical Approach to Investigating Cow Head Chert Acquisition. Master's thesis, Department of Archaeology, Memorial University of Newfoundland, St. John's.
Maxwell, M.S. 1985. *Prehistory of the eastern Arctic*. New York: Academic Press.
Pastore, R.T. 1986. The spatial distribution of late Palaeo-Eskimo sites on the island of Newfoundland. In *Palaeo-Eskimo cultures in Newfoundland, Labrador and Ungava*, 125–134. Reports in Archaeology No. 1, Memorial University of Newfoundland, St. John's.
Pastore, R.T. 1992. *Shanawdithit's people*. St. John's: Atlantic Archaeology.
Penney, G. 1989. Historic Resources Overview Assessment, Daniel's Harbour Transmission Line. On file, Provincial Archaeology Office, Government of Newfoundland and Labrador, St. John's.
Petré, R. 1977. Investigations. In *The discovery of a Norse settlement in America*, ed. A. Stine Ingstad, 49–62. Oslo: Universitetsforlaget.
Pintal, J.-Y. 1998. *Aux frontières de la mer: La préhistoire de Blanc-Sablon*. Les Publications du Québec, Dossiers 102. Collections Patrimoines et Municipalité de Blanc-Sablon, Québec.
Pintal, J.-Y. 2003. Préhistoire de la chasse au phoque dans le Détroit de Belle-Isle. *Recherches Amérindiennes au Québec* 33(1): 35–43.
Rast, T.L. 1998. Burgeo Coast Archaeological Research Project: Interim Report. On file, Provincial Archaeology Office, Government of Newfoundland and Labrador, St. John's.
Rast, T.L. 1999. Investigating a Palaeo-Eskimo and Indian Settlement Patterns along a Submerging Coast at Burgeo, Newfoundland. Master's thesis, Department of Anthropology, Memorial University of Newfoundland, St. John's.
Rast, T.L., M.A.P. Renouf, and T. Bell. 2004. Patterns in precontact site location on the southwest coast of Newfoundland. *Northeast Anthropology* 68(1): 41–55.
Renouf, M.A.P., and T. Bell. 2000. Integrating sea level history and geomorphology in targeted archaeological site survey: The Gould site (EeBi-42), Port au Choix, Newfoundland. *Northeast Anthropology* 59: 47–64.
Renouf, M.A.P., and T. Bell. 1999. Report of the Port au Choix Archaeology and Sea Level History Project: Preliminary Report of the 1998 Field Season. On file, Provincial Archaeology Office, Government of Newfoundland and Labrador, St. John's.
Renouf, M.A.P., T. Bell, and M.A. Teal. 2000. Making contact: Recent Indians and Palaeoeskimos on the island of Newfoundland. In *Identities and cultural contacts in the Arctic*, ed. M. Appelt, J. Berglund, and H.C. Gulløv, 106–119. Copenhagen: The Danish National Museum and Danish Polar Center.
Rowley-Conwy, P.A. 1990. Settlement patterns of the Beothuk Indians of Newfoundland: A view from away. *Canadian Journal of Archaeology* 14: 13–32.
Sanger, C.W. 1977. The evolution of sealing and the spread of settlement in northeastern Newfoundland. In *The peopling of Newfoundland*, ed. J.J. Mannion, 136–152. Social and Economic Papers 8, Institute for Social and Economic Research, St. John's.
Schwarz, F.A. 1984. The Little Passage Complex in Newfoundland: A Comparative Study of Assemblages. Honour's dissertation, Department of Anthropology, Memorial University of Newfoundland, St. John's.
Schwarz, F.A. 1987. An Archaeological Survey of Terra Nova Lake and Gambo Pond. On file, Provincial Archaeology Office, Government of Newfoundland and Labrador, St. John's.
Schwarz, F.A. 1994. Paleo-Eskimo and Recent Indian subsistence and settlement patterns on the island of Newfoundland. *Northeast Anthropology* 47: 55–70.
Stuiver, M., and P.J. Reimer. 1993. Extended 14C database and revised CALIB radiocarbon calibration program. *Radiocarbon* 35: 215–230.
Teal, M.A. 2001. An Archaeological Investigation of the Gould Site (EeBi-42) in Port au Choix, Northwestern Newfoundland: New Insight into the Recent Indian Cow Head Complex. Master's thesis, Department of Anthropology, Memorial University of Newfoundland, St. John's.
Thomas, A.V. 2008. The fire stones carry: Ethnographic records and archaeological expectations for hot-rock cookery in western North America. *Journal of Anthropological Archaeology* 27(4): 443–460.

Thomson, J.C. 1987. Archaeological Survey of Two Interior Remote Cottage Areas at Parsons Pond and Portland Creek Pond. On file, Provincial Archaeology Office, Government of Newfoundland and Labrador, St. John's.
Tuck, J.A. 1978. Excavations at Cow Head, Newfoundland: An interim report. *Études/Inuit/Studies* 2(1): 138–141.
Tuck, J.A. 1988. Prehistory of Atlantic Canada. Manuscript on file, Department of Archaeology, Memorial University of Newfoundland, St. John's.
Wallace, B. 1989. Native occupations of L'Anse aux Meadows. On file, Parks Canada, Archaeology, Atlantic Region, Halifax.
Wallace, B. 1991. L'Anse aux Meadows: Gateway to Vinland. *Acta Archaeologica* 61: 166–195.
Wallace, B. 2003. An Old Ship Plank and Other Lessons Learnt from Archaeology: Report on the excavation at L'Anse aux Meadows National Historic Site, September 3 to 8, 2002. On file, Parks Canada, Archaeology, Atlantic Region, Halifax.
Wallace, B. 2006. *Westward Vikings: The saga of L'Anse aux Meadows*. Historic Sites Association of Newfoundland and Labrador, St. John's.
Wandsnider, L. 1997. The roasted and the boiled: Food composition and heat treatment with special emphasis on pit-hearth cooking. *Journal of Anthropological Archaeology* 16: 1–48.

# Chapter 14
# The Life History of Port au Choix Landscapes

**M.A.P. Renouf**

## Introduction

This chapter synthesizes the cultural themes of Port au Choix's occupational history from a landscape perspective, including the natural and cultural environments. It first addresses the adaptation of Amerindian and Palaeoeskimo populations to the Newfoundland physical environment through a comparison of subsistence, mobility and site location patterns and then compares site location patterns at Port au Choix. This provides the foundation for a discussion of three Port au Choix landscapes from the physical perspective of the changing coastline and the cultural perspective of who occupied these landscapes, how they lived on them, how they may have perceived them, and how they may have ascribed them cultural meaning. Layers of enculturation at sites and among connected sites collectively created the life history of these landscapes. I argue that people acknowledged earlier activities and occupations thereby linking the layers of life history through time.

## Context

### *Origins and Adaptations*

Every precontact culture that lived in Newfoundland had an important connection to Port au Choix (Table 1.1, Renouf, Chap. 1), in large part as a result of the area's rich marine resources. In particular, large herds of migratory harp seals were predictably accessible in the late winter/early spring, that is, the end of March and beginning of April. Although faunal studies (Hodgetts et al. 2003; Wells 2005) show that harp seals were of prime importance to these precontact hunting economies,

M.A.P. Renouf (✉)
Department of Archaeology, Memorial University, St. John's, NL, Canada, A1C 5S7
e-mail: mapr@mun.ca

M.A.P. Renouf (ed.), *The Cultural Landscapes of Port au Choix*, Interdisciplinary Contributions to Archaeology, DOI 10.1007/978-1-4419-8324-4_14,

other resources were available at Port au Choix as well: sea birds, ducks, freshwater fish, near-shore marine fish, small game and caribou. Maritime Archaic Indian (6290–3340 cal BP[1]), Groswater Palaeoeskimo (2950–1820 cal BP), Dorset Palaeoeskimo (1990–1180 cal BP) and Recent Indian (2110–680 cal BP) cultures approached this range of resources in culturally particular ways, reflected in characteristic patterns of site location. These site location preferences were rooted in cultural origins and structured by life in Newfoundland.

The origins of the Maritime Archaic Indians of Newfoundland are unclear, but are likely connected to late Archaic populations in southern and central Labrador that are characterized by side-notched projectile points (Fitzhugh 1972, 2006; Pintal 1998, 2006) and who originated in the Gulf of St. Lawrence region. Shared projectile point styles indicate that subsequent late Archaic communication networks included Maine and the Maritime Provinces (Tuck 1991). The cultural mosaic was complex; in southern Labrador and the Quebec Lower North Shore late Archaic site assemblages characterized either by stemmed or side-notched projectile points chronologically overlap. The presence of stemmed points in late Archaic contexts in Newfoundland which are otherwise characterized by side-notched points indicates regular communication between the two regions. This is also indicated by the presence of Ramah chert, found only in northern Labrador, in Newfoundland Archaic lithic assemblages.

Maritime Archaic Indians occupied Port au Choix as early as 6290 cal BP at the Gould site (EeBi-42), close to a small pond, Field Pond. Reid (2008) interprets this component as a brief occupation linked to contemporaneous Maritime Archaic occupations of southern Labrador (McGhee and Tuck 1975:219, 235), reflecting population movement back and forth across the narrow Strait of Belle Isle. The similarity of this material on both sides of the Strait underscores the mobility of the Maritime Archaic populations and their use of watercraft to travel between Newfoundland and Labrador and along the Northern Peninsula coastline.

The later Maritime Archaic occupation (4800–3340 cal BP) at the Gould site was contemporaneous with Port au Choix-3 (EeBi-2), the Maritime Archaic burial ground near the harbour-front of the modern town of Port au Choix.

The Archaic in Newfoundland are characterized as maritime (Tuck 1976) based on the coastal location of many of their sites (e.g. Austin 1980; Carignan 1975), including Port au Choix-3. In addition, there are maritime elements in Port au Choix-3 burial goods, for example two whale effigies, several carved representations of ducks and auks, and individual bones of gulls, ducks, seals and fish (Tuck 1976). This coastal focus is substantiated by the isotopic signature of a sample of human bones that indicates a marine diet (Jelsma 2000, 2006). At the same time, examination of Maritime Archaic site distribution in Newfoundland suggests a significant riverine-interior dimension to their settlement pattern (Renouf and Bell 2006). This makes sense in light of cognate Archaic populations in the Atlantic Provinces and Gulf of Maine for whom rivers were important settlement areas and

---

[1] All dates are expressed as calendar years before present (cal BP) at the one sigma probability range, calibrated by Calib 6.0html (Stuiver and Reimer 1993). See Appendix for details of radiocarbon dates mentioned in this chapter.

routes of travel (Robinson 2006). It also reinforces the inference that Newfoundland Maritime Archaic populations were highly mobile.

Maritime Archaic Indians were followed by Groswater Palaeoeskimos, a culture whose distant origins centred on the eastern Arctic and who expanded into Newfoundland from Labrador (Anton 2004; Fitzhugh 1972; Loring and Cox 1986), reaching as far south as Saint-Pierre and Miquelon. Groswater Palaeoeskimos are the Newfoundland and Labrador regional expression of what is called the transitional period of Canadian Arctic prehistory, chronologically situated between Pre-Dorset and Dorset Palaeoeskimo occupations (Maxwell 1985; Nagy 1994). It is generally accepted that Groswater populations are an in situ development of Pre-Dorset populations in Labrador but are not the direct ancestors of the Dorset (Tuck and Fitzhugh 1986).

In Newfoundland the Groswater are neither transitory nor transitional. According to the Provincial Archaeology Office (PAO) site database there are 92 known Groswater sites and site components in Newfoundland which collectively span more than 1,000 years (Provincial Archaeology Office 2010). One of the largest and best-preserved Groswater sites is Phillip's Garden East at Port au Choix.

Groswater Palaeoeskimos are characterized as marine specialists who also have a significant terrestrial component to their economy (Pastore 1986; Schwarz 1994). Their sites are found at the exposed outer coast for exploitation of marine resources, in more sheltered inner coastal bays where a wide variety of marine, littoral and terrestrial resources are available (e.g. Loring and Cox 1986), and in the interior for access to terrestrial, riverine and lacustrine resources (Holly and Erwin 2009; Schwarz 1994). LeBlanc (1996, 2000a) characterizes Groswater Palaeoeskimos on the Northern Peninsula as logistically-organized seal hunters, in contrast to Groswater populations on the Quebec Lower North Shore (Pintal 1994, 1998) who she argues employed more forager-like strategies. Possibly these strategies were part of a single annual round that encompassed both sides of the Strait of Belle Isle which, if so, indicates high mobility.

Groswater populations were succeeded by Dorset Palaeoeskimos, a second Arctic-adapted population with origins in the eastern Arctic who followed a similar route of movement from Labrador to Newfoundland and Saint-Pierre and Miquelon (LeBlanc 2000b, 2008). According to the PAO site database there are 254 known Dorset sites and site components in Newfoundland (Provincial Archaeology Office 2010). A few of these sites are quite large (e.g. Fogt 1998; Linnamae 1975) and Phillip's Garden at Port au Choix is by far the largest (Harp 1976; Renouf 1999).

Dorset Palaeoeskimos are characterized as marine specialists whose large sites are predominantly in outer coastal locations for the exploitation of harp seals and other marine resources (Pastore 1986); in many cases these locations were occupied by earlier Groswater groups for similar purposes. A small number of sites are deep in inner bays (Barnable 2008; Schwarz 1993) as well as the interior (Schwarz 1994). Large site size and regionally distinct patterns of raw material and style of harpoon endblade suggest that Dorset populations were less mobile than Groswater groups (Erwin 2001; LeBlanc 2000b, 2008; Robbins 1985). On the Northern Peninsula the preferred raw material was Cow Head chert, a fine-grained stone found at a number of outcrops and as beach cobbles, about 150 km south of Port au Choix.

Recent Indians temporally overlapped with Dorset populations in Newfoundland and their sites are found in the same regions, often at the same locations within regions (Renouf 2003a). In general, Recent Indians were boreal-adapted cultures whose origins are not established but likely lie around the Gulf of St. Lawrence, in interior Quebec and in Atlantic Canada. In Newfoundland there are three Recent Indian complexes: Cow Head (2110–930 cal BP), Beaches (1900–800 cal BP) and Little Passage (1170–300 cal BP). Beaches and Little Passage complexes are thought to comprise a cultural sequence that includes Newfoundland's historic period Amerindians, the Beothuks (Pastore 1992); Cow Head is culturally distinct (Hartery 2007). Since single component Cow Head complex sites are known only from the Northern Peninsula and comparable tool assemblages are known from the Quebec Lower North Shore (Pintal 1998), it seems that the Cow Head complex was a regional phenomenon centred on the Strait of Belle Isle. Recent Indian site locations, tools and faunal data indicate that they had a generalized economic pattern that included seals, birds, caribou, small game, and freshwater and anadromous fish (Holly 2002; Pastore 1986; Rowley-Conwy 1990). Recent Indian sites are small compared to Dorset and Groswater sites, reflecting their shorter stays and greater mobility.

Figure 14.1 shows the distribution of known Maritime Archaic Indian, Groswater Palaeoeskimo, Dorset Palaeoeskimo and Recent Indian sites in Port au Choix. Consistent

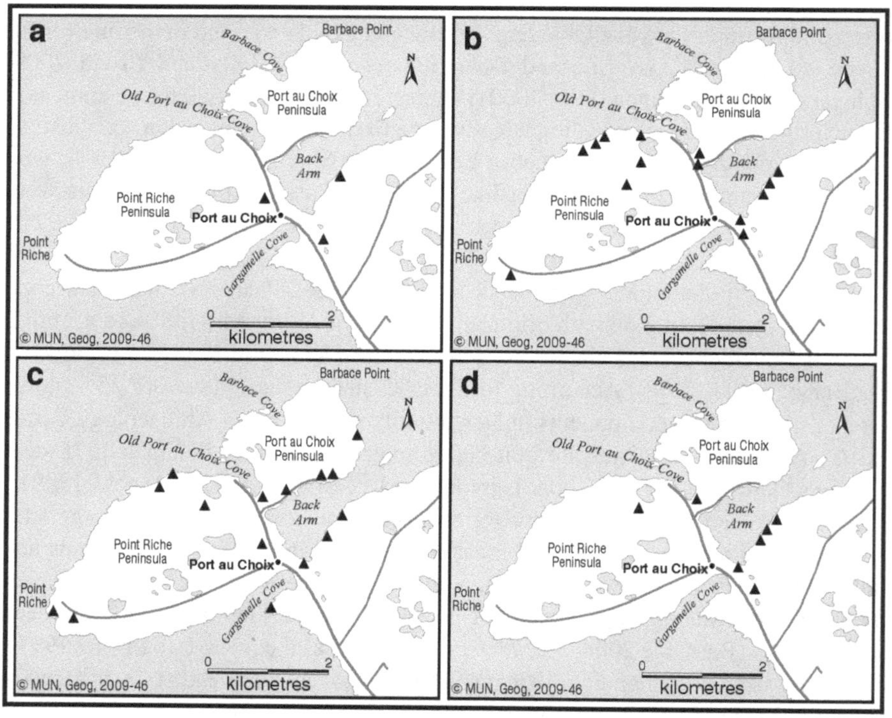


**Fig. 14.1** Distribution of precontact archaeological sites in Port au Choix by culture: (**a**) Maritime Archaic Indian, (**b**) Groswater Palaeoeskimo, (**c**) Dorset Palaeoeskimo, and (**d**) Recent Indian

with the cultural patterns of site location in Newfoundland, the main Palaeoeskimo sites are on exposed headlands in contrast to the Amerindian sites which are in more sheltered inner coastal areas. However, smaller Palaeoeskimo sites also occur in the sheltered head of Back Arm, reflecting other aspects of their subsistence pattern and illustrating their fine-grained use of the Port au Choix landscape.

Zedeño (2000) refers to the life history of places (which she calls landmarks) and the connections among places (which she calls landscape). Pope (2009:136) notes that landmarks and landscapes are recursive whereby a landmark at one spatial scale (e.g. a site) is a landscape at another (e.g. the connections among features within a site). Each landmark and landscape has a life history comprised of layers of different kinds of meaning, for example pertaining to resources, topography, routes of movement, events, stories, myths, and rituals. These life histories evolve as meanings accumulate over time.

Certain areas of Port au Choix stand out as foci of multiple occupations: (1) the isthmus connecting the Point Riche/Port au Choix Peninsulas to the mainland, (2) the southeastern shore of Back Arm and (3) the northwestern area of the Point Riche headland (Fig. 14.2). Each of these areas is a landmark within Port au Choix and is itself a landscape comprised of a network of sites. The next section examines the life history of these landscapes, addressing the sequence of occupation of each, how each became a series of cultural landscapes and how these series were connected over time.

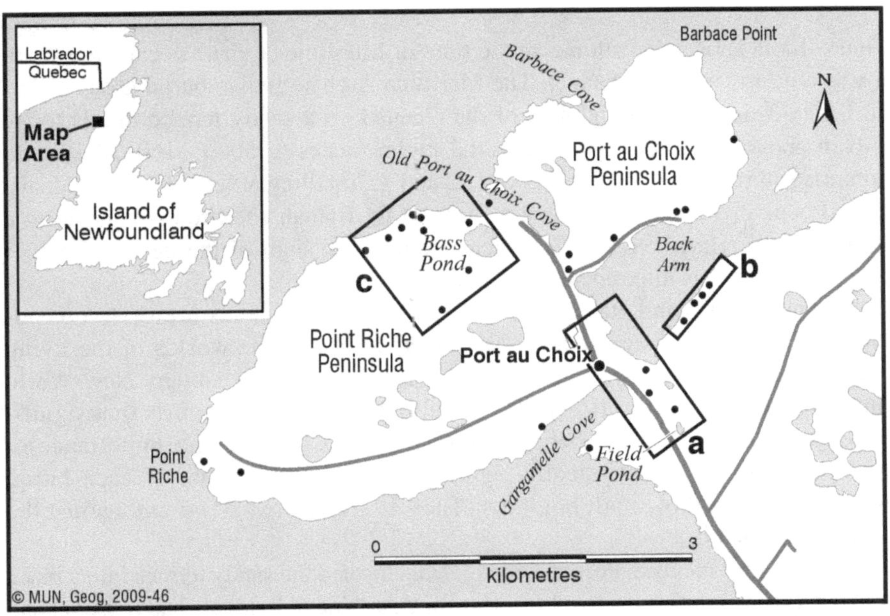


**Fig. 14.2** The three Port au Choix landscapes examined in this chapter, with the precontact archaeological sites that are identified to culture shown as *dots*: (**a**) the isthmus connecting the Point Riche/Port au Choix Peninsulas to the mainland Northern Peninsula, (**b**) the southeastern shore of Back Arm, and (**c**) the northwestern area of the Point Riche headland

# The Life History of Three Port au Choix Landscapes

## *Changing Coastline*

These landscapes must be understood in the context of the changing coastal topography. As summarized in Bell et al. (2005a) and Bell and Renouf (see Chap. 2) the coastline of Port au Choix changed over the past 11,000 years as a result of the adjustment of post-glacial land emergence and sea level rise, resulting in the continuous lowering of relative sea level (RSL) (Liverman 1994; Smith et al. 2005). While the change was dramatic from 11,000 to 7,000 years ago, it began to slow down by 6,000 years ago, reaching present sea level soon after 1,000 years ago (Bell et al. 2005a). At around 5000 cal BP the Point Riche and Port au Choix Peninsulas were two adjacent islands separated from the mainland by a narrow marine channel. By around 3000 cal BP the two islands were connected and the marine channel had narrowed. By around 2000 cal BP the channel had closed (Bell et al. 2005a:21). This changing landscape had repercussions for coastal settlement and travel routes, discussed below.

## *The Isthmus*

Figure 14.3a shows the isthmus at the time of Maritime Archaic occupation when it was a narrow marine channel. The Maritime Archaic Indian burial ground, Port au Choix-3, is on the island side of the channel on a sandy terrace that is today 4–6 m above sea level (asl). The burial ground dates to 5040–3160 cal BP and comprises three burial clusters (Loci 1, 2 and 4), totalling at least 117 individuals. Near Locus 2, a landowner found a cache of 64 Ramah chert tools arranged in a small mound (Harp 1964a:142) which is notable in light of the near absence of chipped stone tools interred with the burials (Tuck 1976).

Cooney (2003) and Rainbird (2007) observe that islands are often considered places of ritual significance because they bridge the worlds of the living and the dead. This is consistent with the island location of many New World hunter–gatherer burials (Holly 2003; Walthall 1999) and suggests that significance was inherent in the geography of Port au Choix-3. The importance of Port au Choix-3 was physically signified by the demarcation of each burial with a low mound of small boulders (Tuck 1976:96) that stood out against the surrounding sand.

The first interments at Port au Choix-3 transformed the sandy terrace into a place with ancestral significance to which Maritime Archaic individuals had personal and community ties. The burials reflect the Maritime Archaic world view which was likely similar to that of other northern and circumpolar hunter-gatherers who have an intimate relationship with animals and land (Boas 1888; Fienup-Riordan 1990; Holly 2003; Jordan 2003; Nuttall 1992; Rasmussen 1929; Sanger 2003;

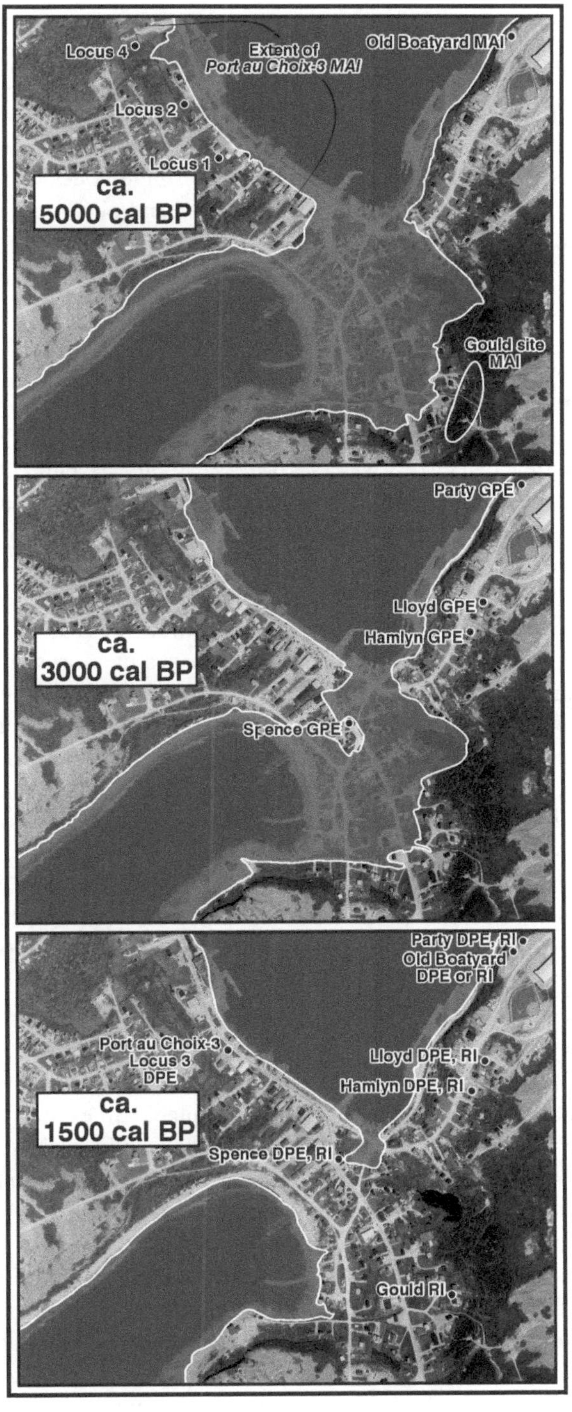


**Fig. 14.3** This figure shows the changing coastal topography of the isthmus area with precontact archaeological sites superimposed on their contemporaneous palaeo-shoreline reconstruction: (**a**) reconstructed around 5000 cal BP, at the time of Maritime Archaic Indian (MAI) occupation, (**b**) reconstructed around 3000 cal BP, at the time of Groswater Palaeoeskimo (GPE) occupation, and (**c**) reconstructed around 1500 cal BP, at the time of Dorset Palaeoeskimo (DPE) and Recent Indian (RI) occupation. Figure is based on Bell et al. (2005a:27)

Speck 1935; Tanner 1979). This relationship requires respectful behaviour towards animals which expresses itself in a variety of ways, including the use of amulets. The Port au Choix-3 grave goods include carved representations of animals, in particular mergansers and auks, and also single bones of a wide range of animal species, including seals, fish, caribou, wolf and pine marten, which likely served as amulets in this context.

Birds were a dominant theme which suggests their particular importance, perhaps connected to their ability to fly and thus mediate between this world and the next (see Balzer 1996; Fienup-Riordan 1996; Mannermaa 2008; Merkur 1987). Heads of mergansers and great auks were carved into the ends of combs and pins; leg bones of trumpeter swans were made into whistles; and wings of harlequin ducks, red-throated loons, gulls and gannets were included in the graves. Eagles, marsh hawks, snowy owls, shearwaters, cormorants, puffins, eider ducks, and common murres were represented by leg bones, bills and mandibles. There were small pebbles in the shape of birds and larger pebbles in the shape of eggs (Renouf 1999:24–25; Tuck 1976:237).

For hunter-gatherers globally, places are associated with stories, events, activities and performances (Morphy 1995; Nuttall 1992; Stewart et al. 2004; Zedeňo 2000) creating what Nuttall (1992:51) calls a memoryscape. The Port au Choix-3 memoryscape was created by acts of burial, no doubt accompanied by storytelling and recollection. The inclusion of special objects in the graves and the burial of a cache of Ramah chert tools were performances that literally placed meaning into the ground. The translucence that is characteristic of Ramah chert (Loring 2002) was likely another material expression of the transition between worlds.

The Gould site is directly across from the burial ground, on a gravel terrace that is today 8–10 m asl. In Chap. 3 Bell and I argue that the younger Maritime Archaic component of this site was an intermittently occupied stop-over camp associated with the burial ground, within direct line of vision and ease of access. Perhaps performances such as storytelling and recollection took place at the stop-over as well as at the burial ground itself.

Groswater Palaeoeskimos succeeded the Maritime Archaic occupations at around 2950 cal BP, at which time the marine channel still existed but in narrower form (Fig. 14.3b). The main focus of Groswater occupation at Port au Choix was elsewhere but nonetheless there are scattered Groswater finds around the isthmus. Groswater individuals and groups would have moved by foot and boat around the island and mainland, passing by and observing the burial grounds. Once seen, this knowledge would have passed into the collective memory of the Groswater population.

At some point after use the burial ground became overlain by a thick layer of peat (Tuck 1976:23), related to increased cooling and expansion of bog plants such as mosses. Although we do not know exactly when that accumulation began it was likely sometime after the Groswater occupation for the simple reason that Groswater sites at Port au Choix are themselves capped by peat. Using the youngest calibrated date ranges at one sigma probability[2], we can infer when peat accumulated at a number of Groswater sites: at Phillip's Garden East after 2350–2130 cal BP, at Phillip's Garden West after 2000–1820 cal BP, and at Area 1 of the Party site after 2850–2770 cal BP. In Area 2 of the Party site there was already 5–10 cm of peat at 2700–2370 cal BP. Additionally, at the Gould site the earliest Recent Indian deposit dating to 2110–2000 cal BP was above 20–30 cm of peat. The general congruence of these dates suggests the onset of peat development at Port au Choix sometime around 2700–1800 cal BP. This is consistent with macrofossil data that indicate cooling of freshwater and marine ecosystems around 3200–2100 cal BP and pollen data that indicate the cooling of terrestrial ecosystems after 3000 cal BP (Bell and Renouf, Chap. 2).

By the time Dorset Palaeoeskimos and Recent Indians occupied the isthmus around 1500 cal BP the marine channel had disappeared and only a shallow inlet remained (Fig. 14.3c). The isthmus was a primary focus of Recent Indian occupation and only a minor focus for the Dorset. The Recent Indian Spence site (EeBi-36), on the banks of the inlet, was primarily a Beaches complex and Little Passage complex occupation but there was also a small Dorset component and a few Groswater artefacts. Not far away, another small Dorset component was found at Port au Choix-3, called Locus 3 (Tuck 1976:22). The sacred landscape of Port au Choix-3 may or may not have been known to Dorset and Recent Indian individuals since some peat certainly had accumulated by then. At the same time, the burials might have been made visible through erosion, frost-heave and rodent burrowing, in much the same way that peat-covered sites in Port au Choix are sometimes exposed today.

Cow Head complex Recent Indians occupied the Gould site, by which time RSL had dropped to the point where the site was set back about 350 m from the shoreline (Fig. 14.3c). This Cow Head complex occupation was a residential base in the shelter of the spruce forest, near the Field Pond outflow stream. Site stratigraphy and Field Pond pollen data indicate that the occupants cut down trees to make a clearing (Renouf et al. 2009). Although this was a protected forest enclave with little or no ocean view, it was within easy access of the shoreline.

A second cultural identity was in the forest clearing. As discussed in Teal (2001) and Renouf et al. (2000) the presence of a small number and narrow range of Dorset tools found in good context at the site indicates the presence of one or more Dorset individuals. We argue that this contact makes sense in light of the complementary landscape experience of Dorset and Cow Head complex people. Whereas the Dorset were knowledgeable about harp seals, the ocean and ice conditions, Cow Head complex Recent Indians were knowledgeable about caribou and the interior. Perhaps each culture was cautious about a landscape or seascape about which they had little expertise (see Grønnow 2009) and therefore had social mechanisms for exchanging meat, fat, hides, antler and information.

---

[2] These ranges exclude dates of uncertain context or with a standard deviation >110 radiocarbon years.

## *Southeastern Shore of Back Arm*

The dominant topographic characteristic of this area (Fig. 14.2) is the almost vertical limestone escarpment set back 50–100 m from the water and paralleling the shore. This rock wall has been fully exposed for a distance of about 150 m by modern house-building on the 6–8 m terrace at its base (Fig. 14.4) and for another 150 m along the shoreline road. Presumably it extends farther along the shore where it is obscured by spruce forest and peat. The rock wall faces west, heating up from the afternoon and evening sun and blocking southeasterly winds; it provides a ready-made wall against which to pitch a tent or construct a dwelling. There is a wide ocean view.

Cultural material was found on the Hamlyn and Lloyd properties in the course of landscaping. Tom and George Hamlyn reported finding four hearths 25 ft (7.6 m) apart along the base of the escarpment. They showed us Groswater and Dorset artefacts and a burned sandstone slab which bore the outline of a Dorset soapstone pot. All but one hearth were destroyed. We excavated this hearth which was a distinct oblong comprised of large fire-cracked rocks surrounding a sandstone slab, or lamp stand (Renouf and Bell 1998:24). We sifted through the landscaping back-dirt, finding Groswater and Dorset artefacts in addition to a few flakes of the

**Fig. 14.4** Limestone escarpment exposed in Tom Hamlyn's back yard where a series of Palaeoeskimo hearths was found

coarse-grained material favoured by Recent Indians. Further along the same rock face, Jim Lloyd found a hearth and several Groswater, Dorset and Recent Indian artefacts in his backyard. A few Dorset artefacts were found between the Hamlyn (EeBi-39) and Lloyd (EeBi-41) sites (Fig. 14.3b).

The Old Boatyard (EeBi-43) and Party (EeBi-30) sites are a short distance farther along the shore (Fig. 14.3c) where the escarpment disappears into the spruce forest. At Old Boatyard we found a Maritime Archaic stemmed slate point just above the rounded gravel beach, dating to 4780–4250 cal BP. In the overlying peat, about 20–30 cm above the beach stratum, we found large hearth stones dating to 1610–1420 cal BP. This hearth continues the pattern of hearths at the escarpment base, but we do not know if it is Dorset or Recent Indian since we did not find any associated artefacts. The Party site, dated to 2850–2370 cal BP, is primarily a Groswater site within which a small number of Dorset and Recent Indian tools and flakes were found.

Stiwich (Chap. 6) argues that the Party site was a warm-weather occupation, based on the paucity of artefacts, two small hearths, and available resources such as harbour seals, ducks, seabirds, freshwater fish and shellfish. If she is correct, it is likely that the above-mentioned cultural traces along this shore are from similar warm-weather occupations. This is consistent with the hearth feature at the Hamlyn site which is unlike anything found at cold-weather Palaeoeskimo sites on the Point Riche headland.

Human occupation on the southeastern shore of Back Arm was structured by the rock escarpment which attracted activity at its base. There was a panoramic view of Back Arm whereby Groswater people could look outwards to the open sea or inwards to the sheltered shore where they could observe the Maritime Archaic burial ground. After the Groswater occupation, Dorset and Recent Indian families camped at the rock face for the same reasons and no doubt saw traces of earlier Groswater people as well as of each other. By then some vegetation had grown over the Maritime Archaic burial ground and Dorset and Recent Indian people may or may not have been aware of its presence.

## *The Point Riche Headland*

The northwestern shore of the Point Riche headland (Fig. 14.2) was an intensively-used landscape, directly related to the availability of harp seals in late winter/early spring. This abundance was predictable because the deep bathymetry close to shore created a regular lead in the sea ice that attracted and concentrated seals (LeBlanc 1996). It was here that Groswater and Dorset Palaeoeskimos left their heaviest imprint on the landscape (Renouf 1994a, 2005, 2009) notably, three habitation sites, Phillip's Garden East (EeBi-1), Phillip's Garden (EeBi-1) and Phillip's Garden West (EeBi-11), in addition to a burial, Crow Head Cave (EeBi-4), an isolated find, David's Site (EeBi-24), and an unregistered cairn (Fig. 14.5).

**Fig. 14.5** Landmarks on the northwest area of the Point Riche headland: *a* Phillip's Garden East; *b* Bass Pond; *c* Crow Head peak; *d* David's Site; *e* Phillip's Garden West; *f* Phillip's Garden; and *g* an unregistered cairn

Phillip's Garden East is a 0.15 ha Groswater site at 12 m asl, close to a small pond, Bass Pond. The site is dated to 2950–2130 cal BP and is characterized by extensive deposits of seal bone and fire-cracked rock, large numbers of flakes and artefacts, the remains of a dwelling and a large pit structure that may or may not be a second dwelling (Renouf 1994a, 2005). Faunal and artefactual data demonstrate a sharp focus on the late winter/early spring harp seal hunt (Kennett 1990; LeBlanc 1996). Pollen and macrofossil evidence from Bass Pond dating to 3000–2200 cal BP indicates Groswater activities at and near the pond, in particular tree clearing, pond-side trampling and a localized fire event (Bell et al. 2005b).

Ingold (1993) describes a taskscape as the spatial, physical, social and experiential context within which tasks are carried out. The Phillip's Garden East taskscape included (but was not limited to) the site area, Bass Pond, the beach directly below where the hunt was launched and landed, and Crow Head peak overlooking the site which was an excellent place from which to monitor ice conditions and seal availability. Near Crow Head peak, at David's site (Fig. 14.5), we found a Groswater preform and a few Palaeoeskimo artefacts and flakes. Pathways must have emanated outward from Phillip's Garden East to include the Phillip's Garden beach and terraces, the second Groswater site at Phillip's Garden West and possibly Crow Head Cave, discussed below.

Phillip's Garden West is a small, well-defined terrace at 13 m asl, dating to 2760–1820 cal BP. There is a panoramic view of the ocean from the terrace which, like Crow Head, makes it an excellent area for monitoring ice, ocean and seals (Fig. 2.6, Bell and Renouf, Chap. 2). The terrace is sheltered from southerly winds by the hillside at its back. Midden deposits were thrown over the terrace edge, accumulating as a palimpsest covering the hillside slope below.

Like Phillip's Garden East, the faunal assemblage from Phillip's Garden West reflects late winter/early spring harp seal hunting (Wells 2005, Chap. 4). However, this seasonality contrasts with the remains of a small and lightly constructed structure that suggests a summer tent. It also contrasts with the absence of fire-cracked rock, which was ubiquitous at Phillip's Garden East. Twenty-one radiocarbon dates from both sites overlap between 2760 and 2150 cal BP, a period of 600 years, indicating that the difference between the sites is not temporal.

The Phillip's Garden West lithic assemblage has been established as a Groswater lithic variant (Renouf 2005). Figure 14.6a shows two particularly fine examples; both endblades have tiny edge serrations, a convex base and narrow side-notches, all of which characterize less finely-made examples of Phillip's Garden West-type endblades and other tool categories. As Ryan discusses in Chap. 5, the most elongated and finely-made endblades from this site (including those in Fig. 14.6a) are the youngest and are from the upper terrace. Ryan notes that single specimens of similar endblades occur sporadically throughout Newfoundland in typical Groswater assemblages as well as in Dorset and Recent Indian contexts. She attributes this wide distribution to exchange rather than movement of people which suggests that these items were curated and therefore of some significance. Since the only known assemblage of these finely-made endblades comes from the upper terrace of Phillip's Garden West, their significance was likely linked to this original landscape context.

Although the abundant seal bones at Phillip's Garden West indicate harp seal hunting, there is only a single harpoon head in contrast to the nine whole and broken specimens from Phillip's Garden East (Renouf 1994a). The Phillip's Garden West specimen shares the partially open socket characteristic of those Groswater harpoon heads; however, its distal end is so thin and rudimentary that it may not have been functional (Fig. 14.6b).

In her comparison of Phillip's Garden East and West faunal assemblages, Wells (Chap. 4) notes a relative under-representation of seal cranial fragments in the Phillip's Garden West assemblages, which from a taphonomic perspective is unusual since certain cranial elements are very dense and therefore are commonly over-represented in seal bone assemblages. She links this under-representation to the significance that many northern people attribute to animal crania as an expression of the relationship between humans and animals. Circumpolar ethnography describes how heads were often carefully treated, for example, placed in trees away from animal gnawing (Tanner 1979:171), brought back to the settlement rather than abandoned at the kill site (Spencer 1957:264), or laid out on the ground outside dwellings and pointing in the direction of the next move (Rasmussen 1931:181). Wells suggests that at Phillip's Garden West the cranial elements were intentionally removed for reasons that are not clearly understood but which, like the Newfoundland-wide distribution of Phillip's Garden West-type specimens, suggests some kind of significance associated with the Phillip's Garden West landscape.

**Fig. 14.6** (**a**) Two finely-made harpoon endblades from the upper terrace of Phillip's Garden West. Note the fine edge serration, narrow side-notches and concave base, all attributes of the Phillip's Garden West lithic variant, (**b**) the only harpoon head found at Phillip's Garden West. This has the basal attributes of other Groswater harpoon heads, but the distal end is unique and possibly non-functional

The Phillip's Garden West taskscape encompassed Phillip's Garden, the large Dorset site between Phillip's Garden East and Phillip's Garden West (Fig. 14.5). Lavers (2006) identified 287 Groswater artefacts from 17 of 24 Dorset dwellings at Phillip's Garden which she argued were in secondary context and represented an earlier Groswater presence at the site. Except for the scrapers which she could not confidently assign to type, she identified all the other Groswater artefacts as Phillip's Garden West-type. Based on the high relative proportion of endblades and bifaces compared to other Groswater sites, she argued that the Groswater component at Phillip's Garden was a seal processing area (Lavers 2006:55). This is consistent with the concentration of these artefacts on a lower terrace that would have been approximately 3–4 m asl at that time (Bell et al. 2005a) and thus close to the beach where the seals were landed.

Phillip's Garden West is puzzling in a number of respects: the lack of seasonal correspondence between the dwelling feature and the faunal assemblage; the significance of the Phillip's Garden West lithic variant in general and its finest examples in particular; the single and singular harpoon head; and the low representation of seal cranial fragments in the faunal assemblage. Pursuing each of these puzzles leads to the same speculation that the significance of the site and associated material culture lies in the ideational aspects of the seal hunt (Renouf 2005). If so, the small terrace that is Phillip's Garden West would have been a symbolically charged place.

Dorset seal hunters followed close on the heels of the Groswater occupation. Based on radiocarbon dates, the first Dorset use of Phillip's Garden (1990–1870 cal BP) began not long after cessation of Groswater activities at Phillip's Garden East (2348–2130 cal BP) and Phillip's Garden West (2000–1820 cal BP). Although Phillip's Garden East and West were eventually covered by a thick layer of peat, it is unlikely that much peat had accumulated in the short period between Groswater abandonment and Dorset arrival. In addition, Phillip's Garden site stratigraphy indicates that it was unvegetated or thinly vegetated when Dorset first arrived. If this is correct, then the pioneer Dorset task groups or families would have seen Groswater artefacts at Phillip's Garden, abundant seal bones and fire-cracked rock at Phillip's Garden East and seal bones cascading down the hillside of Phillip's Garden West. Possibly some vegetation had developed over the organically-enriched middens; however, it is unlikely that they would have been obliterated since bones would have protruded and the vegetation itself would have indicated a culturally created anomaly. Incoming Dorset would have understood these signs on the landscape, which communicated earlier cultural activity and the suitability of the location for harp seal hunting. Dorset occupation continued to 1180 cal BP.

Everything we know about Phillip's Garden suggests that it was a highly enculturated landscape, linked to its subsistence function as a major harp seal hunting site. It had substantially constructed multi-family houses, many thousands of artefacts and seal bones, and a rich build-up of organic matter that created the heavily vegetated meadow of the present day. These indicate that Phillip's Garden was a permanent place on the landscape that was seasonally occupied year after year for approximately eight centuries (Renouf, Chap. 7).

Several multi-family houses, framed with whalebones and covered in seal hides (Renouf 2009), comprised a built environment around which seal processing activities focused. This included the many stages of activity, from preparations for the hunt at the beginning to disposal of food refuse at the end, in addition to hide working which involved many stages of its own (Renouf and Bell 2008). The location of houses and in some cases the houses themselves were re-occupied over long periods, sometimes several centuries (Renouf 2006), which means that they structured indoor and outdoor activities over generations.

The Phillip's Garden taskscape encompassed Bass Pond where there is a distinct disruption across pollen spectra dating to 2200–1400 cal BP.[3] We relate this disturbance and a synchronous increase in pond salinity (Rosenberg et al. 2005) to Dorset pond-side activities, in particular soaking sealskins for depilation (Bell et al. 2005b; Renouf and Bell 2008), similar to modern Northern Peninsula practice (Firestone 1992) (Fig. 7.6b, Renouf, Chap. 7).

Miniature sled shoes are interpreted as toys on the basis of width and/or placement of lashing holes and were found in small numbers ($n=38$) in eight of the house assemblages and outside two houses. These toys are proxy for the presence of children. They are also a reminder of the individual biographies associated with Phillip's Garden, as the children who lived there grew up to be seal hunters and hide processors. As the generations turned over, these biographies comprised the history and identity of Phillip's Garden and those connected to it.

I have earlier suggested that there was a significant ideational dimension to Phillip's Garden connected to the intensity of its seal hunting focus (Renouf 2000). A common theme among circumpolar hunting and gathering societies, related to their world view concerning the respectful treatment of animals, is to separate the products of the land from those of the sea, at least at important hunting sites (McGhee 1977; Søby 1969/1970). For example, among the Iglulik of the central Canadian Arctic, caribou bones could not be broken during the walrus hunting season (Rasmussen 1929:191) and among the coastal Inupiat of Alaska, hunters could not embark on the first whale hunt of the season until all land animal bones that had accumulated during the previous year were gathered up and burned (Spencer 1957:272). As discussed in Renouf (2000) there were only ten caribou bones in a sample of >25,000 bones from five midden features at Phillip's Garden and these were all non-meat bearing. Although the virtual absence of caribou bones might simply be a consequence of the seal hunting specialization whereby there was not the time to exploit caribou, it is hard to imagine ignoring these animals as they wandered about on the Point Riche headland during the winter, spring and early summer, as they do today. A possible explanation is an ideologically-based avoidance of caribou hunting at such an important harp seal hunting location. If so, it is intriguing that most of the Phillip's Garden harpoon heads are made of caribou antler (Wells 2010, personal communication).

I have also suggested that Phillip's Garden was a population aggregation site where groups of related families engaged in communal ritual and social activities that were essential to their cultural identity and social cohesion (Renouf 1994b). As discussed in Chap. 7, the late winter/early spring seal hunt created the necessity for a large labour force which in turn created the opportunity for community activities. If so, Phillip's Garden must have been an important place in the identity of the families who gathered there year after year. It was a fixed place in peoples' idea of who they were which they carried around in their minds, taking Phillip's Garden with them wherever they went.

The Phillip's Garden Dorset further signified the landscape by committing their dead to it. An infant was buried in the central area of House 12, although this seems

[3] A second Bass Pond core showed the same disruption across taxa but dated 200 years earlier (Bambrick 2009). This illustrates the difficulties in establishing chronological correspondence between the same disturbance events in two sediment cores and between sediment cores and archaeological sites.

to have been a single event rather than a general practise (Harp and Hughes 1968). Crow Head Cave was a more formal burial where at least six individuals were sequentially interred (Brown 1988, Chap. 12). Similar to the common hunter-gatherer practice of burying the dead in prominent places from which there is a good view (Charles and Buikstra 2002; Littleton and Allen 2007; Walthall 1999), Crow Head Cave lies beneath a distinctive peak overlooking the north coast of the Point Riche headland (Fig. 14.7a).

**Fig. 14.7** (**a**) The distinctive peak at Crow Head, beneath which is Crow Head Cave, (**b**) Groswater and Maritime Archaic Indian artefacts from Crow Head Cave. From *left* to *right* shows a Groswater harpoon head with partially open socket, a Groswater chipped and ground burin-like tool, the tip fragment of a finely serrated Phillip's Garden West-type endblade, and a cut and ground shell similar to specimens from the Maritime Archaic burial site, Port au Choix-3

Several Groswater artefacts were among the Crow Head burial goods that otherwise comprised a wide range of Dorset items (Fig. 12.2, Brown, Chap. 12). The Groswater artefacts included two Phillip's Garden West-type unnotched endblades (Renouf 2005, Fig. 11), the tip of another (Fig. 14.7b), and a Groswater harpoon head with a slotted tip appropriate for an unnotched endblade (Fig. 14.7b). Brown (1988:51) notes that these Groswater items were in good burial context and suggests the possibility that one of the inhumations was a Groswater individual. If correct, this indicates that the cave had ancestral significance for both Palaeoeskimo cultures. However, it is also possible that the Dorset appropriated Groswater objects for inclusion in Dorset graves. That this was indeed the case is suggested by the inclusion of a ground periwinkle shell (Fig. 14.7b), also in good burial context, which is similar to those from several Maritime Archaic burials from Port au Choix-3 (see Tuck 1976:231) and presumably comes from one of them. If so, not only were the Dorset aware of earlier populations but they also understood the significance of their objects.

The three Phillip's Garden sites lie between two markers on the horizon. To the east is the peak of Crow Head (Fig. 14.7a) and to the west is a cairn that overlooks the sea (Fig. 14.8), Phillip's Garden West (see Fig. 2.6, Bell and Renouf, Chap. 2), and Phillip's Garden. We do not know the date or cultural affiliation of the cairn, but the heavy growth of lichen suggests some antiquity. Together the cairn and the peak effectively function as navigational beacons directing someone at sea to the Phillip's Garden location. We do not know if this was their actual function.

**Fig. 14.8** An undated, lichen-covered cairn that overlooks the sea, Phillip's Garden West, and Phillip's Garden. There is no sightline between the cairn and Phillip's Garden West in contrast to the clear sightline between the cairn and Phillip's Garden

## The Life History of Port au Choix

People did not limit themselves to the three landscapes described above, but moved among and beyond them. The geography of some of these sites suggests paths of movement, as for example, the Maritime Archaic burial ground overlooking a constriction on a major coastal travel route (Renouf and Bell, Chap. 3) and the Groswater sites along the Back Arm escarpment that were similarly positioned. In addition, three Dorset burial caves monitored movement in and around Port au Choix: Crow Head Cave, Eastern Point (EeBi-10) and Gargamelle Rockshelter (EeBi-25) (Fig. 14.9). Each of these burial caves looks out to the sea in a different direction, Crow Head north over the Point Riche headland, Eastern Point east over the entranceway to Back Arm, and Gargamelle Rockshelter east over Gargamelle Cove. Finally, the Recent Indian occupation of the Gould site suggests inland paths of movement in their choice to step back from the ocean view, preferring instead to be oriented in the direction of the interior.

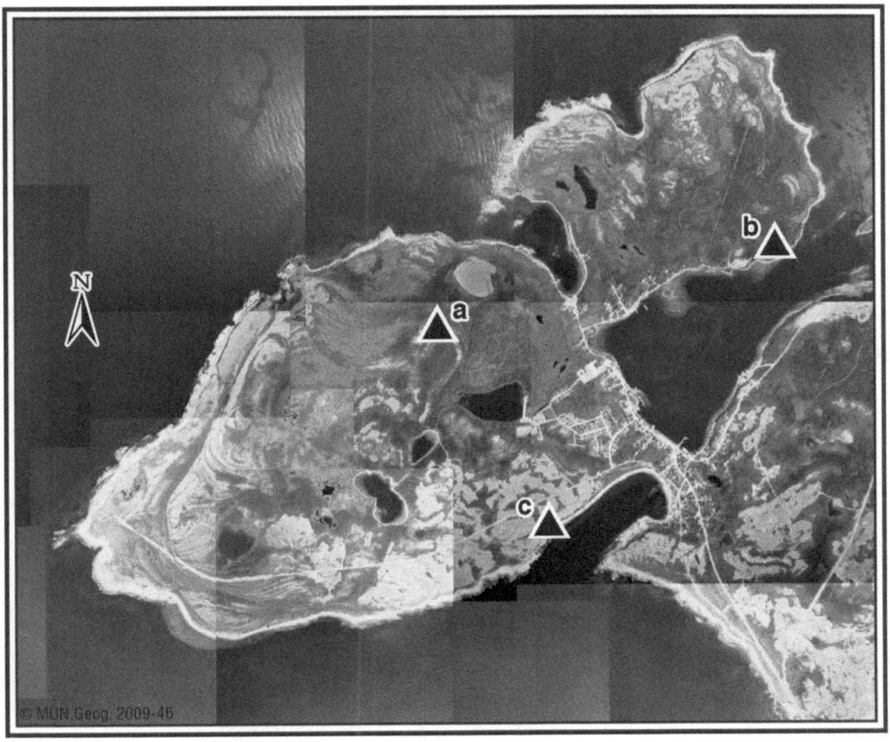


**Fig. 14.9** The location of three Dorset burial caves at Port au Choix, each overlooking the ocean in a different direction. (**a**) Crow Head Cave, (**b**) Eastern Point, and (**c**) Gargamelle Rockshelter

## The Life History of the Northern Peninsula

Culturally, Port au Choix is a deeply stratified location by virtue of 6,000 years of Amerindian and Palaeoeskimo occupation. Other locations were similarly stratified, for example (Fig. 14.10): L'Anse aux Meadows (Wallace 2006), Big Brook

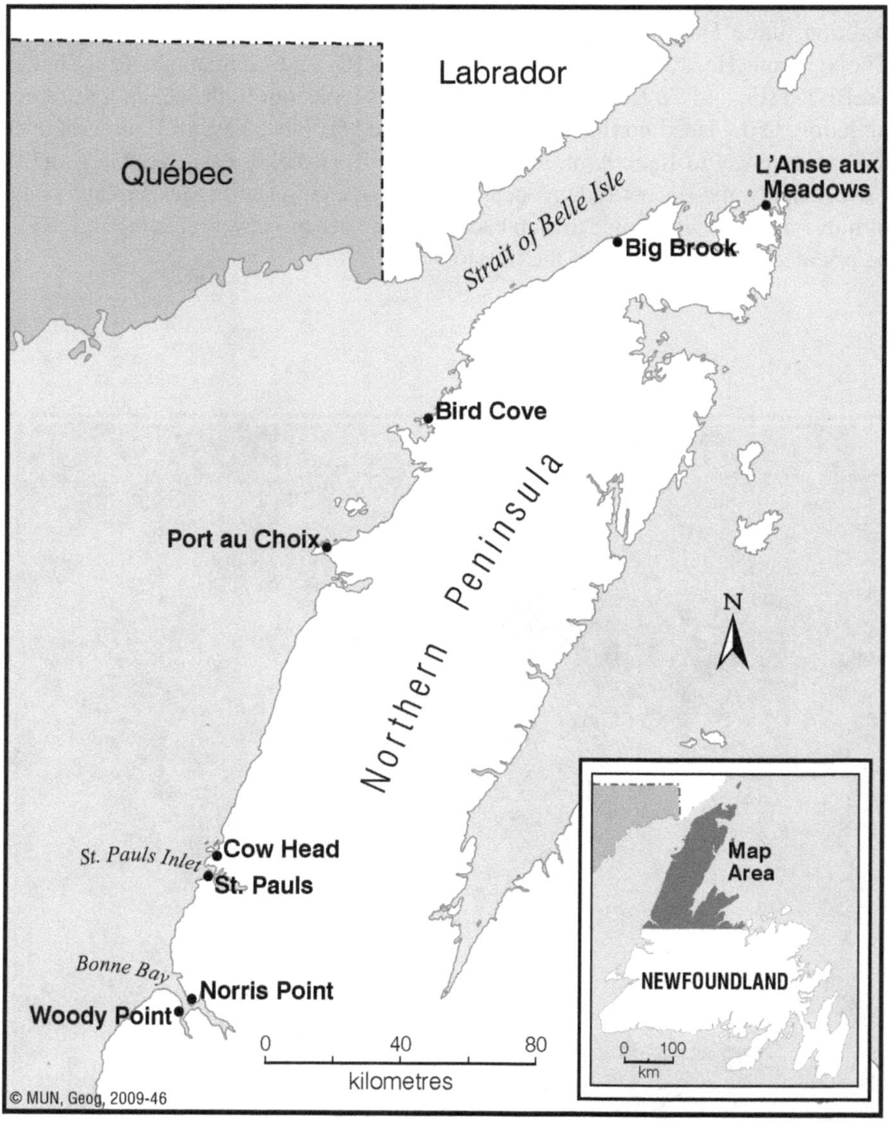


**Fig. 14.10** Culturally stratified locations on the west coast of the Northern Peninsula which are landmarks within the Northern Peninsula landscape

(Beaton 2004), Bird Cove (Hartery and Rast 2001), Cow Head (Auger 1984; Tuck 1978), St. Pauls (Lavers 2010), Norris Point (Harp 1964a, b), and Woody Point (Harp 1964a, b; Schwarz and Skanes 2005). These locations were all landmarks linked together in a network of connections and relations that comprised the cultural landscape of the west coast of the Northern Peninsula.

Port au Choix was a prominent physical landmark within this landscape, projecting out from the otherwise straight Northern Peninsula coastline to serve as a distinctive visual marker for all who travelled up and down the coast. It was also a prominent cultural landmark. Maritime Archaic individuals and groups throughout the Northern Peninsula were connected to Port au Choix through the burial ground at Port au Choix-3. Groswater people throughout the Northern Peninsula knew Phillip's Garden East as a very important harp seal hunting locale. Dorset families from the larger area were drawn to the permanent harp seal hunting settlement at Phillip's Garden where they engaged in economic, social and ritual activities; they would have identified with ancestors buried in Port au Choix caves. We know very little about the Recent Indian populations; however, it is interesting that contact between Cow Head complex Recent Indians and Dorset Palaeoeskimos took place at Port au Choix, a place to which both groups had ties.

## Discussion

These life histories are an example of the fundamental interplay between the physical and cultural dimensions of landscape (Cooney 2000; David and Thomas 2008; Ingold 1993; Knapp and Ashmore 1999; Tilley 1994; Zvelebil 2003). For example, at Port au Choix we can see how the physical landscape structured site locations: an island was considered appropriate for a Maritime Archaic burial ground, particularly since it was a projection that was visible from a long distance; the protective escarpment at the head of Back Arm attracted a sequence of occupations along its base; a persistent lead in the sea ice attracted Palaeoeskimo seal hunters at Phillip's Garden for almost 2,000 years; a high terrace was a good seal and ice observation post; and caves were appropriate for Dorset burials. More generally, Palaeoeskimo populations were attracted to outer coastal areas for access to marine resources and Amerindians were attracted to sheltered inner coasts for access to a broader resource range.

These physical landscapes changed over time (Bell et al. 2005a). The Maritime Archaic burial was proximal to a narrow marine channel that was dry land by the time of Recent Indian occupation. As landscape changed, so did human perception and use of it. Maritime Archaic Indians chose the coastal location of the Gould site in part because it had a direct line of vision to the burial ground whereas Recent Indians chose the same location because it was a sheltered, wooded spot that looked inward to the interior rather than outward to the ocean.

Port au Choix landscapes became enculturated as people experienced them, turning them into places permeated with knowledge, memory, history, identity and emotion. The Maritime Archaic and Dorset burials were places connoting ancestral history and social identity. Phillip's Garden was an important reference point in the life cycle and identity of the families who gathered there. Phillip's Garden East and Phillip's Garden West were enculturated with many seasons of seal harvesting activities including the symbolic dimension of the hunt.

Hunter-gatherers worldwide give meaning to places through naming and storytelling. Nuttall (1992) notes how for the Kangersuatsiarmiut of northwest Greenland, naming places is important because it situates a location in the individual and community memoryscape. Many of the place names refer to landscape use, in particular associated resources. For example, some of the place name translations are "arctic char," "walrus" and "river that belongs to the place of the spotted seal" (Nuttall 1992:51). Often there are additional layers of meaning such as stories about events, for example, the place where a man once made a beautiful harpoon head which he lost when trying to kill a walrus (Nuttall 1992:55). Stewart et al. (2004:191) make similar observations about how the Kivallirmiut of interior Nunavut organize their landscape through place names, referring to geographic features, resources, events and wayfinding. Morphy (1995) notes that for the Yolngu of northern Australia, places are identified by stories that situate a person within social and mythical context. Such place names are an ideational and locational grid on a landscape that, like the Arctic barrens or Australian desert, might seem featureless to the outsider but which is a detailed network of known places to those who live there (Aporta 2009; Nuttall 1992; Stewart et al. 2004; Whitridge 2004). The Northern Peninsula must have been similarly storied and named. We can guess that the names and stories associated with the Phillip's Garden locality might have referenced seals, sea and ice.

Landscape is animated by paths of movement among named places (Aporta 2004; Littleton and Allen 2007; Zedeño 2000). The Maritime Archaic burial ground and Dorset burial caves were positioned with respect to the ocean, overlooking people travelling by sea. Travel to the sources of Cow Head chert, 150 km south of Port au Choix, is visible in the presence of this fine-grained lithic at most Port au Choix sites. At a far greater distance, the landscape of northern Labrador is visible in the cache of Ramah chert tools at Port au Choix-3 and small amounts of Ramah chert at other Port au Choix sites, reflecting travel routes between Newfoundland and Labrador. It may also reflect knowledge of a landscape never visited but nonetheless known, a map of distant places in the imagination (Brody 1981; Whitridge 2004).

Landscapes are complemented by seascapes (Cooney 2003; Niven 2003; Wells 2009), which must have been a prominent element of Port au Choix life, in particular for Palaeoeskimo marine specialists. Two potential navigational beacons, Crow Head peak and an ancient cairn, mark the location of Phillip's Garden from the sea, and Dorset burial caves collectively survey the seascape around Port au Choix. Seascapes become icescapes (Aporta 2002; Nuttall 1992), which would have been an important aspect of the Phillip's Garden experience where the subsistence focus was ice-edge seal hunting. A panoramic view of the ice fields could be had from Phillip's Garden West and Crow Head.

Landscape is cumulative (Zedeño 2008:214), evolving over time with sequences of use and layers of signification that collectively comprise its life history. At Port au Choix those layers were connected through time via knowledge of prior occupations that were part of the collective memory, analogous perhaps to Inuit legends of the Tunniit, giants reputed to be the pioneers of the Arctic (Hallendy 2000; McGhee 1981). Groswater people were aware of the Maritime Archaic burial ground, Dorset and Recent Indian people knew about Groswater activities, and Dorset and Recent Indians were aware of each other. Dorset people may have appropriated objects of symbolic importance to Groswater and Maritime Archaic cultures for inclusion in the Crow Head Cave burials.

The life history of Port au Choix continues to evolve. Early historians and archaeologists (Harp 1964a, b; Howley 1915; Wintemberg 1939) became knowledgeable about its precontact occupations and their investigations set the stage for subsequent archaeological work (Renouf 1999; Tuck 1976). The results of this research have been memorialized in a stone building on a prominent limestone terrace above Gargamelle Rockshelter (Fig. 14.11). The Visitor Centre of the Port au Choix National Historic Site tells the story of the many thousands of years of human occupation of Port au Choix, connecting these lifeways with modern rural settlement (Harp 2003; Renouf 1999, 2003b). The Visitor Centre is a structure on the landscape that incorporates memory, experience and history. It has created a new dynamic of human interaction as people walk from the centre along a network of hiking trails that connect the area's archaeological sites, each of which is a landmark on the trail system. The centre can be seen from many miles away as a

**Fig. 14.11** Port au Choix National Historic Site Visitor Centre, which can be seen as a peak on the horizon for a distance of several kilometres as one travels the coast by road or water

circumflex on the horizon, not unlike the point that defines Crow Head. It is the most recent in a long and connected lineage of cultural landscapes that together make up the long-term history of Port au Choix.

## Conclusions

The Amerindian and Palaeoeskimo populations who occupied Newfoundland viewed the physical landscape differently, based on their distinct cultural origins and historical trajectories. These divergent perspectives converged in Port au Choix and were expressed in different patterns of land use and landscape creation. A close examination of three Port au Choix landscapes shows how successive cultural occupations cumulatively created layers of meaning that comprised each landscape's life history. Each of these landscapes was a landmark within Port au Choix, which was itself an important landmark within the larger landscape of the west coast of the Northern Peninsula. Precontact peoples were aware of landscape life histories thereby linking them through time in a process that continues to this very day.

**Acknowledgements** I would like to thank Parks Canada, the Social Sciences and Humanities Research Council of Canada, and Memorial University, for their sustained funding without which the past 25 years of research at Port au Choix would not have been possible. The Canada Research Chairs Program, the Canada Foundation for Innovation and the Research Council of Newfoundland and Labrador have greatly enhanced the last 10 years of research. Thanks to Charles Conway, Memorial University Geography Department, for drafting the figures and Chris Hammond, Memorial University Photographic Services, for taking the artefact photographs. Thanks also to the Provincial Archaeology Office for their generous sharing of information and expertise.

## References

Anton, E.P. 2004. St. John's Harbour 5 (HeCi-30) and an Examination of Groswater and Early Dorset Relationships in Labrador. Master's thesis, Department of Anthropology, Memorial University of Newfoundland, St. John's.

Aporta, C. 2002. Life on the ice: Understanding the codes of a changing environment. *Polar Record* 30(207): 341–354.

Aporta, C. 2004. Routes, trails and tracks: Trail breaking among the Inuit of Igloolik. *Études/Inuit Studies* 28(2): 9–38.

Aporta, C. 2009. The trail as home: Inuit and their pan-Arctic network of routes. *Human Ecology* 37: 131–146.

Auger, R. 1984. Factory Cove: Recognition and Definition of the Early Palaeo-Eskimo Period in Newfoundland. Master's thesis, Department of Anthropology, Memorial University of Newfoundland, St. John's.

Austin, S.J. 1980. Cape Cove Beach (DhAi-5, 6, 7), Newfoundland: Prehistoric Cultures. Master's thesis, Department of Anthropology, Memorial University of Newfoundland, St. John's.

Balzer, M.M. 1996. Flights of the sacred: Symbolism and theory in Siberian shamanism. *American Anthropologist* 98(2): 305–318.

Bambrick, J. 2009. Palaeoeskimo Sealskin Processing in Port au Choix, Northwestern Newfoundland: A Paleoenvironmental Analysis. Master's thesis, Department of Archaeology, Memorial University of Newfoundland, St. John's.

Barnable, S.K. 2008. Rattling Brook 1 (DgAt-1): An Examination of Middle Dorset Inner Bay Settlements. Master's thesis, Department of Archaeology, Memorial University of Newfoundland, St. John's.

Beaton, G. 2004. A Chip off the Old Block: Investigations of a Maritime Archaic Lithic Workshop/Quarry Site in Big Brook (EjBa-2), Northwestern Newfoundland. Master's thesis, Department of Anthropology, Memorial University of Newfoundland, St. John's.

Bell, T., I.R. Smith, and M.A.P. Renouf. 2005a. Postglacial sea-level history and coastline change at Port au Choix, Great Northern Peninsula, Newfoundland. *Newfoundland and Labrador Studies* 20(1): 9–31.

Bell, T., J.B. Macpherson, and M.A.P. Renouf. 2005b. Late prehistoric impact on Bass Pond, Port au Choix. *Newfoundland and Labrador Studies* 20(1): 107–129.

Boas, F. 1888. *The Central Eskimo*. Sixth Annual Report of the Bureau of American Ethnology. Washington, D.C.: Smithsonian Institution Press.

Brody, H. 1981. *Maps and dreams: Indians and the British Columbia frontier*. Vancouver: Douglas and McIntyre.

Brown, S.C. 1988. Archaeological Investigations at Crow Head Cave and the Gargamelle Rockshelter in the Port au Choix National Historic Park, Newfoundland: Report of 1986 Field Activities. On file, Parks Canada, Archaeology, Atlantic Region, Halifax.

Carignan, P. 1975. Archaeological Survey 1975. Report for the Newfoundland Museum. On file, Provincial Archaeology Office, Government of Newfoundland and Labrador, St. John's.

Charles, D.K., and J.E. Buikstra. 2002. Siting, sighting, and citing the dead. In *The space and place of death*, ed. H. Silverman and D.B. Small, 13–25. Archaeological Papers of the American Anthropological Association 11. American Anthropological Association, Arlington, VA.

Cooney, G. 2000. *Landscapes of neolithic Ireland*. London: Routledge.

Cooney, G. 2003. Introduction: Seeing land from the sea. *World Archaeology* 35(3): 323–328.

David, B., and J. Thomas. 2008. Landscape archaeology: introduction. In *Handbook of landscape archaeology*, ed. B. David and J. Thomas, 27–43. Walnut Creek: Left Coast Press.

Erwin, J. 2001. A Prehistoric Soapstone Quarry in Fleur de Lys, Newfoundland. PhD thesis, Department of Anthropology, University of Calgary, Calgary.

Fienup-Riordan, A. 1990. *Eskimo essays: Yup'ik lives and how we see them*. London: Rutgers University Press.

Fienup-Riordan, A. 1996. *The living tradition of Yup'ik masks*. Seattle: University of Washington Press.

Firestone, M. 1992. Inuit derived culture traits in northern Newfoundland. *Arctic Anthropology* 29(1): 112–128.

Fitzhugh, W.W. 1972. *Environmental archeology and cultural systems in Hamilton Inlet, Labrador: Survey of the central Labrador coast from 3000 B.C. to the present*. Washington, D.C.: Smithsonian Institution Press.

Fitzhugh, W.W. 2006. Settlement, social and ceremonial change in the Labrador Maritime Archaic. In *The Archaic of the far northeast*, ed. D. Sanger and M.A.P. Renouf, 47–81. Orono: Maine University Press.

Fogt, L.M. 1998. The Excavation and Analysis of a Dorset Palaeoeskimo Dwelling at Cape Ray, Newfoundland. Master's thesis, Department of Anthropology, Memorial University of Newfoundland, St. John's.

Grønnow, B. 2009. Blessings and horrors of the interior: Ethno-historical studies of Inuit perceptions concerning the inland region of west Greenland. *Arctic Anthropology* 46(1–2): 191–201.

Hallendy, N. 2000. *Inuksuit: Silent messengers of the Arctic*. Toronto: Douglas and McIntyre.

Harp, E. 1964a. *The cultural affinities of the Newfoundland Dorset Eskimo*. National Museums of Canada Bulletin 200, Ottawa.

Harp, E. 1964b. Evidence of Boreal Archaic culture in southern Labrador and Newfoundland. In *Contributions to anthropology 1961–62: Part I*, 184–260. National Museums of Canada Bulletin 193, Ottawa.

Harp, E. 1976. Dorset settlement patterns in Newfoundland and southeastern Hudson Bay. In *Eastern Arctic prehistory: Paleoeskimo problems*, ed. M.S. Maxwell, 119–138. Washington, D.C.: Society for American Archaeology.

Harp, E. 2003. *Lives and landscapes: A photographic memoir of outport Newfoundland and Labrador, 1949–1963*. Montreal: McGill-Queens University Press.

Harp, E., and D.R. Hughes. 1968. Five prehistoric burials from Port au Choix. *Polar Notes* 8: 1–47.

Hartery, L. 2007. *The Cow Head complex and the Recent Indian period in Newfoundland, Labrador and the Quebec Lower North Shore*. Occasional Papers in Northeastern Archaeology 17. St. John's: Copetown Press.

Hartery, L., and T. Rast. 2001. Bird Cove Archaeology Project 2000 Field Season: Final report. On file, Provincial Archaeology Office, Government of Newfoundland and Labrador, St. John's.

Hodgetts, L.M., M.A.P. Renouf, M.S. Murray, D. McCuaig-Balkwill, and L. Howse. 2003. Changing subsistence practices at the Dorset Paleoeskimo site of Phillip's Garden, Newfoundland. *Arctic Anthropology* 40(1): 106–120.

Holly, D.H. 2002. From Space to Place: An Archaeology and Historical Geography of the Recent Indian Period in Newfoundland. PhD dissertation, Department of Anthropology, Brown University, Providence.

Holly, D.H. 2003. Places of the living, places of the dead: Situating a sacred geography. *Northeast Anthropology* 66: 57–75.

Holly, D.H., and J.C. Erwin. 2009. Terra incognita, still: Archaeological investigations in the interior of the island of Newfoundland. *Archaeology of Eastern North America* 37: 65–84.

Howley, J.P. 1915. *The Beothucks or Red Indians: The aboriginal inhabitants of Newfoundland*. Cambridge: Cambridge University Press.

Ingold, T. 1993. The temporality of landscape. *World Archaeology* 25(2): 152–174.

Jelsma, J. 2000. A Bed of Ochre: Mortuary Practices and Social Structure of a Maritime Archaic Indian Society at Port au Choix, Newfoundland. PhD dissertation, University of Groningen, Groningen.

Jelsma, J. 2006. Three social status groups at Port au Choix: Maritime Archaic mortuary practices and social structure. In *The Archaic of the far northeast*, ed. D. Sanger and M.A.P. Renouf, 83–103. Orono: University of Maine Press.

Jordan, P. 2003. *Material culture and sacred landscape: The anthropology of the Siberian Khanty*. New York: Rowman and Littlefield Publishers.

Kennett B.L. 1990. Phillip's Garden East: An Examination of the Groswater Palaeoeskimo Phase. Master's thesis, Department of Anthropology, Memorial University of Newfoundland, St. John's.

Knapp, A.B., and W. Ashmore. 1999. Archaeological landscapes: Constructed, conceptualized, ideational. In *Archaeologies of landscape: Contemporary perspectives*, ed. A.B. Knapp and W. Ashmore, 1–30. Oxford: Blackwell Publishers.

Lavers, D. 2006. The Groswater Palaeoeskimo Component at Phillip's Garden (EeBi-1), Port au Choix, Newfoundland. Honour's dissertation, Department of Anthropology and Archaeology, Memorial University of Newfoundland, St. John's.

Lavers, D. 2010. The Recent Indian Cow Head Complex Occupation of the Northern Peninsula, Newfoundland: A Geochemical Approach to Investigating Cow Head Chert Acquisition. Master's thesis, Department of Archaeology, Memorial University of Newfoundland, St. John's.

LeBlanc, S. 1996. A Place with a View: Groswater Subsistence-Settlement Patterns in the Gulf of St. Lawrence. Master's thesis, Department of Anthropology, Memorial University of Newfoundland, St. John's.

LeBlanc, S. 2000a. Groswater technological organization: A decision-making approach. *Arctic Anthropology* 37(2): 23–37.

LeBlanc, S. 2000b. Middle Dorset (1900–1100 BP) regional variability on the island of Newfoundland and in Saint-Pierre et Miquelon. In *Identities and cultural contacts in the Arctic*, ed. M. Appelt, J. Berglund, and H.C. Gulløv, 97–105. Copenhagen: The Danish National Museum and Danish Polar Center.

LeBlanc, S. 2008. Middle Dorset Variability and Regional Culture Traditions: A Case Study from Newfoundland and Saint-Pierre and Miquelon. PhD thesis, Department of Anthropology, University of Alberta, Edmonton.

Linnamae, U. 1975. *The Dorset culture: A comparative study in Newfoundland and the Arctic*. Technical Papers of the Newfoundland Museum 1, St. John's.

Littleton, J., and H. Allen. 2007. Hunter-gatherer burials and the creation of persistent places in southeastern Australia. *Journal of Anthropological Archaeology* 26(2): 282–298.

Liverman, D.G.E. 1994. Relative sea-level history and isostatic rebound in Newfoundland, Canada. *Boreas* 23: 217–230.

Loring, S. 2002. And they took away the stones from Ramah: Lithic raw material sourcing and eastern Arctic archaeology. In *Honoring our elders: A history of eastern Arctic archaeology*, ed. W.W. Fitzhugh, S. Loring, and D. Odess, 153–185. Washington, D.C.: Arctic Studies Center.

Loring, S., and S. Cox. 1986. The Postville Pentecostal Groswater site, Kaipokok Bay, Labrador. In *Palaeo-Eskimo cultures in Newfoundland, Labrador and Ungava*, 65–93. Reports in Archaeology No. 1, Memorial University of Newfoundland, St. John's.

Mannermaa, K. 2008. Birds and burials at Ajvide (Gotland, Sweden) and Zvejnieki (Latvia) about 8000–3900 BP. *Journal of Anthropological Archaeology* 27: 201–225.

Maxwell, M.S. 1985. *Prehistory of the eastern Arctic*. New York: Academic Press.

McGhee, R.J. 1977. Ivory for the sea woman: The symbolic attributes of a prehistoric technology. *Canadian Journal of Archaeology* 1: 141–149.

McGhee, R.J. 1981. *The Tunit: First explorers of the high Arctic*. National Museums of Canada, Ottawa.

McGhee, R.J., and J.A. Tuck. 1975. *An Archaic sequence from the Strait of Belle Isle, Labrador*. Mercury Series Archaeology Paper 34, National Museums of Canada, Ottawa.

Merkur, D. 1987. Eagle, the hunter's helper: The cultic significance of Inuit mythological tales. *History of Religions* 27(2): 171–188.

Morphy, H. 1995. Landscape and the reproduction of the ancestral past. In *The anthropology of landscape: Perspectives on place and space*, ed. E. Hirsch and M. O'Hanlon, 184–209. Oxford: Clarendon Press.

Nagy, M. 1994. A critical review of the Pre-Dorset/Dorset transition. In *Threads of Arctic prehistory: Papers in honour of William E. Taylor, Jr.*, ed. D. Morrison and J.-L. Pilon, 1–14. Mercury Series Archaeology Paper 149, Canadian Museum of Civilization, Ottawa.

Niven, I.J. 2003. Saltwater people: Spiritscapes, maritime rituals and the archaeology of Australian indigenous landscapes. *World Archaeology* 35(3): 329–349.

Nuttall, M. 1992. *Arctic homeland: Kinship, community and development in northwest Greenland*. Toronto: University of Toronto Press.

Pastore, R.T. 1986. The spatial distribution of late Palaeo-Eskimo sites on the island of Newfoundland. In *Palaeo-Eskimo cultures in Newfoundland, Labrador and Ungava*, 125–134. Reports in Archaeology No. 1, Memorial University of Newfoundland, St. John's.

Pastore, R.T. 1992. *Shanawdithit's people*. St. John's: Atlantic Archaeology.

Pintal, J.-Y. 1994. A Groswater site at Blanc-Sablon, Quebec. In *Threads of Arctic prehistory: Papers in honour of William E. Taylor, Jr*, ed. D. Morrison and J.-L. Pilon, 145–164. Mercury Series Archaeology Paper 149, Canadian Museum of Civilization, Ottawa.

Pintal, J.-Y. 1998. *Aux frontières de la mer: La préhistoire de Blanc-Sablon*. Les Publications du Québec, Dossiers 102. Collections Patrimoines et Municipalité de Blanc-Sablon, Québec.

Pintal, J.-Y. 2006. The Archaic sequence of the St. Lawrence Lower North Shore, Quebec. In *The Archaic of the far northeast*, ed. D. Sanger and M.A.P. Renouf, 105–138. Orono: Maine University Press.

Pope, P.E. 2009. Transformation of the maritime cultural landscape of Atlantic Canada by migratory fishermen, 1500–1800. In *Fisheries of the North Atlantic, the North Sea and the Baltic, 900–1850*, ed. L. Sicking and D. Abreu-Ferreira, 123–154. Leiden: Brill.

Provincial Archaeology Office. 2010. Archaeological Site Database. Department of Tourism, Culture and Recreation, Government of Newfoundland and Labrador, St. John's.

Rainbird, P. 2007. *The archaeology of islands*. Cambridge: Cambridge University Press.

Rasmussen, K. 1929. *Intellectual culture of the Iglulik Eskimos*. Report of the Fifth Thule Expedition 1921–24 , Vol. 7(1). Guldendalske Boghandel, Nordisk Forlag, Copenhagen.
Rasmussen, K. 1931. *The Netsilik Eskimos: Social life and spiritual culture*. Report of the Fifth Thule Expedition 1921–24 , Vol. 8(1–2). Guldendalske Boghandel, Nordisk Forlag, Copenhagen.
Reid, H.S. 2008. A Study of Southern Variant, Maritime Archaic Sites from the Northern Peninsula and Strait of Belle Isle, Newfoundland and Labrador. Master's thesis, Department of Anthropology, Memorial University of Newfoundland, St. John's.
Renouf, M.A.P. 1994a. Two transitional sites at Port au Choix, northwestern Newfoundland. In *Threads of Arctic prehistory: Papers in honour of William E. Taylor, Jr.*, ed. D. Morrison and J.-L. Pilon, 165–196. Mercury Series Archaeology Paper 149, Canadian Museum of Civilization, Ottawa.
Renouf, M.A.P. 1994b. Hunter-gatherer population aggregations in prehistory: A case study from Port au Choix, northwestern Newfoundland. In *Hunters and gatherers in the modern context*, Proceedings of the Seventh International Conference on Hunting and Gathering Societies, ed. L. Ellanna, 529–539. Department of Anthropology, University of Alaska at Fairbanks, Fairbanks.
Renouf, M.A.P. 1999. *Ancient cultures, bountiful seas: The story of Port au Choix*. Historic Sites Association of Newfoundland and Labrador, St. John's.
Renouf, M.A.P. 2000. Symbolism and subsistence: Seals and caribou at Port au Choix, northwestern Newfoundland. In *Animal bones, human societies*, ed. P. Rowley-Conwy, 65–73. Oxford: Oxbow Books.
Renouf, M.A.P. 2003a. Hunter-gatherer interactions: Mutualism and resource partitioning on the island of Newfoundland. *Before Farming* 1(4): 1–16.
Renouf, M.A.P. 2003b. Editor's introduction: Newfoundland outports. In *Lives and landscapes: a photographic memoir of outport Newfoundland and Labrador, 1949–1963*, ed. E. Harp Jr., xv–xxi. Montreal: McGill-Queens University Press.
Renouf, M.A.P. 2005. Phillip's Garden West: a Newfoundland Groswater variant. In *Contributions to the study of the Dorset Palaeo-Eskimos*, ed. P.D. Sutherland, 57–80. Mercury Series Archaeology Paper 167, Canadian Museum of Civilization, Gatineau.
Renouf, M.A.P. 2006. Re-investigating a middle phase Dorset dwelling at Phillip's Garden, Port au Choix, Newfoundland. In *Dynamics of northern societies*, Proceedings of the SILA/NABO Conference on Arctic and North Atlantic Archaeology, Copenhagen, May 10th–14th, 2004, ed. J. Arneborg and B. Grønnow, 119–128. Publications from the National Museum Studies in Archaeology and History Vol. 10, Copenhagen.
Renouf, M.A.P. 2009. Dorset Palaeoeskimo whalebone use at Phillip's Garden, Port au Choix. In *On the track of the Thule culture, from Bering Strait to east Greenland*, Proceedings of the SILA Conference The Thule Culture – New Perspectives in Inuit Prehistory, ed. B. Grønnow, 91–104. Publications from the National Museum Studies in Archaeology and History Vol. 15, Copenhagen.
Renouf, M.A.P., and T. Bell. 1998. Searching for the Maritime Archaic Indian Habitation Site at Port au Choix, Newfoundland: An Integrated Approach using Archaeology, Geomorphology and Sea Level History. On file, Provincial Archaeology Office, Government of Newfoundland and Labrador, St. John's.
Renouf, M.A.P., and T. Bell. 2006. Maritime Archaic site location patterns on the island of Newfoundland. In *The Archaic of the far northeast*, ed. D. Sanger and M.A.P. Renouf, 1–46. Orono: Maine University Press.
Renouf, M.A.P., and T. Bell. 2008. Dorset Palaeoeskimo sealskin processing at Phillip's Garden, Port au Choix, northwestern Newfoundland. *Arctic* 61(1): 35–47.
Renouf, M.A.P., T. Bell, and M.A. Teal. 2000. Making contact: Recent Indians and Palaeoeskimos on the island of Newfoundland. In *Identities and cultural contacts in the Arctic*, ed. M. Appelt, J. Berglund, and H.-C. Gulløv, 106–119. Copenhagen: The Danish National Museum and Danish Polar Center.
Renouf, M.A.P., T. Bell, and J. Macpherson. 2009. Hunter-gatherer impact on Subarctic vegetation: Amerindian and Palaeoeskimo occupations of Port au Choix, northwestern Newfoundland. *Arctic Anthropology* 46(1–2): 176–190.

Robbins, D.R. 1985. Stock Cove Trinity Bay: The Dorset Eskimo Occupation of Newfoundland from a Southeastern Perspective. Master's thesis, Department of Anthropology, Memorial University of Newfoundland, St. John's.

Robinson, B.S. 2006. Burial ritual, technology, and cultural landscape in the far northeast: 8600–3700 B.P. In *The Archaic of the far northeast*, ed. D. Sanger and M.A.P. Renouf, 341–381. Orono: Maine University Press.

Rosenberg, S., I. Walker, and J. Macpherson. 2005. Environmental changes at Port au Choix as reconstructed from fossil midges. *Newfoundland and Labrador Studies* 20(1): 57–73.

Rowley-Conwy, P.A. 1990. Settlement patterns of the Beothuk Indians of Newfoundland: A view from away. *Canadian Journal of Archaeology* 14: 13–32.

Sanger, D. 2003. Who lived in pre-European Maine?: A cosmology approach to social patterning on the landscape. *Northeast Anthropology* 66: 29–39.

Schwarz, F.A. 1993. Archaeological Investigations on the Bay of Exploits, June–August 1993. On file, Exploits Valley Tourism Association, Grand Falls-Windsor, Newfoundland.

Schwarz, F.A. 1994. Paleo-Eskimo and Recent Indian subsistence and settlement patterns on the island of Newfoundland. *Northeast Anthropology* 47: 55–70.

Schwarz, F., and R. Skanes. 2005. Stage 2 Assessment of DjBl-04, Woody Point, Bonne Bay, Newfoundland: Interim Report on Site Significance. On file, Provincial Archaeology Office, Government of Newfoundland and Labrador, St. John's.

Smith, I.R., T. Bell, and M.A.P. Renouf. 2005. Testing a proposed late Holocene sea-level oscillation using the isolation basin approach, Great Northern Peninsula, Newfoundland. *Newfoundland and Labrador Studies* 20(1): 33–55.

Søby, R.M. 1969/1970. The Eskimo animal cult. *Folk* 11–12: 43–78.

Speck, F.G. 1935. *Naskapi: The savage hunters of the Labrador Peninsula*. Norman: University of Oklahoma Press.

Spencer, R.F. 1957. *The North Alaskan Eskimo: A study in ecology and society*. Bureau of American Ethnology Bulletin 171. Washington, D.C.: Smithsonian Institution Press.

Stewart, A.M., D. Keith, and J. Scottie. 2004. Caribou crossings and cultural meanings: Placing traditional knowledge and archaeology in context in an Inuit landscape. *Journal of Archaeological Method and Theory* 11(2): 183–211.

Stuiver, M., and P.J. Reimer. 1993. Extended 14C database and revised CALIB radiocarbon calibration program. *Radiocarbon* 35: 215–230.

Tanner, A. 1979. *Bringing home animals: Religious ideology and mode of production of the Mistassini Cree hunters*. Newfoundland Social and Economic Studies 23, Institute of Social and Economic Research, Memorial University of Newfoundland, St. John's.

Teal, M.A. 2001. An Archaeological Investigation of the Gould Site (EeBi-42) in Port au Choix, Northwestern Newfoundland: New Insight into the Recent Indian Cow Head Complex. Master's thesis, Department of Anthropology, Memorial University of Newfoundland, St. John's.

Tilley, C. 1994. *A phenomenology of landscape: Places*. Oxford: Berg Publishers.

Tuck, J.A. 1976. *Ancient people of Port au Choix*. Newfoundland Social and Economic Studies 17, Institute of Social and Economic Research, Memorial University of Newfoundland, St. John's.

Tuck, J.A. 1978. Excavations at Cow Head, Newfoundland: An interim report. *Études/Inuit/Studies* 2(1): 138–141.

Tuck, J.A. 1991. The Archaic period in the maritime provinces. In *Prehistoric archaeology in the maritime provinces: Past and present research*, ed. M. Deal and S. Blair, 29–57. The Council of Maritime Premiers, Reports in Archaeology 8, Fredericton.

Tuck, J.A., and W.W. Fitzhugh. 1986. Palaeo-Eskimo traditions of Newfoundland and Labrador: A re-appraisal. In *Palaeo-Eskimo cultures in Newfoundland, Labrador and Ungava*, 161–167. Reports in Archaeology No. 1, Memorial University of Newfoundland, St. John's.

Wallace, B.L. 2006. *Westward vikings: The saga of L'Anse aux Meadows*. Historic Sites Association of Newfoundland and Labrador, St. John's, St. John's.

Walthall, J.A. 1999. Mortuary behaviour and early Holocene land use in the North American midcontinent. *North American Archaeologist* 20(1): 1–30.

Wells, P. 2005. Animal exploitation and season of occupation at the Groswater Palaeoeskimo site of Phillip's Garden West. *Newfoundland and Labrador Studies* 20(1): 75–90.

Wells, P. 2009. Constructed seascapes of Newfoundland inshore fishers: A model for precontact small scale societies. In *On the track of the Thule culture, from Bering Strait to east Greenland*, Proceedings of the SILA Conference The Thule Culture – New Perspectives in Inuit Prehistory, ed. B. Grønnow, 105–117. Publications from the National Museum Studies in Archaeology and History Vol. 15, Copenhagen.

Whitridge, P. 2004. Landscapes, houses, bodies, things: "Place" and the archaeology of Inuit imaginaries. *Journal of Archaeological Method and Theory* 11(2): 213–250.

Wintemberg, W.J. 1939. Eskimo sites of the Dorset culture in Newfoundland, Part I. *American Antiquity* 5(1): 83–102.

Zedeño, M.N. 2000. On what people make of places: A behavioural cartography. In *Social theory in archaeology*, ed. M.B. Schiffer, 97–226. Salt Lake City: The University of Utah Press.

Zedeño, M.N. 2008. The archaeology of territory and territoriality. In *Handbook of landscape archaeology*, ed. B. David and J. Thomas, 210–217. Walnut Creek: Left Coast Press.

Zvelebil, M. 2003. Enculturation of mesolithic landscapes. In *Mesolithic on the move*, ed. L. Larsson, H. Kindgren, K. Knutsson, D. Loeffler, and A. Åkerlund, 65–73. Oxford: Oxbow Books.

# Appendix
# Radiocarbon Dates from Port au Choix

This appendix is an archive of radiocarbon dates from Port au Choix archaeological sites and pond sediments. They are presented in alphabetical order by site in two tables, Table 1 comprising provenance and radiometric data and Table 2 presenting detailed contextual information.

The measured radiocarbon age before present is presented for each date. Where data are available the normalized, or conventional, age before present is given in addition to the isotopic fractionation of the sample (δ13C value). Isotopic fractionation was not determined for samples analysed before 1990. All dates were calibrated using Calib 6.0html (Stuiver and Reimer 1993); for the sake of consistency, measured age rather than normalized age was the basis of calibration. Calibration ranges are presented as calendar years before present at one and two sigma probability in addition to the median date. All dates are rounded to the nearest decade.

These dates are the basis for the chronological ranges given for precontact cultures in Port au Choix and for the temporal span of Port au Choix sites. However, certain dates were not used in our analyses, namely: dates based on seal fat; dates with a standard deviation >110 BP; dates of uncertain archaeological context; and dates that are anomalously young or old in the context of existing dates for that site or culture.

Phillip's Garden dates include Elmer Harp's (1976), which can be recognized by the lab designation of the University of Pennsylvania (P), in contrast to the dates from the Port au Choix Archaeology Project which are identified by the Beta Analytic lab designation (Beta). The contextual information for Harp's dates is based on his research notes which he kindly shared with us.

The following is a list of abbreviations:

| | |
|---|---|
| AA | National Science Foundation |
| AMS | Accelerator mass spectrometry |
| BASS | Bass Pond |
| Beta | Beta Analytic |
| BP | Before present |
| BP | Bass Pond (where it appears in the sample number field) |
| Cal BP | Calendar years before present based on calibrated dates |
| CP | Control Pond |
| δ13C | sample isotopic fractionation |
| DPE | Dorset Palaeoeskimo |

M.A.P. Renouf (ed.), *The Cultural Landscapes of Port au Choix*, Interdisciplinary Contributions to Archaeology, DOI 10.1007/978-1-4419-8324-4,

| | |
|---|---|
| Ext. | Extended count |
| F | Feature |
| FCR | Fire-cracked rock |
| FP | Field Pond |
| GPE | Groswater Palaeoeskimo |
| GrA | Groningen Accelerator |
| GSC | Geological Survey of Canada |
| I | Teledyne Isotopes |
| L | Level |
| LL | Lower level |
| L2A | Soil level with very greasy texture |
| L2WC | Second layer of rocks on inside slope of dwelling wall area |
| MAI | Maritime Archaic Indian |
| N/A | Not applicable |
| P | University of Pennsylvania |
| PAC | Port au Choix |
| PGE1, PGE2 | Earlier and later period of Phillip's Garden East occupation |
| PGW1, PGW2 | Earlier and later period of Phillip's Garden West occupation |
| RI | Recent Indian |
| RL | Rock level, in cases where there is more than one layer of rocks |
| SkP | Skin Pond |
| SP | Stove Pond |
| TO | IsoTrace Laboratory |
| UNK | Unknown cultural affiliation |
| Y | Yale University |
| 1 | Port au Choix Archaeology and Sea Level Project |
| 2 | Hodgetts 2002:35 |
| 3 | Harp 1976:137 |
| 4 | Tuck 1976:162 |
| 5 | Robinson 2006:363 |
| 6 | Jelsma 2000:191 |
| 7A | Port au Choix National Historic Site; subsequent numbers are part of the Parks Canada provenience system and refer to locations within sites |

## References


Harp, E. 1976. Dorset settlement patterns in Newfoundland and southeastern Hudson Bay. In *Eastern Arctic prehistory: Paleoeskimo problems*, ed. M.S. Maxwell, 119–138. Washington, D.C.: Society for American Archaeology.

Hodgetts, L.M. 2002. Report of the 2001 Excavations at Phillip's Garden, Port au Choix National Historic Site. On file, Parks Canada, Atlantic Region, Halifax.

Jelsma, J. 2000. A Bed of Ochre: Mortuary Practices and Social Structure of a Maritime Archaic Indian Society at Port au Choix, Newfoundland. PhD dissertation, University of Groningen, Netherlands.

Robinson, B.S. 2006. Burial ritual, technology, and cultural landscape in the far northeast: 8600–3700 B.P. In *The Archaic of the far northeast*, ed. D. Sanger and M.A.P. Renouf, 341–381. Orono: Maine University Press.

Stuiver, M. and P.J. Reimer. 1993. Extended 14C database and revised CALIB radiocarbon calibration program. Radiocarbon 35(215): 215–230.

Tuck, J.A. 1976. Ancient people of Port au Choix. Newfoundland Social and Economic Studies 17, Institute of Social and Economic Research, Memorial University of Newfoundland, St. John's.

Table 1

| Site Name | Borden number | Sample number | Lab number | Analysis type | Measured age (BP) | δ13C | Normalized/ conventional age (BP) | ± | cal BP median age | cal BP 1σ older | cal BP 1σ younger | cal BP 2σ older | cal BP 2σ younger |
|---|---|---|---|---|---|---|---|---|---|---|---|---|---|
| Bass Pond | | BP-02-06 | Beta 238618 | AMS | 190 | −28.1 | 140 | 40 | 180 | 290 | 0 | 310 | 0 |
| Bass Pond | | BASS19310.5 | Beta 115780 | AMS | 1270 | −27.4 | 1230 | 60 | 1200 | 1200 | 1140 | 1290 | 1060 |
| Bass Pond | | BP-02-50 | Beta 238620 | AMS | 1730 | −25.8 | 1720 | 40 | 1640 | 1700 | 1570 | 1730 | 1540 |
| Bass Pond | | BP1-27/29 | TO-8744 | AMS | 1980 | | | 50 | 1930 | 1990 | 1880 | 2060 | 1820 |
| Bass Pond | | BP-02-18 | Beta 238619 | AMS | 2040 | −24.7 | 2040 | 40 | 2000 | 2050 | 1930 | 2120 | 1900 |
| Bass Pond | | BPI-40 | TO-9162 | AMS | 2230 | | | 60 | 2130 | 2330 | 2160 | 2350 | 2070 |
| Bass Pond | | BP-04-21 | Beta 238549 | AMS | 2180 | −24.9 | 2180 | 40 | 2220 | 2310 | 2130 | 2330 | 2060 |
| Bass Pond | | BP-02-78 | Beta 238621 | AMS | 3000 | −26.8 | 2970 | 40 | 3200 | 3320 | 3080 | 2240 | 3070 |
| Bass Pond | | BP-01-60 | TO-9163 | AMS | 3420 | | | 60 | 3680 | 3820 | 3580 | 3840 | 3490 |
| Bass Pond | | BP-03-44.5 | Beta 249548 | AMS | 3660 | −29.2 | 3590 | 40 | 3990 | 4080 | 3910 | 4140 | 3870 |
| Bass Pond | | BP-04-63 | Beta 249550 | AMS | 4580 | −28.9 | 4520 | 40 | 5290 | 5440 | 5080 | 5450 | 5050 |
| Bass Pond | | BASS19382 | Beta 115782 | AMS | 5100 | −24.8 | 5100 | 50 | 5820 | 5910 | 5750 | 5940 | 5720 |
| Bass Pond | | BASS193200/210 | GSC 5661 | Radiometric | 9210 | −14.5 | 9210 | 75 | 10370 | 10480 | 10270 | 10500 | 10250 |
| Control Pond | | CP-02-53a | Beta 249553 | AMS | 2520 | −25.9 | 2510 | 40 | 2590 | 2730 | 2500 | 2750 | 2470 |
| Control Pond | | CP-02-20a | Beta 249551 | AMS | 3440 | −25.5 | 3430 | 40 | 3700 | 3820 | 3640 | 3830 | 3590 |
| Control Pond | | CP-02-32a | Beta 249552 | AMS | 4320 | −24.1 | 4330 | 40 | 4890 | 4960 | 4841 | 5030 | 4830 |
| Field Pond | | FP2:5–5.5 | Beta 155668 | AMS | 670 | −25 | 670 | 40 | 630 | 560 | 670 | 560 | 680 |
| Field Pond | | FP2:16 | Beta 149993 | AMS | 700 | −26.7 | 700 | 50 | 660 | 570 | 690 | 560 | 730 |
| Field Pond | | FP2:34-35 | Beta 155669 | AMS | 1730 | −25 | 1730 | 50 | 1640 | 1570 | 1700 | 1530 | 1810 |
| Field Pond | | FP2:54.5 | Beta 151259 | AMS | 3460 | −26.7 | 3460 | 40 | 3730 | 3640 | 3830 | 3630 | 3840 |
| Gould Site | EeBi-42 | EeBi-42:1531 | Beta 134154 | AMS | 70 | −27.6 | 70 | 40 | 110 | 250 | 0 | 270 | 0 |
| Gould Site | EeBi-42 | EeBi-42:886 | TO-7856 | AMS | modern | | 138.04 | 0.92 | 150 | 270 | 20 | 270 | 10 |
| Gould Site | EeBi-42 | EeBi-42:872 | Beta 121859 | AMS | modern | −29 | 144.6 | 1.3 | 190 | 270 | 20 | 270 | 10 |
| Gould Site | EeBi-42 | EeBi-42:1357 | Beta 134150 | Radiometric | 1480 | | | 70 | 1380 | 1480 | 1300 | 1520 | 1290 |
| Gould Site | EeBi-42 | EeBi-42:1585 | Beta 134155 | AMS | 1510 | −27.2 | 1510 | 40 | 1390 | 1490 | 1340 | 1520 | 1310 |
| Gould Site | EeBi-42 | EeBi-42:1738 | Beta 134156 | AMS | 1510 | −25.7 | 1500 | 40 | 1390 | 1490 | 1340 | 1520 | 1310 |
| Gould Site | EeBi-42 | EeBi-42:63 | Beta 108552 | Radiometric | 1520 | | | 60 | 1410 | 1520 | 1350 | 1530 | 1310 |
| Gould Site | EeBi-42 | EeBi-42:1254 | Beta 134149 | Radiometric | 1870 | | | 60 | 1810 | 1870 | 1730 | 1950 | 1630 |

(continued)

**Table 1** (continued)

| Site Name | Borden number | Sample number | Lab number | Analysis type | Measured age (BP) | δ13C | Normalized/ conventional age (BP) | ± | cal BP median age | cal BP 1σ older | cal BP 1σ younger | cal BP 2σ older | cal BP 2σ younger |
|---|---|---|---|---|---|---|---|---|---|---|---|---|---|
| Gould Site | EeBi-42 | EeBi-42:500 | Beta 120796 | Radiometric | 1950 | | | 60 | 1900 | 1990 | 1830 | 2040 | 1730 |
| Gould Site | EeBi-42 | EeBi-42:977 | Beta 134147 | AMS | 2080 | −25.1 | 2080 | 40 | 2053 | 2110 | 2000 | 2150 | 1950 |
| Gould Site | EeBi-42 | EeBi-42:1968 | Beta 148517 | Radiometric | 2820 | −23 | 2850 | 40 | 2930 | 2970 | 2870 | 3060 | 2800 |
| Gould Site | EeBi-42 | Eebi-42:923 | Beta 132364 | Radiometric | 3200 | | | 100 | 3430 | 3560 | 3340 | 3690 | 3160 |
| Gould Site | EeBi-42 | EeBi-42:33 | Beta 108099 | AMS | 3270 | −25.7 | 3260 | 50 | 3500 | 3560 | 3450 | 3620 | 3390 |
| Gould Site | EeBi-42 | EeBi-42:1481 | Beta 134153 | Radiometric | 3420 | | | 60 | 3680 | 3820 | 3580 | 3840 | 3490 |
| Gould Site | EeBi-42 | EeBi-42:414 | Beta 120795 | Radiometric | 3450 | | | 70 | 3720 | 3830 | 3640 | 3900 | 3490 |
| Gould Site | EeBi-42 | EeBi-42:1219 | Beta 134148 | AMS | 3460 | −25.7 | 3450 | 50 | 3730 | 3830 | 3640 | 3850 | 3590 |
| Gould Site | EeBi-42 | EeBi-42:55 | Beta 107795 | Radiometric | 3740 | −25.8 | 3720 | 50 | 4100 | 4210 | 3990 | 4240 | 3930 |
| Gould Site | EeBi-42 | EeBi-42:2213 | Beta 121295 | Radiometric ext. | 3850 | | | 100 | 4260 | 4420 | 4100 | 4520 | 3980 |
| Gould Site | EeBi-42 | EeBi-42:828 | TO-8520 | AMS | 4010 | | | 160 | 4490 | 4810 | 4260 | 4860 | 3990 |
| Gould Site | EeBi-42 | EeBi-42:134151 | Beta 146081 | Radiometric | 4060 | | | 80 | 4570 | 4800 | 4430 | 4830 | 4300 |
| Gould Site | EeBi-42 | EeBi-42:1767 | TO-8519 | AMS | 4140 | | | 80 | 5390 | 5570 | 5300 | 5580 | 5060 |
| Gould Site | EeBi-42 | EeBi-42:860 | TO-8581 | AMS | 4670 | −25.7 | | 120 | 5390 | 5590 | 5290 | 5640 | 4980 |
| Gould Site | EeBi-42 | EeBi-42:656 | Beta 148519 | Radiometric | 5390 | −24.1 | 5440 | 40 | 6210 | 6280 | 6130 | 6290 | 6020 |
| Gould Site | EeBi-42 | EeBi-42:2385 | Beta 134151 | AMS | 5420 | −23.9 | 5440 | 50 | 6230 | 6280 | 6200 | 6310 | 6020 |
| Gould Site | EeBi-42 | EeBi-42:2213 | Beta 148518 | Radiometric | 5430 | -24.1 | 5440 | 50 | 6240 | 6290 | 6200 | 6310 | 6020 |
| Old Boatyard Site | EeBi-43 | EeBi-43:1 | Beta 121296 | Radiometric ext. | 1640 | | | 70 | 1540 | 1610 | 1420 | 1710 | 1390 |
| Old Boatyard Site | EeBi-43 | EeBi-43:2 | Beta 121297 | Radiometric ext. | 3980 | | | 110 | 4450 | 4780 | 4250 | 4820 | 4150 |
| Party Site | EeBi-30 | EeBi-30:311 | Beta 183604 | Radiometric | 2460 | | | 70 | 2540 | 2700 | 2370 | 2720 | 2360 |
| Party Site | EeBi-30 | EeBi-30:142 | Beta 146666 | Radiometric | 2570 | | | 60 | 2630 | 2760 | 2510 | 2790 | 2370 |
| Party Site | EeBi-30 | EeBi-30:268 | Beta 183603 | AMS | 2710 | −25.1 | 2710 | 40 | 2810 | 2850 | 2770 | 2920 | 2750 |
| Phillip's Garden | EeBi-1 | 7A284C92 | Beta 15639 | Radiometric | 1250 | | | 60 | 1180 | 1270 | 1090 | 1290 | 1020 |
| Phillip's Garden | EeBi-1 | | P-737 | Radiometric | 1321 | | | 49 | 1250 | 1300 | 1180 | 1330 | 1100 |
| Phillip's Garden | EeBi-1 | 7A368D338 | Beta 160977 | Radiometric | 1360 | | | 80 | 1280 | 1350 | 1180 | 1410 | 1080 |
| Phillip's Garden | EeBi-1 | | P-676 | Radiometric | 1502 | | | 49 | 1290 | 1480 | 1320 | 1520 | 1310 |
| Phillip's Garden | EeBi-1 | 7A368D79 | Beta 66436 | Radiometric | 1370 | | | 90 | 1290 | 1370 | 1180 | 1510 | 1070 |
| Phillip's Garden | EeBi-1 | 7A368C743 | Beta 66435 | Radiometric | 1410 | | | 100 | 1330 | 1410 | 1180 | 1530 | 1090 |
| Phillip's Garden | EeBi-1 | 7A279A136 | Beta 238479 | AMS | 1450 | −25.8 | 1440 | 40 | 1340 | 1370 | 1310 | 1400 | 1300 |

| | | | | | | | | | | | | | |
|---|---|---|---|---|---|---|---|---|---|---|---|---|---|
| Phillip's Garden | EeBi-1 | | P-734 | Radiometric | 1465 | | | 51 | 1360 | 1390 | 1310 | 1510 | 1290 |
| Phillip's Garden | EeBi-1 | 7A368D223 | Beta 160976 | AMS | 1480 | −24.4 | 1490 | 40 | 1370 | 1400 | 1330 | 1510 | 1300 |
| Phillip's Garden | EeBi-1 | 7A281D104 | Beta 238481 | AMS | 1480 | −24.4 | 1490 | 40 | 1370 | 1400 | 1330 | 1510 | 1300 |
| Phillip's Garden | EeBi-1 | 7A348C142 | Beta 211271 | AMS | 1510 | −25.9 | 1500 | 40 | 1390 | 1490 | 1240 | 1520 | 1310 |
| Phillip's Garden | EeBi-1 | | P-696 | Radiometric | 1509 | | | 47 | 1400 | 1510 | 1340 | 1520 | 1310 |
| Phillip's Garden | EeBi-1 | 7A324D1118 | Beta 19084 | Radiometric | 1520 | | | 90 | 1420 | 1520 | 1340 | 1610 | 1280 |
| Phillip's Garden | EeBi-1 | | P-729 | Radiometric | 1538 | | | 55 | 1440 | 1520 | 1370 | 1540 | 1320 |
| Phillip's Garden | EeBi-1 | 7A348C102 | Beta 211270 | AMS | 1550 | −24 | 1570 | 40 | 1450 | 1520 | 1400 | 1530 | 1360 |
| Phillip's Garden | EeBi-1 | 7A323A211 | Beta 15381 | Radiometric | 1570 | | | 70 | 1460 | 1530 | 1390 | 1610 | 1320 |
| Phillip's Garden | EeBi-1 | | P-733 | Radiometric | 1565 | | | 53 | 1460 | 1520 | 1410 | 1560 | 1350 |
| Phillip's Garden | EeBi-1 | 7A259D881 | Beta 211268 | AMS | 1570 | −23.9 | 1590 | 40 | 1460 | 1520 | 1410 | 1540 | 1380 |
| Phillip's Garden | EeBi-1 | | P-727 | Radiometric | 1580 | | | 54 | 1470 | 1520 | 1410 | 1600 | 1350 |
| Phillip's Garden | EeBi-1 | 7A279A472 | Beta 238480 | AMS | 1580 | −23.5 | 1600 | 40 | 1470 | 1520 | 1420 | 1550 | 1380 |
| Phillip's Garden | EeBi-1 | | P-694 | Radiometric | 1602 | | | 49 | 1480 | 1540 | 1420 | 1610 | 1380 |
| Phillip's Garden | EeBi-1 | 7A348D137 | Beta 211272 | AMS | 1600 | −26.3 | 1580 | 40 | 1480 | 1530 | 1420 | 1570 | 1390 |
| Phillip's Garden | EeBi-1 | 7A270C505 | Beta 238478 | AMS | 1610 | −24.8 | 1610 | 40 | 1490 | 1540 | 1420 | 1600 | 1400 |
| Phillip's Garden | EeBi-1 | 7A259A329 | Beta 211267 | AMS | 1630 | −26.2 | 1610 | 40 | 1520 | 1570 | 1420 | 1680 | 1410 |
| Phillip's Garden | EeBi-1 | 7A338A40 | Beta 211269 | AMS | 1630 | −27.8 | 1580 | 40 | 1520 | 1570 | 1420 | 1680 | 1410 |
| Phillip's Garden | EeBi-1 | | P-679 | Radiometric | 1632 | | | 47 | 1530 | 1600 | 1420 | 1690 | 1410 |
| Phillip's Garden | EeBi-1 | | P-693 | Radiometric | 1659 | | | 48 | 1560 | 1690 | 1450 | 1690 | 1420 |
| Phillip's Garden | EeBi-1 | 7A349D557 | Beta 160975 | Radiometric | 1640 | | | 70 | 1540 | 1610 | 1420 | 1710 | 1290 |
| Phillip's Garden | EeBi-1 | 7A249C794 | Beta 211266 | AMS | 1680 | −26.1 | 1660 | 40 | 1590 | 1690 | 1540 | 1700 | 1450 |
| Phillip's Garden | EeBi-1 | | P-736 | Radiometric | 1683 | | | 49 | 1590 | 1690 | 1530 | 1710 | 1420 |
| Phillip's Garden | EeBi-1 | | P-695 | Radiometric | 1712 | | | 40 | 1620 | 1690 | 1560 | 1710 | 1540 |
| Phillip's Garden | EeBi-1 | 7A269B208 | Beta 238476 | AMS | 1730 | −26.2 | 1710 | 50 | 1640 | 1700 | 1570 | 1810 | 1530 |
| Phillip's Garden | EeBi-1 | | P-692 | Radiometric | 1736 | | | 48 | 1650 | 1710 | 1570 | 1810 | 1540 |
| Phillip's Garden | EeBi-1 | 7A270C433 | Beta 238477 | AMS | 1750 | −25.1 | 1750 | 40 | 1660 | 1710 | 1610 | 1810 | 1550 |
| Phillip's Garden | EeBi-1 | 7A295D301 | Beta 42968 | Radiometric | 1770 | | | 120 | 1690 | 1820 | 1550 | 1950 | 1410 |
| Phillip's Garden | EeBi-1 | 7A284D284 | Beta 15379 | Radiometric | 1850 | | | 110 | 1780 | 1920 | 1630 | 2040 | 1530 |
| Phillip's Garden | EeBi-1 | 7A250A47 | Beta 42967 | Radiometric | 1890 | | | 90 | 1830 | 1930 | 1720 | 2040 | 1570 |
| Phillip's Garden | EeBi-1 | 7A324D1058 | Beta 23978 | Radiometric | 1900 | | | 110 | 1840 | 1990 | 1710 | 2110 | 1570 |
| Phillip's Garden | EeBi-1 | 7A323A540 | Beta 15638 | Radiometric | 1920 | | | 110 | 1870 | 1990 | 1730 | 2130 | 1570 |

(continued)

**Table 1** (continued)

| Site Name | Borden number | Sample number | Lab number | Analysis type | Measured age (BP) | δ13C | Normalized/ conventional age (BP) | ± | cal BP median age | cal BP 1σ older | cal BP 1σ younger | cal BP 2σ older | cal BP 2σ younger |
|---|---|---|---|---|---|---|---|---|---|---|---|---|---|
| Phillip's Garden | EeBi-1 | 7A294A535 | Beta 23977 | Radiometric | 1970 | | | 60 | 1920 | 1990 | 1870 | 2110 | 1740 |
| Phillip's Garden | EeBi-1 | 7A294A142 | Beta 23976 | Radiometric | 2140 | | | 100 | 2130 | 2310 | 2000 | 2340 | 1900 |
| Phillip's Garden | EeBi-1 | | P-683 | Radiometric | 1593 | | | 49 | 1480 | 1530 | 1420 | 1600 | 1370 |
| Phillip's Garden | EeBi-1 | | P-735 | Radiometric | 1837 | | | 60 | 1770 | 1860 | 1710 | 1920 | 1610 |
| Phillip's Garden | EeBi-1 | | P-682 | Radiometric | 1879 | | | 60 | 1820 | 1880 | 1740 | 1970 | 1630 |
| Phillip's Garden | EeBi-1 | | P-730 | Radiometric | 1906 | | | 55 | 1850 | 1930 | 1740 | 1990 | 1710 |
| Phillip's Garden | EeBi-1 | | P-731 | Radiometric | 1911 | | | 65 | 1850 | 1930 | 1740 | 2000 | 1700 |
| Phillip's Garden | EeBi-1 | | P-678A | Radiometric | 2006 | | | 65 | 1970 | 2040 | 1880 | 2130 | 1820 |
| Phillip's Garden | EeBi-1 | | P-732 | Radiometric | 2314 | | | 60 | 2330 | 2430 | 2160 | 2670 | 2150 |
| Phillip's Garden East | EeBi-1 | 7A383D475 | Beta 23980 | Radiometric | 1730 | | | 200 | 1660 | 1860 | 1420 | 2130 | 1280 |
| Phillip's Garden East | EeBi-1 | 7A383D555 | Beta 19088 | Radiometric | 1910 | | | 150 | 1860 | 2050 | 1630 | 2310 | 1520 |
| Phillip's Garden East | EeBi-1 | 7A382C66 | Beta 19085 | Radiometric | 1930 | | | 140 | 1880 | 2050 | 1710 | 2300 | 1540 |
| Phillip's Garden East | EeBi-1 | 7A385B1026 | Beta 49755 | Radiometric ext. | 2240 | | | 100 | 2230 | 2348 | 2130 | 2680 | 1950 |
| Phillip's Garden East | EeBi-1 | 7A385B1193 | Beta 50022 | Radiometric ext. | 2260 | | | 70 | 2240 | 2340 | 2160 | 2460 | 2060 |
| Phillip's Garden East | EeBi-1 | 7A384C41 | Beta 42970 | Radiometric | 2310 | | | 90 | 2330 | 2460 | 2160 | 2700 | 2120 |
| Phillip's Garden East | EeBi-1 | 7A383D539 | Beta 19087 | Radiometric | 2320 | | | 100 | 2350 | 2650 | 2160 | 2710 | 2120 |
| Phillip's Garden East | EeBi-1 | 7A393D384 | Beta 50023 | Radiometric | 2350 | -24.2 | 2360 | 90 | 2420 | 2690 | 2180 | 2710 | 2150 |
| Phillip's Garden East | EeBi-1 | 7A394A727 | Beta 42972 | Radiometric | 2350 | | | 100 | 2420 | 2690 | 2180 | 2710 | 2150 |
| Phillip's Garden East | EeBi-1 | 7A383D613 | Beta 19089 | Radiometric | 2370 | | | 160 | 2430 | 2710 | 2180 | 2760 | 2000 |
| Phillip's Garden East | EeBi-1 | 7A394A426 | Beta 42971 | Radiometric ext. | 2420 | | | 110 | 2500 | 2700 | 2350 | 2750 | 2160 |
| Phillip's Garden East | EeBi-1 | 7A385A173 | Beta 50021 | Radiometric | 2500 | | | 60 | 2570 | 2720 | 2490 | 2740 | 2360 |
| Phillip's Garden East | EeBi-1 | 7A383D403 | Beta 19086 | Radiometric | 2510 | | | 90 | 2570 | 2740 | 2490 | 2750 | 2360 |
| Phillip's Garden East | EeBi-1 | 7A382B2 | Beta 15375 | Radiometric | 2660 | | | 70 | 2780 | 1850 | 2740 | 2950 | 2510 |
| Phillip's Garden East | EeBi-1 | 7A383D377+371 | Beta 23979 | Radiometric | 2760 | | | 90 | 2880 | 2950 | 2770 | 3140 | 2740 |
| Phillip's Garden West | EeBi-11 | 7A711D253 | Beta 66438 | Radiometric | 1960 | | | 80 | 1910 | 2000 | 1820 | 2110 | 1720 |
| Phillip's Garden West | EeBi-11 | 7A701B302 | Beta 49757 | Radiometric | 2090 | | | 70 | 2070 | 2150 | 1950 | 2300 | 1890 |
| Phillip's Garden West | EeBi-11 | 7A701B236 | Beta 49756 | Radiometric ext. | 2190 | | | 100 | 2190 | 2330 | 2070 | 2430 | 1900 |
| Phillip's Garden West | EeBi-11 | 7A702A79 | Beta 42973 | Radiometric ext. | 2200 | | | 110 | 2200 | 2340 | 2060 | 2460 | 1900 |
| Phillip's Garden West | EeBi-11 | 7A711A842+843 | Beta 66437 | Radiometric | 2240 | | | 70 | 2230 | 2340 | 2160 | 2360 | 2050 |

| | | | | | | | | | | | | | |
|---|---|---|---|---|---|---|---|---|---|---|---|---|---|
| Phillip's Garden West | EeBi-11 | 7A714C30+35 | Beta 66439 | Radiometric | 2340 | | | 70 | 2390 | 2650 | 2180 | 2700 | 2150 |
| Phillip's Garden West | EeBi-11 | 7A711D166 | Beta 49760 | Radiometric ext. | 2340 | | | 100 | 2400 | 2680 | 2160 | 2720 | 2150 |
| Phillip's Garden West | EeBi-11 | 7A701B386 | Beta 49758 | Radiometric | 2350 | | | 80 | 2410 | 2680 | 2210 | 2706 | 2160 |
| Phillip's Garden West | EeBi-11 | 7A711D177 | Beta 49761 | Radiometric ext. | 2460 | | | 120 | 2540 | 2700 | 2360 | 2780 | 2180 |
| Phillip's Garden West | EeBi-11 | 7A711A650 | Beta 49759 | Radiometric | 2540 | | | 160 | 2600 | 2760 | 2370 | 3000 | 2160 |
| Point Riche | EeBi-20 | 7A525B113 | Beta 15377 | Radiometric | 1546 | | | 80 | 1450 | 1520 | 1370 | 1610 | 1300 |
| Point Riche | EeBi-20 | 7A555A146+203 | Beta 160980 | AMS | 1590 | −21.5 | 1650 | 40 | 1470 | 1530 | 1420 | 1560 | 1390 |
| Point Riche | EeBi-20 | 7A547B499 | Beta 15382 | Radiometric | 1750 | | | 90 | 1670 | 1810 | 1550 | 1880 | 1420 |
| Point Riche | EeBi-20 | 7A547D380 | Beta 15376 | Radiometric | 1750 | | | 80 | 1670 | 1770 | 1560 | 1870 | 1420 |
| Point Riche | EeBi-20 | 7A543C511 | Beta 50025 | Radiometric ext. | 1760 | | | 150 | 1680 | 1870 | 1530 | 2000 | 1350 |
| Point Riche | EeBi-20 | 7A557D11 | Beta 50026 | Radiometric ext. | 1800 | −26.9 | 1770 | 70 | 1730 | 1820 | 1630 | 1880 | 1560 |
| Point Riche | EeBi-20 | 7A542A59 | Beta 160978 | AMS | 1810 | −23.9 | 1830 | 40 | 1750 | 1810 | 1710 | 1860 | 1620 |
| Point Riche | EeBi-20 | 7A543C466+471 | Beta 50024 | Radiometric | 1830 | | | 90 | 1760 | 1870 | 1630 | 1950 | 1540 |
| Port au Choix-3 | EeBi-2 | | I-4380 | Radiometric | 3230 | | | 220 | 3460 | 3820 | 3160 | 3980 | 2870 |
| Port au Choix-3 | EeBi-2 | | Y-2608 | Radiometric | 3370 | | | 80 | 3620 | 3700 | 3480 | 3830 | 3440 |
| Port au Choix-3 | EeBi-2 | | I-4677 | Radiometric | 3410 | | | 100 | 3670 | 3830 | 3560 | 3910 | 3410 |
| Port au Choix-3 | EeBi-2 | | I-4682 | Radiometric | 3690 | | | 90 | 4040 | 4150 | 3900 | 4380 | 3730 |
| Port au Choix-3 | EeBi-2 | | AA 33919 | AMS | 3777 | −21 | | 44 | 4150 | 4230 | 4090 | 4340 | 3990 |
| Port au Choix-3 | EeBi-2 | | AA 29482 | AMS | 3825 | −19.7 | | 60 | 4230 | 4380 | 4100 | 4420 | 4110 |
| Port au Choix-3 | EeBi-2 | | AA 33920 | AMS | 3826 | −17.4 | | 38 | 4230 | 4290 | 4150 | 4410 | 4090 |
| Port au Choix-3 | EeBi-2 | | I-4678 | Radiometric | 3930 | | | 130 | 4370 | 4570 | 4150 | 4820 | 3990 |
| Port au Choix-3 | EeBi-2 | | GrA-6525 | AMS | 4000 | | | 50 | 4480 | 4520 | 4420 | 4780 | 4300 |
| Port au Choix-3 | EeBi-2 | | GrA-6496 | AMS | 4000 | | | 50 | 4480 | 4520 | 4420 | 4790 | 4300 |
| Port au Choix-3 | EeBi-2 | | GrA-6527 | AMS | 4110 | | | 50 | 4640 | 4810 | 4530 | 4820 | 4450 |
| Port au Choix-3 | EeBi-2 | | GrA-6666 | AMS | 4130 | | | 50 | 4670 | 4810 | 4570 | 4830 | 4530 |
| Port au Choix-3 | EeBi-2 | | GrA-6501 | AMS | 4130 | | | 50 | 4670 | 4810 | 4570 | 4830 | 4530 |
| Port au Choix-3 | EeBi-2 | | GrA-6526 | AMS | 4150 | | | 50 | 4690 | 4820 | 4590 | 4830 | 4530 |
| Port au Choix-3 | EeBi-2 | | GrA-6495 | AMS | 4150 | | | 50 | 4690 | 4820 | 4590 | 4830 | 4530 |
| Port au Choix-3 | EeBi-2 | | GrA-6479 | AMS | 4160 | | | 50 | 4700 | 4820 | 4630 | 4840 | 4530 |
| Port au Choix-3 | EeBi-2 | | GrA-6478 | AMS | 4220 | | | 50 | 4740 | 4850 | 4650 | 4860 | 4580 |
| Port au Choix-3 | EeBi-2 | | I-3788 | Radiometric | 4290 | | | 110 | 4870 | 5040 | 4630 | 5280 | 4530 |
| Port au Choix-3 | EeBi-2 | | GSC-1403 | Radiometric | 4690 | | | 130 | 5400 | 5590 | 5290 | 5660 | 4980 |

(continued)

**Table 1** (continued)

| Site Name | Borden number | Sample number | Lab number | Analysis type | Measured age (BP) | δ13C | Normalized/ conventional age (BP) | ± | cal BP median age | cal BP 1σ older | cal BP 1σ younger | cal BP 2σ older | cal BP 2σ younger |
|---|---|---|---|---|---|---|---|---|---|---|---|---|---|
| Port au Choix-3 | EeBi-2 | | Y-2609 | Radiometric | 5120 | | | 120 | 5870 | 5990 | 5670 | 6180 | 5610 |
| Sea Level Project | | PAC-15-S-9631 | Beta 107797 | Radiometric | 1520 | +2.7 | | 80 | 1420 | 1520 | 1340 | 1570 | 1290 |
| Sea Level Project | | PAC-03-S-9718 | Beta 107796 | Radiometric | 7750 | +2.0 | | 90 | 8540 | 8600 | 8430 | 8930 | 8380 |
| Skin Pond | | SkP-01-32 | Beta 238622 | AMS | 340 | −25.3 | 340 | 40 | 400 | 460 | 320 | 490 | 310 |
| Spence Site | EeBi-36 | EeBi-36:1773 | Beta 66440 | Radiometric | 840 | | 840 | 90 | 780 | 900 | 680 | 930 | 660 |
| Spence Site | EeBi-36 | EeBi-36:2042/6 | Beta 66441 | AMS | 1020 | −25.2 | 1020 | 60 | 940 | 1050 | 800 | 1060 | 790 |
| Spence Site | EeBi-36 | EeBi-36:2544 | Beta 66442 | Radiometric | 1340 | | 1340 | 80 | 1250 | 1330 | 1180 | 1380 | 1070 |
| Spence Site | EeBi-36 | 7A77A183 | Beta 49753 | Radiometric ext. | 1360 | | | 80 | 1280 | 1350 | 1180 | 1410 | 1080 |
| Spence Site | EeBi-36 | 7A77A383 | Beta 49754 | Radiometric ext. | 1420 | | | 70 | 1330 | 1390 | 1290 | 1520 | 1180 |
| Stove Pond | | SP-01-50 | Beta 36436 | AMS | 4400 | | | 90 | 5020 | 5270 | 4860 | 5290 | 4840 |
| Stove Pond | | SP-01-90 | Beta 32595 | AMS | 5420 | | | 70 | 6220 | 6300 | 6030 | 6390 | 6000 |
| Stove Pond | | SP-01-120 | Beta 32596 | AMS | 7180 | | | 140 | 8010 | 8170 | 7870 | 8310 | 7720 |
| Stove Pond | | SP-01-139 | Beta 32597 | AMS | 7420 | | | 130 | 8230 | 8380 | 8060 | 8440 | 7970 |
| Stove Pond | | SP-01-180 | Beta 32598 | AMS | 7920 | | | 130 | 8780 | 8980 | 8610 | 9090 | 8430 |

**Table 2**

| Site Name | Borden number | Sample number | Lab number | Measured age BP | Sample material | Context | Cultural affiliation | Data source |
|---|---|---|---|---|---|---|---|---|
| Bass Pond | | BP-02-06 | Beta 238618 | 190±40 | Wood | 5.5 cm depth | N/A | 1 |
| Bass Pond | | BASS19310.5 | Beta 115780 | 1270±60 | *Abies balsamea* twig | 10.5 cm depth | N/A | 1 |
| Bass Pond | | BP-02-50 | Beta 238620 | 1730±40 | Wood | 26.5 cm depth | N/A | 1 |
| Bass Pond | | BP1-27/29 | TO-8744 | 1980±50 | *Nemopanthus* bark | 27–29 cm depth | N/A | 1 |
| Bass Pond | | BP-02-18 | Beta 238619 | 2040±40 | Wood | 11 cm depth | N/A | 1 |
| Bass Pond | | BP1-40 | TO-9162 | 2230±60 | *Nemopanthus* twig? | 40 cm depth | N/A | 1 |
| Bass Pond | | BP-04-21 | Beta 238549 | 2180±40 | Wood | 21 cm depth | N/A | 1 |
| Bass Pond | | BP-02-78 | Beta 238621 | 3000±40 | Wood | 40 cm depth | N/A | 1 |
| Bass Pond | | BP1-60 | TO-9163 | 3420±60 | *Picea* bark | 60 cm depth | N/A | 1 |
| Bass Pond | | BP-03-44.5 | Beta 249548 | 3660±40 | Wood | 44.5 cm depth | N/A | 1 |
| Bass Pond | | BP-04-63 | Beta 249550 | 4580±40 | Wood | 63 cm depth | N/A | 1 |
| Bass Pond | | BASS19382 | Beta 115782 | 5100±50 | Conifer bark | 82 cm depth | N/A | 1 |
| Bass Pond | | BASS193200/210 | GSC 5661 | 9210±75 | Marine organic sandy mud | 200–210 cm depth | N/A | 1 |
| Control Pond | | CP-02-53a | Beta 249553 | 2520±40 | Wood | 53 cm depth | N/A | 1 |
| Control Pond | | CP-02-20a | Beta 249551 | 3440±40 | Wood | 20 cm depth | N/A | 1 |
| Control Pond | | CP-02-32a | Beta 249552 | 4320±40 | Wood | 32 cm depth | N/A | 1 |
| Field Pond | | FP2:5–5.5 | Beta 155668 | 470±40 | Needles | 5.5 cm core depth | N/A | 1 |
| Field Pond | | FP2:16 | Beta 149993 | 700±50 | Wood Fragment | 16 cm core depth | N/A | 1 |
| Field Pond | | FP2:34-35 | Beta 155669 | 1730±50 | Needles | 34–35 cm core depth | N/A | 1 |
| Field Pond | | FP2:54.5 | Beta 151259 | 3460±40 | Plant material | 54.5 cm core depth | N/A | 1 |
| Gould Site | EeBi-42 | EeBi-42:1531 | Beta 134154 | 70±40 | Wood | Area 99–1, L5, near large natural T-shaped pit | modern | 1 |
| Gould Site | EeBi-42 | EeBi-42:886 | TO-7856 | modern | Caribou bone | Area 99–2, level unknown, Ben Ploughman's putting green | modern | 1 |
| Gould Site | EeBi-42 | EeBi-42:872 | Beta 121859 | modern | Wood | Area 99–1, L6, from natural pit | modern | 1 |
| Gould Site | EeBi-42 | EeBi-42:1357 | Beta 134150 | 1480±70 | Charcoal | Area 98–8, L2, refuse pit F216 | RI | 1 |
| Gould Site | EeBi-42 | EeBi-42:1585 | Beta 134155 | 1510±40 | Charcoal | Area 99–12, O'Keefe property, L6 | UNK | 1 |
| Gould Site | EeBi-42 | EeBi-42:1738 | Beta 134156 | 1510±40 | Charcoal | Area 99–11, L2, dwelling F280 | RI | 1 |
| Gould Site | EeBi-42 | EeBi-42:63 | Beta 108552 | 1520±60 | Charcoal | Area 98–8, L2, hearth F3 | RI | 1 |

(continued)

**Table 2** (continued)

| Site Name | Borden number | Sample number | Lab number | Measured age BP | Sample material | Context | Cultural affiliation | Data source |
|---|---|---|---|---|---|---|---|---|
| Gould Site | EeBi-42 | EeBi-42:1254 | Beta 134149 | 1870±60 | Charcoal | Area 98–7, L2, hearth F26 | RI | 1 |
| Gould Site | EeBi-42 | EeBi-42:500 | Beta 120796 | 1950±60 | Charcoal | Area 98–1, L2, bone & charcoal F17 associated with hearth F2 | RI | 1 |
| Gould Site | EeBi-42 | EeBi-42:977 | Beta 134147 | 2080±40 | Charcoal | Area 98–1, L2, concentration of FCR F53 associated with hearth F2 | RI | 1 |
| Gould Site | EeBi-42 | EeBi-42:1968 | Beta 148517 | 2820±40 | Charcoal | Area 99-2C, Ploughman property, L4, associated with FCR and flakes | UNK | 1 |
| Gould Site | EeBi-42 | Eebi-42:923 | Beta 132364 | 3200±100 | Charcoal | Area 99–1, L6, beneath dark soil F46, associated with a dark flake | MAI | 1 |
| Gould Site | EeBi-42 | EeBi-42:33 | Beta 108099 | 3270±50 | Charcoal | Area 98–1 (Area 1 in Chapter 3), L4, small dump F1 | MAI | 1 |
| Gould Site | EeBi-42 | EeBi-42:1481 | Beta 134153 | 3420±60 | Charcoal | Area 3 in Chapter 3, L6, hearth F264 in woods near small pond | MAI | 1 |
| Gould Site | EeBi-42 | EeBi-42:414 | Beta 120795 | 3450±70 | Wood (*Picea* sp.) | Area 98–1 (Area 1 in Chapter 3), L4, culturally modified spruce log F11a | MAI | 1 |
| Gould Site | EeBi-42 | EeBi-42:1219 | Beta 134148 | 3460±50 | Charcoal | Area 99–1 (Area 2 in Chapter 3), L5/6, black soil F46 with MAI points | MAI | 1 |
| Gould Site | EeBi-42 | EeBi-42:55 | Beta 107795 | 3740±50 | Charcoal | Area 99–6 (Area 4 in Chapter 3) near Boyd property, L4, beneath hearth | MAI | 1 |
| Gould Site | EeBi-42 | EeBi-42:2213 | Beta 121295 | 3850±100 | Charcoal | Area 98–1 (Area 1 in Chapter 3) Level 5, hearth F24 | MAI | 1 |
| Gould Site | EeBi-42 | EeBi-42:828 | TO-8520 | 4010±160 | Leaf and bark | Main site area, 70 cm depth | MAI | 1 |
| Gould Site | EeBi-42 | EeBi-42:134151 | Beta 146081 | 4060±80 | Charcoal | TT00-16 (Area 1 in Chapter 3), L3X, charcoal, FCR & flakes F515 | MAI | 1 |
| Gould Site | EeBi-42 | EeBi-42:1767 | TO-8519 | 4140±80 | Bark fragments | Main site area 85–86 cm depth | MAI | 1 |
| Gould Site | EeBi-42 | EeBi-42:860 | TO-8581 | 4670±120 | Stem fragments | Near Boyd property line (Area 4 in Chapter 3) 122–123 cm depth | MAI | 1 |
| Gould Site | EeBi-42 | EeBi-42:656 | Beta 148519 | 5390±40 | Charcoal | Area 99-2E, Ploughman property, L4 | MAI | 1 |
| Gould Site | EeBi-42 | EeBi-42:2385 | Beta 134151 | 5420±50 | Charcoal | Area 99-2B, Ploughman property, L5 | MAI | 1 |
| Gould Site | EeBi-42 | EeBi-42:146081 | Beta 148518 | 5430±50 | Charcoal | Area 99-2E, Ploughman property, flake concentration F502 | MAI | 1 |

| | | | | | | | | |
|---|---|---|---|---|---|---|---|---|
| Old Boatyard Site | EeBi-43 | EeBi-43:1 | Beta 121296 | 1640±70 | Charcoal | Associated with FCR in upper level of peat | RI | 1 |
| Old Boatyard Site | EeBi-43 | EeBi-43:2 | Beta 121297 | 3980±110 | Charcoal | Associated with flakes, on top of rounded gravel; stemmed point found nearby | MAI | 1 |
| Party Site | EeBi-30 | EeBi-30:311 | Beta 183604 | 2460±70 | Charcoal | Area 2, L5, midden F4 | GPE | 1 |
| Party Site | EeBi-30 | EeBi-30:142 | Beta 146666 | 2570±60 | Charcoal | Upper terrace, near Area 2, associated with animal skin and flakes | RI | 1 |
| Party Site | EeBi-30 | EeBi-30:268 | Beta 183603 | 2710±40 | Charcoal | Area 1, L2, hearth F1 | GPE | 1 |
| Phillip's Garden | EeBi-1 | 7A284C92 | Beta 15639 | 1250±60 | Charcoal | House F1, charcoal-stained deposit F4 on top of rock perimeter | DPE? | 1 |
| Phillip's Garden | EeBi-1 | | P-737 | 1321±49 | Charcoal | House 20, sterile sand | DPE | 3 |
| Phillip's Garden | EeBi-1 | 7A368D338 | Beta 160977 | 1360±80 | Charred material | House F55, L3, midden F73 | DPE | 2 |
| Phillip's Garden | EeBi-1 | | P-676 | 1502±49 | Charcoal | House 5, main layer in centre of house | DPE | 3 |
| Phillip's Garden | EeBi-1 | 7A368D79 | Beta 66436 | 1370±90 | Charcoal | House F55, L2A, midden F73 | DPE | 1 |
| Phillip's Garden | EeBi-1 | 7A368C743 | Beta 66435 | 1410±100 | Charcoal | House 55, L2, house centre | DPE | 1 |
| Phillip's Garden | EeBi-1 | 7A279A136 | Beta 238479 | 1450±40 | Charcoal | House 17, L2A, midden F167 | DPE | 1 |
| Phillip's Garden | EeBi-1 | | P-734 | 1465±51 | Charcoal | House 17, central fire pit | DPE | 3 |
| Phillip's Garden | EeBi-1 | 7A368D223 | Beta 160976 | 1480±40 | Charred material | House F55, L2, midden F73 | DPE | 1 |
| Phillip's Garden | EeBi-1 | 7A281D104 | Beta 238481 | 1480±40 | Charcoal | House 17, L2A, midden F164 | DPE | 1 |
| Phillip's Garden | EeBi-1 | 7A348C142 | Beta 211271 | 1510±40 | Charcoal | House 2, LL2, fill from storage pit F87b | DPE | 1 |
| Phillip's Garden | EeBi-1 | | P-696 | 1509±47 | Charcoal | House 11 | DPE | 3 |
| Phillip's Garden | EeBi-1 | 7A324D1118 | Beta 19084 | 1520±90 | Charcoal | Midden F2, L2A, dump episode 2T | DPE | 1 |
| Phillip's Garden | EeBi-1 | | P-729 | 1538±55 | Charcoal | House 12, infant burial | DPE | 3 |
| Phillip's Garden | EeBi-1 | 7A348C102 | Beta 211270 | 1550±40 | Charcoal | House 2, LL2, fill from post-hole F87d | DPE | 1 |
| Phillip's Garden | EeBi-1 | 7A323A211 | Beta 15381 | 1570±70 | Charcoal | Midden F2, L2A | DPE | 1 |
| Phillip's Garden | EeBi-1 | | P-733 | 1565±53 | Charcoal | House 16 midden | DPE | 3 |
| Phillip's Garden | EeBi-1 | 7A259D881 | Beta 211268 | 1570±40 | Charcoal | House 18, LL2, bottom of post-hole F119 | DPE | 1 |
| Phillip's Garden | EeBi-1 | | P-727 | 1580±54 | Charcoal | House 4, house centre | DPE | 3 |
| Phillip's Garden | EeBi-1 | 7A279A472 | Beta 238480 | 1580±40 | Charcoal | House 17, L2/3, benearth rock comprising post-hole F170 | DPE | 1 |
| Phillip's Garden | EeBi-1 | | P-694 | 1602±49 | Charcoal | House 10, house centre | DPE | 3 |
| Phillip's Garden | EeBi-1 | 7A348D137 | Beta 211272 | 1600±40 | Charcoal | House 2, west platform, L3A beneath RL1 | DPE | 1 |

(continued)

**Table 2** (continued)

| Site Name | Borden number | Sample number | Lab number | Measured age BP | Sample material | Context | Cultural affiliation | Data source |
|---|---|---|---|---|---|---|---|---|
| Phillip's Garden | EeBi-1 | 7A270C505 | Beta 238478 | 1610±40 | Charcoal | House 17, exterior area, L3, area of mottled soil F197 | DPE | 1 |
| Phillip's Garden | EeBi-1 | 7A259A329 | Beta 211267 | 1630±40 | Charcoal | House 18, L2A, midden F118 | DPE | 1 |
| Phillip's Garden | EeBi-1 | 7A338A40 | Beta 211269 | 1630±40 | Charcoal | House 10, east platform, L2, RL1 | DPE | 1 |
| Phillip's Garden | EeBi-1 | | P-679 | 1632±47 | Charcoal | House 6, second layer in southwest quadrant | DPE | 3 |
| Phillip's Garden | EeBi-1 | | P-693 | 1659±48 | Charcoal | House 2 midden | DPE | 3 |
| Phillip's Garden | EeBi-1 | 7A349D557 | Beta 160975 | 1640±70 | Charred material | House 2, L2, midden F77 | DPE | 2 |
| Phillip's Garden | EeBi-1 | 7A249C794 | Beta 211266 | 1680±40 | Charcoal | House 18, LL2, storage pit, F123 | DPE | 1 |
| Phillip's Garden | EeBi-1 | | P-736 | 1683±49 | Charcoal | House 18 | DPE | 3 |
| Phillip's Garden | EeBi-1 | | P-695 | 1712±40 | Charcoal | House 10 | DPE | 3 |
| Phillip's Garden | EeBi-1 | 7A269B208 | Beta 238476 | 1730±50 | Charcoal | House 17, L3, beneath RL2 in midden F167 | DPE | 1 |
| Phillip's Garden | EeBi-1 | | P-692 | 1736±48 | Charcoal | House 2 midden | DPE | 2 |
| Phillip's Garden | EeBi-1 | 7A270C433 | Beta 238477 | 1750±40 | Charcoal | House 17, exterior area, L2/3, pot stand | DPE | 1 |
| Phillip's Garden | EeBi-1 | 7A295D301 | Beta 42968 | 1770±20 | Charcoal | House F14, L2A, midden F52 | DPE | 1 |
| Phillip's Garden | EeBi-1 | 7A284D284 | Beta 15379 | 1850±110 | Charcoal | House F1, L2A, bone-filled pit F6, house centre | DPE | 1 |
| Phillip's Garden | EeBi-1 | 7A250A47 | Beta 42967 | 1890±90 | Charcoal | Midden F49, L2A | DPE | 1 |
| Phillip's Garden | EeBi-1 | 7A324D1058 | Beta 23978 | 1900±110 | Charcoal | Midden F2, L2A, dump episode F2U, lowest level | DPE | 1 |
| Phillip's Garden | EeBi-1 | 7A323A540 | Beta 15638 | 1920±110 | Charcoal | Midden F2, L2A, hearth dump in lower level of midden | DPE | 1 |
| Phillip's Garden | EeBi-1 | 7A294A535 | Beta 23977 | 1970±60 | Charcoal | House F14, L4, near pit F18 | DPE | 1 |
| Phillip's Garden | EeBi-1 | 7A294A142 | Beta 23976 | 2140±100 | Charcoal | House F14, possible hearth F19, L3/4 | DPE | 1 |
| Phillip's Garden | EeBi-1 | | P-683 | 1593±49 | Seal fat | House 2, main layer, southwest quadrant | DPE | 3 |
| Phillip's Garden | EeBi-1 | | P-735 | 1837±60 | Seal fat | House 17, central fire pit | DPE | 3 |
| Phillip's Garden | EeBi-1 | | P-682 | 1879±60 | Seal fat | House 2, main layer, centre of house | DPE | 3 |
| Phillip's Garden | EeBi-1 | | P-730 | 1906±55 | Seal fat | House 12, central pit | DPE | 3 |
| Phillip's Garden | EeBi-1 | | P-731 | 1911±65 | Seal fat | House 13 | DPE | 3 |
| Phillip's Garden | EeBi-1 | | P-678A | 2006±65 | Seal fat | House 6, main layer, house centre | DPE | 3 |
| Phillip's Garden | EeBi-1 | | P-732 | 2314±60 | Seal fat | House 15 | DPE | 3 |

| | | | | | | | | |
|---|---|---|---|---|---|---|---|---|
| Phillip's Garden East | EeBi-1 | 7A383D475 | Beta 23980 | 1730±200 | Charcoal | PGE1, floor of dwelling/pit F2, L3, may be associated with DPE | GPE/DPE? | 1 |
| Phillip's Garden East | EeBi-1 | 7A383D555 | Beta 19088 | 1910±150 | Charcoal | PGE1, near dwelling/pit F2, L3A, may be associated with DPE | GPE/DPE? | 1 |
| Phillip's Garden East | EeBi-1 | 7A382C66 | Beta 19085 | 1930±140 | Charcoal | PGE1, secondary deposit of FCR F1, L2, may be associated with DPE | GPE/DPE? | 1 |
| Phillip's Garden East | EeBi-1 | 7A385B1026 | Beta 49755 | 2240±100 | Charcoal | PGE2, wall area of dwelling F12, L2 | GPE | 1 |
| Phillip's Garden East | EeBi-1 | 7A385B1193 | Beta 50022 | 2260±70 | Charcoal | PGE2, storage pit F53 likely associated with dwelling F12 | GPE | 1 |
| Phillip's Garden East | EeBi-1 | 7A384C41 | Beta 42970 | 2310±90 | Charcoal | PGE2, central area of dwelling F12, L2 | GPE | 1 |
| Phillip's Garden East | EeBi-1 | 7A383D539 | Beta 19087 | 2320±100 | Charcoal | PGE1 immediately southwest of dwelling/ pit F2, L3A | GPE | 1 |
| Phillip's Garden East | EeBi-1 | 7A393D384 | Beta 50023 | 2350±90 | Charcoal | PGE2 dwelling F12, L3 | GPE | 1 |
| Phillip's Garden East | EeBi-1 | 7A394A727 | Beta 42972 | 2350±100 | Charcoal | PGE2 beneath refuse perimeter of dwelling F12, L3 | GPE | 1 |
| Phillip's Garden East | EeBi-1 | 7A383D613 | Beta 19089 | 2370±160 | Charcoal | PGE1, dwelling/pit F2, L3A | GPE | 1 |
| Phillip's Garden East | EeBi-1 | 7A394A426 | Beta 42971 | 2420±110 | Charcoal | PGE2, F12, L2, exterior area associated with wide side-notched endblades | GPE | 1 |
| Phillip's Garden East | EeBi-1 | 7A385A173 | Beta 50021 | 2500±60 | Charcoal | PGE2, storage pit F55, likely associated with dwelling F12 | GPE | 1 |
| Phillip's Garden East | EeBi-1 | 7A383D403 | Beta 19086 | 2510±90 | Charcoal | PGE1, dwelling/pit F2 wall, L3A, associated with side-notched endblade | GPE | 1 |
| Phillip's Garden East | EeBi-1 | 7A382B2 | Beta 15375 | 2660±70 | Charcoal | PGE1, secondary deposit of FCR, flakes and artefacts F1, L2 | GPE | 1 |
| Phillip's Garden East | EeBi-1 | 7A383D377+371 | Beta 23979 | 2760±90 | Charcoal | PGE1, northwest of dwelling/pit F2, L3 | GPE | 1 |
| Phillip's Garden West | EeBi-11 | 7A711D253 | Beta 66438 | 1960±80 | Charcoal | PGW1, midden F5E, small dump at the bottom of main hillside midden F5 | GPE | 1 |
| Phillip's Garden West | EeBi-11 | 7A701B302 | Beta 49757 | 2090±70 | Charcoal | PGW2, hearth F14, L2, upper terrace | GPE | 1 |
| Phillip's Garden West | EeBi-11 | 7A701B236 | Beta 49756 | 2190±100 | Charcoal | PGW2, stone spiral F23, L2, upper terrace | GPE | 1 |
| Phillip's Garden West | EeBi-11 | 7A702A79 | Beta 42973 | 2200±110 | Charcoal | PGW2, hearth F11, L3, associated with finely-made artefacts | GPE | 1 |
| Phillip's Garden West | EeBi-11 | 7A711A842+843 | Beta 66437 | 2240±70 | Charcoal | PGW1, midden F5C & 5D, small dump bottom of hillside midden F5 | GPE | 1 |

(continued)

**Table 2** (continued)

| Site Name | Borden number | Sample number | Lab number | Measured age BP | Sample material | Context | Cultural affiliation | Data source |
|---|---|---|---|---|---|---|---|---|
| Phillip's Garden West | EeBi-11 | 7A714C30+35 | Beta 66439 | 2340±70 | Charcoal | PGW1, activity area at lower terrace, F28, L3A-B | GPE | 1 |
| Phillip's Garden West | EeBi-11 | 7A711D166 | Beta 49760 | 2340±100 | Charcoal | PGW1, midden F18, dump in midden F5 thrown down from the top, L3 | GPE | 1 |
| Phillip's Garden West | EeBi-11 | 7A701B386 | Beta 49758 | 2350±80 | Charcoal | PGW2, hearth F16, centre of dwelling F25, L2, upper terrace | GPE | 1 |
| Phillip's Garden West | EeBi-11 | 7A711D177 | Beta 49761 | 2460±120 | Charcoal | PGW1, midden F18, dump episode in hillside midden F5, L3 | GPE | 1 |
| Phillip's Garden West | EeBi-11 | 7A711A650 | Beta 49759 | 2540±160 | Charcoal | PGW1, midden F21, a dump within hillside midden F5, L2 | GPE | 1 |
| Point Riche | EeBi-20 | 7A525B113 | Beta 15377 | 1546±80 | Charcoal | Exterior dwelling/midden F1, L2 | DPE | 1 |
| Point Riche | EeBi-20 | 7A555A146+203 | Beta 160980 | 1590±40 | Charred material | Within occupation deposit F59, floor of dwelling F30 | DPE | 1 |
| Point Riche | EeBi-20 | 7A547B499 | Beta 15382 | 1750±90 | Charcoal | Bone-filled pit F2, L3, dwelling(?) F1 | DPE | 1 |
| Point Riche | EeBi-20 | 7A547D380 | Beta 15376 | 1750±80 | Charcoal | Faunal deposit in dwelling(?) F1, L3 | DPE | 1 |
| Point Riche | EeBi-20 | 7A543C511 | Beta 50025 | 1760±150 | Charcoal | Inside slope of wall F29, L2WC, dwelling F8 | DPE | 1 |
| Point Riche | EeBi-20 | 7A557D11 | Beta 50026 | 1800±70 | Charcoal | Lower level of hearth F24 outside dwelling F8 | DPE | 1 |
| Point Riche | EeBi-20 | 7A542A59 | Beta 160978 | 1810±40 | Charred material | On top of burned stones F49 associated with GPE artefacts | GPE | 1 |
| Point Riche | EeBi-20 | 7A543C466+471 | Beta 50024 | 1830±90 | Charcoal | Dwelling F8, L2 | DPE | 1 |
| Port au Choix-3 | EeBi-2 | | I-4380 | 3230±220 | Charcoal | Locus 4, Burial 2 | MAI | 4 |
| Port au Choix-3 | EeBi-2 | | Y-2608 | 3370±80 | Charcoal | Locus 2, Burial 22 | MAI | 4 |
| Port au Choix-3 | EeBi-2 | | I-4677 | 3410±100 | Human bone | Human bone recovered from construction activities in Locus 1 | MAI | 4 |
| Port au Choix-3 | EeBi-2 | | I-4682 | 3690±90 | Charcoal | Locus 2, Burial 22 | MAI | 4 |
| Port au Choix-3 | EeBi-2 | | AA 33919 | 3777±44 | Beaver incisor | Locus 2, Burial 35B | MAI | 5 |
| Port au Choix-3 | EeBi-2 | | AA 29482 | 3825±60 | Caribou bone | Locus 2, Burial 35B | MAI | 5 |
| Port au Choix-3 | EeBi-2 | | AA 33920 | 3826±38 | Caribou bone | Locus 2, Burial 35B | MAI | 5 |
| Port au Choix-3 | EeBi-2 | | I-4678 | 3930±130 | Human bone | Locus 2, Burial 50 | MAI | 4 |

| | | | | | | | | |
|---|---|---|---|---|---|---|---|---|
| Port au Choix-3 | EeBi-2 | | GrA-6525 | 4000±50 | Human bone | Locus 2, Burial B18A | MAI | 6 |
| Port au Choix-3 | EeBi-2 | | GrA-6496 | 4000±50 | Human bone | Locus 2, Burial C40A | MAI | 6 |
| Port au Choix-3 | EeBi-2 | | GrA-6527 | 4110±50 | Human bone | Locus 2, Burial C35A | MAI | 6 |
| Port au Choix-3 | EeBi-2 | | GrA-6666 | 4130±50 | Human bone | Locus 2, Burial A10 | MAI | 6 |
| Port au Choix-3 | EeBi-2 | | GrA-6501 | 4130±50 | Human bone | Locus 2, Burial B30C | MAI | 6 |
| Port au Choix-3 | EeBi-2 | | GrA-6526 | 4150±50 | Human bone | Locus 2, Burial B25 | MAI | 6 |
| Port au Choix-3 | EeBi-2 | | GrA-6495 | 4150±50 | Human bone | Locus 2, Burial C36A | MAI | 6 |
| Port au Choix-3 | EeBi-2 | | GrA-6479 | 4160±50 | Human bone | Locus 2, Burial A12 | MAI | 6 |
| Port au Choix-3 | EeBi-2 | | GrA-6478 | 4220±50 | Human bone | Locus 2, Burial A1B | MAI | 6 |
| Port au Choix-3 | EeBi-2 | | I-3788 | 4290±110 | Charcoal | Locus 2, Burial 22 | MAI | 4 |
| Port au Choix-3 | EeBi-2 | | GSC-1403 | 4690±130 | Long clam shells | From the sand of the 6.1 m Port au Choix-3 beach terrace | MAI | 4 |
| Port au Choix-3 | EeBi-2 | | Y-2609 | 5120±120 | Wood inside bog iron | Locus 2, Burial 42 | MAI | 4 |
| Sea Level Project | | PAC-15-S-9631 | Beta 107797 | 1520±80 | Shell | *Nucella lapillus* at 3 m asl beach gravel | N/A | 1 |
| Sea Level Project | | PAC-03-S-9718 | Beta 107796 | 7750±90 | Shell | *Nucella lapillus* at 9 m asl beach gravel | N/A | 1 |
| Skin Pond | | SkP-01-32 | Beta 238622 | 340±40 | Wood | 16 cm depth | N/A | 1 |
| Spence Site | EeBi-36 | EeBi-36:1773 | Beta 66440 | 840±90 | Charcoal and peat | Main excavation area, core & flake concentration F16, L2-4 | RI | 1 |
| Spence Site | EeBi-36 | EeBi-36:2042/6 | Beta 66441 | 1020±60 | Charred material | Shed area trench, associated with faunal, L3 | RI | 1 |
| Spence Site | EeBi-36 | EeBi-36:2544 | Beta 66442 | 1340±80 | Charcoal | Main excavation area, lithic, bone & charcoal concentration F27, L2-4 | RI | 1 |
| Spence Site | EeBi-36 | 7A77A183 | Beta 49753 | 1360±80 | Charcoal | Main excavation area, hearth F2, L3 | RI | 1 |
| Spence Site | EeBi-36 | 7A77A383 | Beta 49754 | 1420±70 | Charcoal | Main excavation area, hearth F9, L3 | RI | 1 |
| Stove Pond | | SP-50-60 | Beta 36436 | 4400±90 | Gyttja | 50–60 cm depth | N/A | 1 |
| Stove Pond | | SP-90-95 | Beta 32595 | 5420±70 | Bulk organics | 90–95 cm depth | N/A | 1 |
| Stove Pond | | SP-120-127.5 | Beta 32596 | 7180±140 | Bulk organics | 120–127.7 cm depth | N/A | 1 |
| Stove Pond | | SP-139-146 | Beta 32597 | 7420±130 | Bulk organics | 139–146 cm depth | N/A | 1 |
| Stove Pond | | SP-180-187 | Beta 32598 | 7920±130 | Bulk organics | 180–187 cm depth | N/A | 1 |

# Index

**P**

MIX
Papier | Fördert
gute Waldnutzung
FSC® C083411

Kolibri 360 GmbH
Ferdinand-Jühlke-Straße 7
99095 Erfurt, Deutschland
produktsicherheit@kolibri360.de